AF361940

Kinetoplastid Genomics and Beyond

Kinetoplastid Genomics and Beyond

Editor

Jose M. Requena

MDPI • Basel • Beijing • Wuhan • Barcelona • Belgrade • Manchester • Tokyo • Cluj • Tianjin

Editor
Jose M. Requena
Molecular Biology
Universidad Autonoma de
Madrid
Madrid
Spain

Editorial Office
MDPI
St. Alban-Anlage 66
4052 Basel, Switzerland

This is a reprint of articles from the Special Issue published online in the open access journal *Genes* (ISSN 2073-4425) (available at: www.mdpi.com/journal/genes/special_issues/kinetoplastid_ genomics).

For citation purposes, cite each article independently as indicated on the article page online and as indicated below:

LastName, A.A.; LastName, B.B.; LastName, C.C. Article Title. *Journal Name* **Year**, *Volume Number*, Page Range.

ISBN 978-3-0365-1582-3 (Hbk)
ISBN 978-3-0365-1581-6 (PDF)

Contents

About the Editor

Jose M. Requena

Jose M. Requena is a Ph.D. in Molecular Biology (1990). He has been based at the Department of Molecular Biology (Autonomous University of Madrid, Spain) since 1998, where he teaches courses on biochemistry, molecular biology techniques, and microbiology. Since 2003, Dr. Requena is a group leader at Centro de Biologia Molecular Severo Ochoa (Madrid, Spain). His research is focused on the field of trypanosomatids, mainly *Leishmania* species, and includes topics like gene expression regulation, heat shock proteins, vaccines, genomics, transcriptomics, proteomics, phylogenetics, and mechanisms of drug resistance. He serves as an editorial board member for several scientific journals, and is currently Editor-in-Chief for the 'Microbial Genetics and Genomics' section of *Genes*.

Preface to "Kinetoplastid Genomics and Beyond"

Kinetoplastids are a clade of protists located among the earliest-branching eukaryotes. They are mainly recognized because the medical and economic importance of some of their members, e.g. *Trypanosoma* ssp., causing Chagas disease and sleeping sickness in humans, and *Leishmania* spp., causing kala-azar and other types of leishmaniases. Nevertheless, many parasites for plants (*Phytomonas* spp.) and insects (*Leptomonas*, *Crithidia*, and other genera) also belong to this class. In addition, free-living kinetoplastids (bodonids) are abundant and active microbial predators in terrestrial and aquatic ecosystems. Kinetoplastids, together with other two major clades (euglenids and diplonemids), constitute a monophyletic group of flagellates: the Euglenozoa.

Apart from their medical and veterinary relevance, these organisms generate a considerable basic scientific interest due to their bizarre cytology, genome organization, and mechanisms of gene expression regulation. In recent years, the incorporation of "omics"methodologies to the study of these organisms has allowed assembly of the genomes for a growing number of both parasitic and free-living kinetoplastids to analyze changes in gene expression, determine the proteome compendium, establish metabolic pathways, etc.

The aim of this Special Issue was to bring together a set of reviews and research articles on recent and cutting-edge advances in topics related to genome organization, mechanisms of gene expression, and experimental and bioinformatics methodologies, among others. I am grateful to those colleagues who decided contributing to this objective, submitting eleven excellent articles that are now compiled in this book.

In the following paragraphs, with the objective of guiding readers, the contents of the different chapters are briefly summarized, and their connections are highlighted.

As mentioned above, within kinetoplastids, the parasites referred to as trypanosomatids cause severe diseases in humans. Unfortunately, there are no vaccines to prevent these infections, and the available drugs to control these diseases are far from ideal due to host toxicity, limited access, and increasing rates of drug resistance. Here, in the first chapter, Bhattacharya and colleagues present a comprehensive review on current chemotherapy against tripanosomatids and, more importantly, describe the technological advances in parasitology, chemistry, and genomics that have brought improved compound screening technologies and incorporated novel drug concepts. As documented in this review, these new approaches are uncovering new lead compounds and, consequently, more effective treatments are envisioned for the near future [1].

The need for new drugs for treatment and the problem of drug resistance is also illustrated in the article by Ghosh et al [2]. Artemisinin, a drug used for malaria treatment, is being explored as a candidate drug for combating leishmaniasis. Nevertheless, apart from its efficacy, it is mandatory to establish the easiness by which parasites might create resistance. In this study, by comparative genomics and transcriptomics analyses of in vitro-adapted artesunate-resistant Leishmania donovani parasites, the authors have outlined the molecular basis underlying artemisinin resistance in *Leishmania* parasites.

Trypanosomatids exhibit a number of highly peculiar molecular features. Among them, a remarkable peculiarity is the genome structure: genes are organized into large collinear clusters, but contrary to prokaryotic polycistronic units, the genes present have no common nor akin function. Other modulators of genome structure are retroposons and gene families comprised of abundant and sequence variable members. These genomic peculiarities are illustrated in the chapter by Herreros-Cabello et al. [3], who reviewed current knowledge on *Trypanosoma cruzi* genome architecture and plasticity. Within the study of the *T. cruzi* genome, the chapter by Bernardo et al [4] shows an in-depth analysis of the Retrotransposon Hot Spot (RHS) gene family. The RHS family is the largest gene family existing in the *T. cruzi* genome, but, at the same time, the most enigmatic regarding their cellular functions. Based on their nuclear location, the authors suggest that RHS proteins might be involved in the control of the chromatin structure and gene expression along the parasite life cycle. Apart from multigenic families, the *T. cruzi* genome is populated by interspersed repetitive DNA elements that amount for a significant fraction of its genomic content. As suggested by Calderano et al. [5], these repeated elements, often located at the 3′-untranslated regions (3′-UTRs) of genes, may be essential players for the mechanisms regulating gene expression. The kinetoplastids have a unique reliance on post-transcriptional control of gene expression, and the uncovering of cis- and trans-acting regulators is a basic step towards the understanding of how these organisms regulate mRNA and proteins levels.

Crucial components of genome architecture are origins of replication (ORIs), i.e., the places in which DNA replication initiates. The activation (firing) of ORIs is an extremely regulated process, as the cell viability depends on the complete replication of every chromosome within a precise phase (S-phase) of the cell cycle. In kinetoplastids, our knowledge about the number of ORIs required to replicate their genomes is limited. In the chapter by da Silva et al. [6], a bioinformatic tool designed to calculate the minimum number of ORIs required to duplicate an entire chromosome within the S-phase duration in trypanosomatids (*T. cruzi*, *Leishmania major*, and *T. brucei*) and yeasts (*Saccharomyces cerevisiae* and *Schizosaccharomyces pombe*) is described.

Complete and well-annotated genomes represent the ultimate resource for genome-wide scale studies, such as transcriptomic and proteomic analyses. However, as documented in the chapter by Sanchiz et al. [7], a proteogenomic approach should be considered as a first choice when determining the experimental proteome for a given organism, *Leishmania infantum* in this instance. This strategy would allow the uncovering of new protein-coding genes and, consequently, to improve gene annotations.

As mentioned above, kinetoplastids depend on post-transcriptional mechanisms for gene expression regulation. This includes post-transcriptional protein modifications, which contribute to cellular phenotypes by altering protein abundance, function, and localization. In the chapter by Bea et al. [8], it is documented that the machinery of modification of polypeptides by the covalent attachment of small ubiquitin-like modifier (SUMO) moiety (SUMOylation of proteins) is essential for *Leishmania donovani* viability and infectivity. In this study, the CRISPR-Cas9-mediated gene edition system was used. The CRISPR (Clustered Regularly Interspaced Short Palindromic Repeats)–Cas9 (CRISPR-associated protein 9) methodology is revolutionizing in vivo studies aimed to decipher gene function in many organisms. The chapter by Adaui et al. [9] illustrated the use of this technique in *Leishmania braziliensis*. The authors successfully applied a cloning-free, PCR-based CRISPR–Cas9 technology to inactivation of the two alleles of two well-characterized heat-shock genes, HSP23 and HSP100. The detailed description of the technique and the compendium of methods used in the study for the characterization of the mutant lines make this chapter a methodological reference article.

The evolutionary and biogeographical history of kinetoplastids is fascinating, even though it remains a hotly debated topic, and multiple hypotheses have been proposed. The chapter by Cantanhêde et al. [10] reviewed our current knowledge on the origin of *Leishmania* parasites, adding a new player, the RNA viruses identified in many species of this genus. Phylogenetic analyses of the endosymbiotic Leishmania viruses and the *Leishmania* species harbouring them suggest a long coevolutionary relationship, which would enhance parasite survival and virus fitness during leishmaniasis.

As stated above, kinetoplastids belong to the Euglenozoa group, an evolutionary ancient phylum of flagellate eukaryotes. In the final chapter of this book, Cordoba and co-workers [11] present a study aimed to generate a comprehensive transcriptome for *Euglena gracilis*, a known photosynthetic microeukaryote considered as the product of a secondary endosymbiosis between a green alga and a phagotrophic unicellular protist, an evolutionary relative of kinetoplastids. Thus, analysing *E. gracilis* genomic and transcriptomic information is a way to approach the evolution of parasitism. In this regard, the authors of this study show evidence that trans-splicing mechanisms (typical of trypanosomatids) are also occurring in a large percentage of the *E. gracilis* transcripts.

In summary, this collection of eight original research articles and three reviews covers a wide range of topics in the field of kinetoplastids. In addition, readers can find a compendium of experimental methods and bioinformatics tools.

Finally, I would like to express my gratitude to the contributing authors and thanks to Maggie Miao for her invaluable editorial assistance. This book is also dedicated to my daughter Carmen.

References

1. Bhattacharya, A.; Corbeil, A.; Do Monte-Neto, R.L.; Fernandez-Prada, C. Of drugs and trypanosomatids: New tools and knowledge to reduce bottlenecks in drug discovery. Genes (Basel). 2020, 11, 1–24.

2. Ghosh, S.; Verma, A.; Kumar, V.; Pradhan, D.; Selvapandiyan, A.; Salotra, P.; Singh, R. Genomic and transcriptomic analysis for identification of genes and interlinked pathways mediating artemisinin resistance in leishmania donovani. Genes (Basel). 2020, 11, 1362, doi:10.3390/genes11111362.

3. Herreros-Cabello, A.; Callejas-Hernández, F.; Gironès, N.; Fresno, M. Trypanosoma cruzi genome: Organization, multi-gene families, transcription, and biological implications. Genes (Basel). 2020, 11, 1196.

4. Bernardo, W.P.; Souza, R.T.; Costa-Martins, A.G.; Ferreira, E.R.; Mortara, R.A.; Teixeira, M.M.G.; Ramirez, J.L.; Da Silveira, J.F. Genomic organization and generation of genetic variability in the RHS (Retrotransposon hot spot) protein multigene family in Trypanosoma cruzi. Genes (Basel). 2020, 11, 1–19, doi:10.3390/genes11091085.

5. Calderano, S.G.; Nishiyama Junior, M.Y.; Marini, M.; Nunes, N. de O.; Reis, M. da S.; Patané, J.S.L.; da Silveira, J.F.; da Cunha, J.P.C.; Elias, M.C. Identification of novel interspersed DNA repetitive elements in the trypanosoma cruzi genome associated with the 3UTRs of surface multigenic families. Genes (Basel). 2020, 11, 1235, doi:10.3390/genes11101235.

6. da Silva, M.S.; Vitarelli, M.O.; Souza, B.F.; Elias, M.C. Comparative analysis of the minimum number of replication origins in trypanosomatids and yeasts. Genes (Basel). 2020, 11, 523, doi:10.3390/genes11050523.

7. Sanchiz, Á.; Morato, E.; Rastrojo, A.; Camacho, E.; González-de la Fuente, S.; Marina, A.; Aguado, B.; Requena, J.M. The Experimental Proteome of Leishmania infantum Promastigote and Its Usefulness for Improving Gene Annotations. Genes (Basel). 2020, 11, E1036, doi:10.3390/genes11091036.

8. Bea, A.; Kröber-Boncardo, C.; Sandhu, M.; Brinker, C.; Clos, J. The leishmania donovani SENP protease is required for SUMO processing but not for viability. Genes (Basel). 2020, 11, 1–16,

doi:10.3390/genes11101198.

9. Adaui, V.; Kröber-Boncardo, C.; Brinker, C.; Zirpel, H.; Sellau, J.; Arévalo, J.; Dujardin, J.C.; Clos, J. Application of crispr/cas9-based reverse genetics in leishmania braziliensis: Conserved roles for hsp100 and hsp23. Genes (Basel). 2020, 11, 1–24, doi:10.3390/genes11101159.

10. Cantanhêde, L.M.; Mata-Somarribas, C.; Chourabi, K.; Pereira da Silva, G.; Dias Das Chagas, B.; de Oliveira R. Pereira, L.; Côrtes Boité, M.; Cupolillo, E. The maze pathway of coevolution: A critical review over the leishmania and its endosymbiotic history. Genes (Basel). 2021, 12, 657.

11. Cordoba, J.; Perez, E.; Van Vlierberghe, M.; Bertrand, A.R.; Lupo, V.; Cardol, P.; Baurain, D. De Novo Transcriptome Meta-Assembly of the Mixotrophic Freshwater Microalga Euglena gracilis. Genes (Basel). 2021, 12, 842, doi:10.3390/genes12060842.

Jose M. Requena

Editor

 *genes*

Review

Of Drugs and Trypanosomatids: New Tools and Knowledge to Reduce Bottlenecks in Drug Discovery

Arijit Bhattacharya [1], **Audrey Corbeil** [2], **Rubens L. do Monte-Neto** [3] and **Christopher Fernandez-Prada** [2,*]

1   Department of Microbiology, Adamas University, Kolkata, West Bengal 700 126, India; arijbhatta@gmail.com
2   Department of Pathology and Microbiology, Faculty of Veterinary Medicine, Université de Montréal, Saint-Hyacinthe, QC J2S 2M2, Canada; audrey.corbeil@umontreal.ca
3   Instituto René Rachou, Fundação Oswaldo Cruz, Belo Horizonte MG 30190-009, Brazil; rubens.monte@fiocruz.br
*   Correspondence: christopher.fernandez.prada@umontreal.ca; Tel.: +1-450-773-8521 (ext. 32802)

Received: 4 June 2020; Accepted: 26 June 2020; Published: 29 June 2020

Abstract: Leishmaniasis (*Leishmania* species), sleeping sickness (*Trypanosoma brucei*), and Chagas disease (*Trypanosoma cruzi*) are devastating and globally spread diseases caused by trypanosomatid parasites. At present, drugs for treating trypanosomatid diseases are far from ideal due to host toxicity, elevated cost, limited access, and increasing rates of drug resistance. Technological advances in parasitology, chemistry, and genomics have unlocked new possibilities for novel drug concepts and compound screening technologies that were previously inaccessible. In this perspective, we discuss current models used in drug-discovery cascades targeting trypanosomatids (from in vitro to in vivo approaches), their use and limitations in a biological context, as well as different examples of recently discovered lead compounds.

Keywords: trypanosomatids; neglected tropical diseases; *Leishmania*; *Trypanosoma cruzi*; *Trypanosoma brucei*; drug discovery; in vitro models; in vivo models; genomics; drug resistance

---

1. Introduction: Status and Impact of Trypanosomatid-Borne Infections

In 1970, the Rockefeller Foundation coined the term "Neglected Tropical Diseases" (NTDs), which still applies to three major, chronic, debilitating, and poverty-promoting diseases caused by trypanosomatid parasites: human African trypanosomiasis (HAT or sleeping sickness), caused by *Trypanosoma brucei* and transmitted by tsetse flies; Chagas disease (South American trypanosomiasis) caused by *T. cruzi* and transmitted by blood-sucking triatomine bugs; and leishmaniasis, caused by various species of the genus *Leishmania* and transmitted by sand flies. At present, the therapeutic arsenal to combat these infections is ineffective and highly toxic. Progressively over the last two decades, this situation has been aggravated by the emergence and spread of drug-resistant strains [1].

Although the WHO has targeted the elimination of HAT as a public health problem by 2020 (and interruption of transmission for 2030), Chagas disease and leishmaniasis are global threats in continuous expansion [2–6]. Chagas disease affects an estimated 8–10 million people worldwide, approximately 30% of which will develop chronic Chagas cardiac disease, leading to 14,000 deaths per year [1,6]. The cost of Chagas disease was estimated in 2013 at more than US$ 7 B/year, including lost productivity [7]. However, and despite these alarming numbers, only two toxic, old-fashioned compounds, benznidazole and nifurtimox (Figure 1), are approved for the treatment of Chagas disease [6,8]. While benznidazole is only FDA-approved for pediatric and acute cases of *T. cruzi* infection, nifurtimox is still only available under compassionate-use directives from the CDC [9,10]. Moreover, the efficacy of benznidazole treatment in chronic Chagas patients is controversial [10,11].

1

In addition to the unacceptable side effects of these drugs, drug resistance has emerged as a major concern in terms of treatment failure [1,12,13].

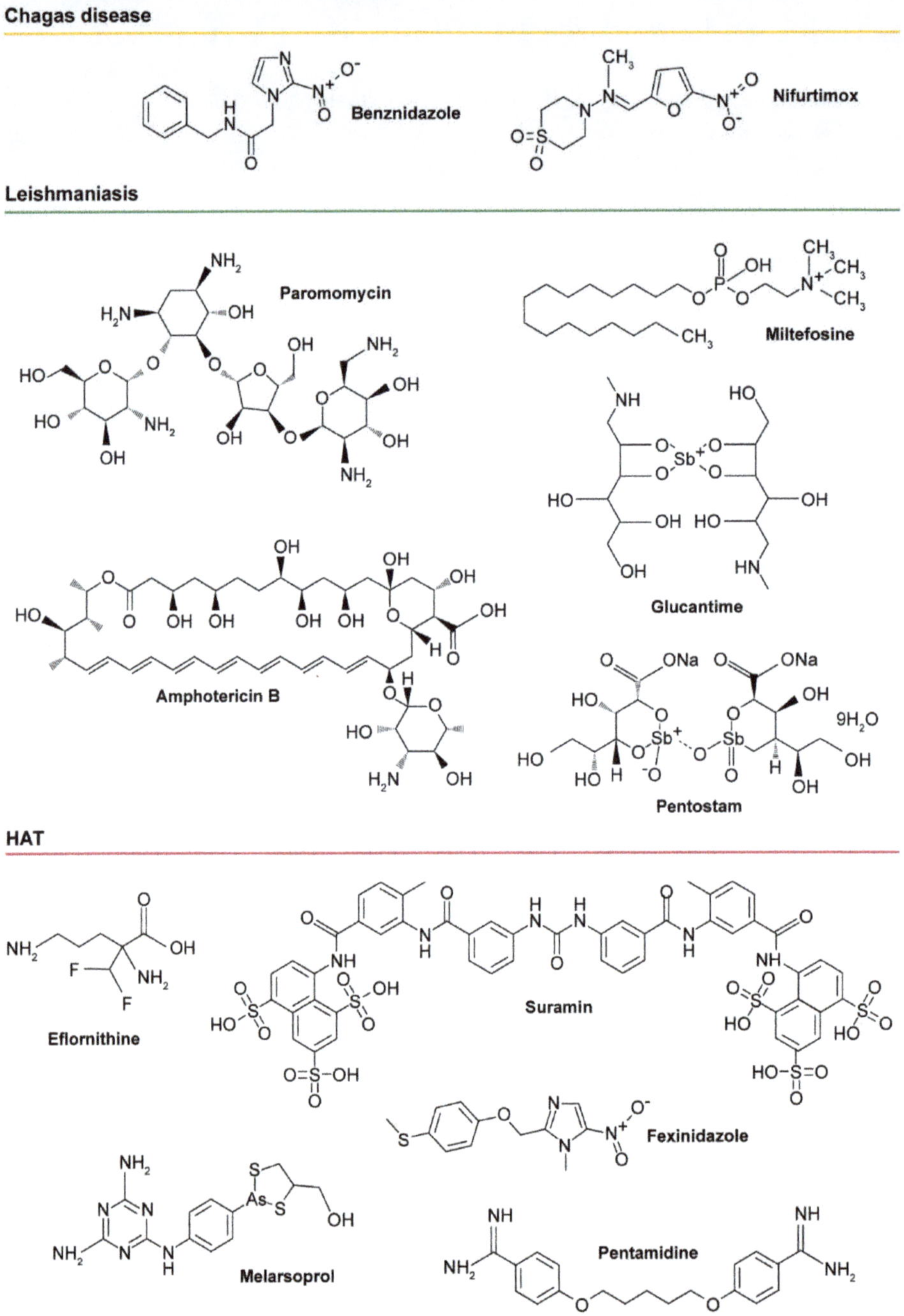

Figure 1. Drugs in clinical use against Chagas disease, leishmaniasis, and human African trypanosomiasis (HAT).

Leishmaniasis is estimated to be the ninth largest disease burden among individual infectious diseases, and the most dangerous of the NTDs. Leishmaniasis currently infects around 12 million people worldwide, and it is spreading with ca. 0.7–1 million new cases per year [14]. Dramatically, its visceral form (also referred as VL) has a 95% fatality rate among the poorest people in the world. The control of leishmaniasis relies on old-fashioned, highly toxic chemotherapy using a very limited number of registered molecules (Figure 1). In addition to toxicity, significant drawbacks such as complex route of administration, length of treatment, emergence of drug resistance, and costs limit their use in endemic areas [1,14]. Furthermore, NTDs are becoming emergent diseases in non-tropical countries, triggering vast socioeconomic consequences. The absence of investment to combat NTDs is likely due to their traditional cause of misfortune to poor, rural, and otherwise marginalized populations. However, their impact has shifted because of resistant strains and globalization. Without effective new drugs, the incidence of Chagas disease and leishmaniasis is expected to spread owing to climate change, global urbanization, immunosuppressive disease, etc. [15,16].

Traditionally, pharmaceutical companies have shown a very limited interest in improving current therapeutics against trypanosomatid parasites because of the expected low return on investment when targeting communities with little to no purchasing power [17,18]. In order to alleviate the costs and accelerate the marketing process [19–21] (e.g., to avoid obstacles during clinical trials, such as drug toxicity or unfavorable pharmacokinetics) [22], many initiatives are trying to find new indications for already-existing drugs, also known as drug repurposing (or drug repositioning) [1]. On the other hand, other initiatives—especially those stemming from academia—are targeted for identifying new points of intervention and to conceive novel drugs. In both cases, interdisciplinary research between experts in parasitology and chemistry is required, such that the former focus primarily on established drugs to treat infection due to limited access to novel molecules. Markedly, the critical situation with NTDs calls for the urgent development of high-throughput approaches for assessing drug efficacy and resistance, as well as novel therapeutics to avoid the emergence and spread of drug-resistant strains. Through this review, we aim to bring together these two major fields of knowledge and shed some light on the different models that are currently available, in order to build a drug-discovery pipeline targeting trypanosomatids (from in vitro to in vivo approaches), their use and limitations, as well as recent endeavors for discovering lead compounds.

2. Trypanosomatids' Life Cycle in the Context of In Vitro Screening Assays

Pathogenic trypanosomatids have complex, digenetic lifecycles, which require the presence of both invertebrate and vertebrate hosts (summarized in Figure 2). In this way, various developmental stages throughout trypanosomatids' lifecycle are required to guarantee their survival and spread.

These diverse stages encompass many metabolic, biochemical, and cell biological adaptations, including a significant variation of cell morphology [23–25]. Because of these changes, it is hard, and sometimes impossible, to establish a correlation between compounds selected in assays targeting different forms of the same parasite (e.g., extracellular vs. intracellular). In the current lack of methodology standardization, this section will discuss the mains aspects to be considered to choose the most adapted in vitro screening assay to start a drug discovery cascade.

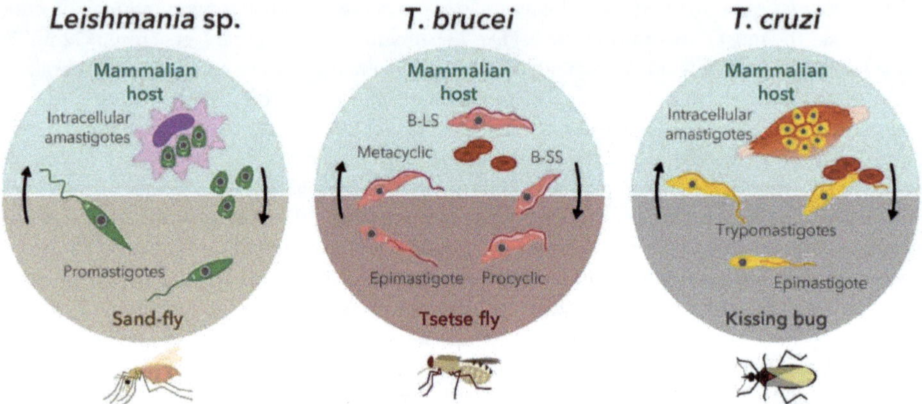

Figure 2. Life cycles of pathogenic trypanosomatid parasites. The clinically relevant life-cycle stages that are targets for drug intervention are intracellular amastigotes in *Leishmania* sp.; bloodstream forms (bloodstream long slender form (B-LS) and bloodstream short stumpy form (B-SS)) in *Trypanosoma brucei*; and infective trypomastigotes and intracellular amastigotes in *Trypanosoma cruzi*.

2.1. Leishmania Parasites

Leishmania parasites cycle between the motile promastigote form in the gut of the sand-fly vector and the intracellular amastigote stage within the macrophages and other types of mononuclear phagocytic cells of the mammalian host. In this way, when invading macrophages, *Leishmania* promastigotes block the phagosome maturation process and create an environment that is propitious to amastigote differentiation. Subsequent divisions and later infection of other mononuclear phagocytic cells, as well as different tissues, leads to the setup and progression of the clinical manifestations related to these diseases [26]. Traditionally, compounds have been evaluated by means of cell-free assays using axenic promastigotes and amastigotes, which allow high-throughput screening and high reproducibility, while relying on a limited number or parasites per evaluation. However, these two parasite forms present several important caveats that can lead to the selection of false candidates. On the one hand, promastigotes are not the mammalian form, and they show significant differences in their metabolic profile when compared to intracellular amastigotes. Moreover, their growth and sensitivity are influenced by different parameters, such as cell culture density, medium composition, and compound mode of action (MoA), among others, so care must be taken in interpreting the data [27]. While closer to the mammalian form, axenic amastigotes retain some promastigote traits, leading to a lack of correlation between axenic forms screenings and intracellular amastigote assays, which increases the false-positive rate of hit discovery when using this artificial form [28]. Consequently, models using the intracellular amastigote infecting mammalian host cells remain the gold standard in the determination of drug sensitivity. These models have great advantages such as the direct evaluation of drug penetration in the host cell, as well as drug activity in the phagolysosome milieu, among others [29,30]. Moreover, intracellular amastigotes are generally more sensitive than promastigotes against most of the drugs currently used in clinic, such as antimony or miltefosine [31,32], which could be a consequence of genes differentially regulated in the two developmental stages of the parasite [31,33,34]. The activity of candidate compounds against intracellular amastigotes is determined by microscopic automatic/manual counting of infected macrophages and the number of parasites per macrophage (parasitic index) or by spectrophotometric (e.g., optical density or staining) and fluorometric methods. These latter include the automated detection and quantification of genetically engineered amastigotes that express fluorescent and bioluminescent reporters, which enables faster read-outs and higher throughput [35]. Nonetheless, determination of the cidal and static effects of candidate compounds against intracellular forms can be very challenging, in part because of the slow replication rate of amastigotes when compared to promastigotes [36–38]. Moreover, this determination could be biased by many confounding factors that can reduce lab-to-lab reproducibility and lead to false hit discoveries. These factors could include macrophage infection rate, incomplete amastigogenesis, impact of distinct culture media, as well as the intrinsic pathogenicity of the strain selected for the assay [39–41].

Despite these potential limitations, in vitro amastigote assays (infecting THP-1 and primary mouse macrophages (PMM cells)) have led to the discovery and optimization of a novel series of amino-pyrazole ureas with potent antileishmanial activity [42]. Likewise, more recently, Van den Kerkhof et al. (2018) evaluated three antileishmanial leads series (nitroimidazoles, oxaboroles and aminopyrazoles) using intracellular *L. donovani* and *L. infantum* amastigotes infecting PMM, and showed a good in vitro to in vivo correspondence, with high efficacy and negligible side effects in vivo [43]. Tunes et al. (2020) found that gold(I)-derived complexes were very active against *L. infantum* and *L. braziliensis* intracellular amastigotes infecting THP-1 cells, including antimony-resistant strains (SbR), and they were potent inhibitors of trypanothione reductase. Moreover, two of these complexes presented very favorable pharmacokinetic and safety profiles in vivo after oral administration [44]. In the search of more robust, scalable, and reproducible models, Melby's team developed an ex vivo splenic explant assay that allows the identification of new compounds active against *Leishmania* within the pathophysiologic environment [45,46]. In this way, they recovered the spleens of hamsters infected with a luciferase-transfected *L. donovani* strain, and used amastigote-harboring splenocytes to evaluate the antileishmanial activity of more than 4000 molecules. This medium-throughput screen revealed 84 small molecules with good antileishmanial activity and an acceptable toxicity evaluation [45]. Similarly, in a drug repurposing initiative, Fernandez-Prada et al. (2013) used BALB/c-derived splenic explants infected with *L. infantum* amastigotes expressing the infrared fluorescent protein IFP1.4 to evaluate the antileishmanial effect of anticancer-drug camptothecin and several analogues [37]. Markedly, and despite their many advantages, engineered parasites are not flawless, and different mitigation strategies should be taken into account in order to avoid any compensatory change in parasite metabolism or virulence (e.g., prioritize the use of integrative strategies to generate the strain) [35]. A final important remark is that, as has been recently demonstrated, there could be different compound efficiencies linked to the drug susceptibility background of the *Leishmania* strains used in the screening process (especially in the case of antimony susceptibility), which shows the potential value of including clinical isolates (and resistant strains) in the drug discovery cascade [47].

2.2. Trypanosoma brucei

Contrary to *Leishmania*, the *T. brucei* life cycle does not require the intracellular environment for any of its developmental forms. *T. brucei* is transmitted between mammalian hosts by *Glossina* spp. (tsetse fly), in which the bloodstream short stumpy form (B-SS) differentiates into the replicative procyclic form (PFs). PFs migrate to the proventriculus were they subsequently differentiate into epimastigotes and into cycle-arrested metacyclics (infective form) in the salivary glands of the tsetse fly. Parasites colonize the mammalian host during the blood meal of the fly and differentiate into bloodstream long slender form (B-LS), which eventually evolves to the B-SS form by a quorum-sensing mechanism [48,49]. Consequently, drug-screening assays targeting *T. brucei* rely on the bloodstream form of the parasite. Different approaches for whole-cell, high-throughput screening have recently been successfully developed. Mackey et al. (2006) screened 2160 FDA-approved drugs, bioactive compounds, and natural products to identify hits that were cytotoxic to *T. brucei* at a concentration of 1 μM or less. This approach led to the identification of 35 new hits from seven different drug categories, which included two approved trypanocidal drugs, suramin and pentamidine [50]. Similar to *Leishmania*, bioluminescent-engineered *T. brucei* have recently been developed and implemented in whole-cell high-throughput screens. Sykes et al. (2009) developed a luciferase-based viability assay for ATP detection in a 384-well format, making high-throughput whole-cell screening in *T. brucei* very reproducible, sensitive, and cost effective [51]. Later, Sykes et al. (2012) described the application of an

Alamar Blue (resazurin)-based, 384-well high-throughput screening (HTS) assay to screen a library of 87,296 compounds, leading to 6 hits from 5 new chemical classes displaying great activity against *T.b. rhodesiense* [52]. As an alternative to luciferase and Alamar Blue, Faria et al. (2015) developed a whole-cell assay in 384-well plates based on the quantitative detection of double-stranded DNA bound to cyanine dye SYBR Green. The assay was a validated screening of a kinase-focused library composed of 4000 compounds, leading to the discovery of novel scaffolds with potent antitrypanosomal activity [53]. In the recent years, thanks to different screening initiatives, several new leads such as diamidine derivatives, fexinidazole, oxaborole SCYX-7158, quinolone amide GHQ168, and acoziborole are now in various stages of the development pipeline for treating HAT [54–56].

2.3. Trypanosoma cruzi

Infective trypomastigotes and intracellular replicative amastigotes are the clinically relevant life-cycle stages of *T. cruzi* that are targets for drug intervention [57]. Briefly, non-dividing *T. cruzi* metacyclic trypomastigotes are transmitted to humans in the feces of infected triatomine bugs at the bite site of these hematophagous insects. Trypomastigotes invade various cell types and transform into intracellular amastigotes, which multiply by binary fission until the host cell is overwhelmed, and then transform into bloodstream trypomastigotes and spread to distant sites through the lymphatics and bloodstream. Once back in the insect vector, trypomastigotes transform into epimastigotes and then differentiate into infective metacyclic trypomastigotes [58]. Despite many efforts, only two compounds, benznidazole (since 1972) and nifurtimox (since 1967), are currently used for the treatment of certain forms of Chagas disease [59]. Markedly, drug discovery in *T. cruzi* is handicapped by the small number of well-established targets (e.g., the sterol biosynthetic pathway, cruzipain, cytochrome b, trypanothione reductase, cyclophilin, or carbonic anhydrases [57]), which explains the wide use of phenotypic approaches that have become the main pillar of Chagas R&D [60]. Drug screening against *T. cruzi* can be performed in cell-free axenic amastigotes and epimastigotes, as well as in intracellular amastigotes, with similar advantages and caveats to those previously discussed for *Leishmania*. In terms of tools for measuring the trypanocidal effect of the compounds, screening systems have evolved from manual microscopic counting of parasite growth; the use of colorimetric substrates (e.g., chlorophenol-red-β-D-galactopyranoside); bioluminescent (e.g., parasites expressing the firefly luciferase) and fluorescent reporters (e.g., tdTomato-expressing lines); and high-content imaging approaches that do not require the incorporation of any reporter molecule [35,61,62]. Engel et al. (2010) developed a cell-based HTS assay that can be used with untransfected *T. cruzi* isolates and host cells that can simultaneously measure efficacy against the parasite and host cell toxicity. This approach was used to screen a library of 909 bioactive compounds, leading to the identification of 55 hits [63]. Using NIH-3T3 fibroblasts infected with a recombinant *T. cruzi* strain expressing beta-galactosidase as an intracellular reporter, Peña et al. (2015) screened the GlaxoSmithKline diversity set of 1.8 million compounds. A total of 2310 compounds were identified with great potency against *T. cruzi* (pIC_{50} > 5) and a selectivity index > 10 [64]. The resulting lead compounds were further validated by Alonso-Padilla et al. (2015) using a novel, highly reproducible, high-content, high-throughput assay using myoblasts [65]. De Rycker et al. (2016) developed a new hit discovery screening cascade designed combining a primary imaging-based assay followed by newly developed and appropriately scaled secondary assays to predict the cidality and rate-of-kill of the compounds. This cascade was used to profile the SelleckChem set (421 FDA-approved drugs) and the NIH Clinical Collection set (727 compounds that have been used in clinical trials), leading to the identification of several known clinical compounds as candidates for a repurposing strategy for Chagas disease [66]. This cascade was further improved by the inclusion of three distinct in vitro assays: the slow replicating/cycling strain potency assay, the trypomastigote assay, and the extended duration washout assay [67]. Recently, Bernatchez et al. (2020) screened 7680 compounds from the Repurposing, Focused Rescue, and Accelerated Medchem library, and identified seven lead compounds with potent in vitro activity against *T. cruzi* and good therapeutic index [68].

3. Animal Models in Drug Discovery and Development against Trypanosomatids

Animal models are expected to mimic the pathophysiological features and immunological responses observed in the human host. A good experimental model for parasitic infections allows estimation of the specificity of drug action in relation to absorption, distribution, metabolism, excretion, and toxicity. Experimental models like rodents, dogs, and monkeys have been developed in order to identify and profile novel drugs against trypanosomatids, though mimicking the pathogenesis of disease and the impact of natural transmission is difficult to emulate under laboratory conditions [69]. The genotypic feature of laboratory models also augments hindrances due to restricted genotypic variations compared to infection with wild varieties. Hence, animal models developed and practiced for *T. brucei*, *T. cruzi*, or *Leishmania* infections do not accurately reproduce the consequences in human hosts, though several of these models exhibit an acceptable degree of proficiency for drug and vaccine development, particularly for the in vivo testing of trial compounds and libraries [70]. Important among them are BALB/c mice and Syrian golden hamster (primary tests), dogs (secondary tests), and monkeys (tertiary screens) as models for VL alongside athymic and SCID mice, which serve as a model for the treatment of VL in immunosuppressed conditions [69,71]. The genetic basis of the degree of susceptibility of mice to *Leishmania* has been linked to the Sc11 1a1 locus, based on which the outcome can be either self-healing or fatal [72]. The widely used (BALB/c and C57BL/6) mice breeds are mutated in the locus. In BALB/c mice, the immunopathology does not actually resemble human infection; instead, after around four weeks of infection, a strong Th1 response results in clearance of the parasite from the liver [72]. BALB/c is also highly susceptible to infection by *L. major*, with severe lesions and parasite-specific Th2 response with the enhanced expression of deactivating macrophage cytokines—particularly interleukin 4 (IL-4), interleukin 10 (IL-10), and transforming growth factor-β (TGF-β) [73]. On the contrary, the majority of inbred mouse strains like CBA and C57BL/6 are resistant to infection by *L. major*, and lesions spontaneously heal in 10–12 weeks [73]. The situation is bit different for the new-world *L. mexicana* and *L. amazonensis*, for which BALB/c, C57BL/6, and CBA/J mice are susceptible to infection [70]. On the contrary, for *L. braziliensis*, majority of mouse strains are resistant as the parasite does not induce protective Th2 response in the host [74]. However, for BALB/c, co-administration with salivary gland exudates of the vector promotes infection by altering the cytokine milieu [74]. Genetic susceptibility studies identified that the scl-1 locus controls the healing versus non-healing responses to *L. major* and the scl-2 is ascribed to the development of *L. mexicana*-induced cutaneous lesions. Around 30 loci have been identified as involved in the complex control of cutaneous leishmaniasis (CL) in mice [75]. BALB/c mice have been exploited as a model to profile metabolic changes during infection by *T. brucei* [72]. Mouse models including BALB/c, SCID, C57BL/6, and CH3 are the most widely used animal models in Chagas disease research [76]. However, the outcome was different in terms of Chagasic cardiomyopathy based on the strain of parasite and mouse line chosen for infection. Among alternative rodent models, guinea pigs have also been used as a model for experimental *T. cruzi* infection for acute and chronic Chagas disease [77–79]. For *T. brucei*, Wistar rats have been exploited as a preclinical model for HAT-associated cardiomyopathy [80]. The cotton rat (*Sigmodon hispidus*) represents one of the most susceptible animal hosts for *L. donovani*. The infection remains for 3–4 months, and after the appearance of initial clinical signs, the disease progresses rapidly, leading to death of the host [81]. Among various hamster species that are susceptible to *L. donovani*, the Syrian golden hamster (*Mesocricetus auratus*) represents a good model for VL with synchronous infection in the liver and spleen that culminates into a chronic non-cure infection with immune responses similar to human VL [81]. However, optimization of this model for drug screening is also effectively achieved through an ex-vivo splenic explant [45]. The only model that shows true potential for the evaluation of potential drugs targeting *L. braziliensis*, with low virulence for mice, is the golden hamster. Disease progression can be monitored over longer periods due to the chronic nature of the disease in the hamster [82]. For *L. infantum*, dogs are the natural reservoir. The natural infection of domestic dogs with *L. braziliensis*, *L. panamensis* and *L. mexicana* has been reported in Latin America. The infection of dogs with *L. infantum* is a pertinent laboratory model because it reproduces the natural infection with considerable similarity to human infections. The use of dogs as experimental models to study VL actually elucidated the role of immune cells, cytokines, and signaling events mediating immune response during *Leishmania* infection, offering crucial clues for developing immunotherapy. Canine models of *L. mexicana* infection have been established with Beagle dogs [83].

Non-human primates are exploited as the first experimental model for evaluating safety and efficacy of drugs and vaccines. For VL, *Macaca* sp. developed low and/or inconsistent infections. However, *Presbytis entellus* showed substantial susceptibility to hamster-derived amastigotes of *L. donovani* with all the clinical-immunopathological features as observed in kala-azar characterized by consistent and progressive acute fatal infection, leading to death between 110 to 150 days post-infection. The *L. major*–rhesus monkey model emulates self-limiting human cutaneous leishmaniasis that resolves within three months [73,84,85]. The model also shows promise in deciphering the intricacies of immune function and granuloma formation by *L. braziliensis*, rendering it as a useful model for drug and vaccine development [86]. Non-human primates have been explored as models for Chagas disease, but in most of the studied cases only a limited number of animals develop typical cardiomyopathy signifying *T. cruzi* infection [87]. Recent analysis of circulating leukocytes from naturally infected non-human primate cynomolgus macaque revealed a strong resemblance with immune-pathological biomarkers of Chagas disease in humans, projecting the prospect of this model in preclinical studies for new drugs for Chagas disease [87].

4. Cheminformatics in Drug Discovery

After the identification of several important and prospective drug targets like reductases of folate metabolic cascade, kinases, cAMP-phosphodiesterases, and enzymes for trypanothione synthesis and purine salvage, cheminformatics studies to identify structure–activity relationships for the design of optimized compounds have been prioritized. In recent times, combinatorial chemistry and HTS have enabled tests on large compound libraries, which encompass a significant chemical diversity, in short time scales [88,89]. Cheminformatics tools are broadly classified into structure- and ligand-based drug design (SBDD and LBDD) approaches. SBDD exploits the 3D coordinates of target structures for favorable ligand interactions. Potential ligands can be screened by molecular docking or structure-based virtual screening of potential ligands. High-affinity interactions between the binding site and ligand can be achieved by exploring binding site attributes like electronic distribution. The establishment of structure–activity relationships (SARs) can be achieved through experiments to further optimize ligand–receptor affinity [90]. Alternatively, ligand-based drug design studies can be performed without the receptor 3D structure. Instead, they require information on the structure, activity, and molecular properties of small molecules [91]. Chemometric models based on quantitative structure–activity and structure–property relationships (QSAR and QSPR, respectively) can be built in order to identify molecular descriptors complementing the target property [92].

Pteridine reductase (PTR1), an enzyme of the folate biosynthetic pathway, was one of the prominent candidates for drug targeting since no homologue of that protein is detectable in mammalian hosts. The crystal structure of LmjPTR1 was determined [93]. Implementing an SBDD strategy, Rasid et al. (2016) identified a number of dihydropyrimidine- and chalcone-based inhibitors for *Leishmania* PTR1 [94]. Using homology model for type 2 NADH dehydrogenase, Stevanovic et al. (2018) conducted a pharmacophore-based virtual screening to identify several hits [95]. A 6-methoxy-quinalidine derivative showed potential inhibition of the recombinant protein and inhibition of amastigotes with an EC_{50} of nanomolar range. Tryparedoxin peroxidase, a parasite-specific enzyme and a key component for parasitic survival under macrophage oxidative stress, has been considered as a key drug target. By performing deep molecular docking analysis with the crystal structure of PTR1 from *L. major*, a series of *N,N*-disubstituted 3-aminomethyl quinolones was identified which might serve as a worthy starting point for a suitable drug. SAR analysis of benzimidazole inhibitors against cysteine proteases cruzain and rhodesain from *T. brucei* and *T. cruzi*, followed by detailed cheminformatic analysis was conducted to find scaffold novelty and favorable physicochemical properties. Distinct endopeptidases like cathepsin-L-like CPB2.8 have emerged as exploitable drug targets in leishmaniasis. De Luca et al. (2018) identified a group of substituted benzimidazole derivatives that displayed strong (nanomolar) affinity for the protease from *L. mexicana* [96]. One of the compounds demonstrated a good bioavailability profile with ADMET analysis, implying it is a good future drug candidate. Carbonic anhydrases (CAs) have recently been identified from trypanosomatids. Cheminformatics analysis targeting this enzyme identified *N*-nitrosulfonamides as prospective inhibitors for CA from *Trypanosoma* and *Leishmania* over mammalian homologues. Being comparable with existing drugs in terms of EC_{50} and cytotoxicity, these compounds might serve as interesting leads for drug development.

Using the ligand-based approach, aminophosphonates have been studied with QSAR modelling [97]. The authors took the gathered data for the whole compound series to build comparative molecular field analysis (CoMFA) models that suggested that several modifications can enhance the anti-leishmanial potential of α–aminophosphonates. Similar approaches identified 1,2,3-triazole and thiosemicarbazone hybrids and tetrahydro-β carboline derivatives as candidate anti-leishmanial drugs [98]. Novel quinazoline and arylimidamide derivatives have been identified using 3D QSAR-based analysis against *T. cruzi* [99]. The structure-guided discovery of a compound (compound 7) from the pyrazolopyrimidine series against a known protein kinase scaffold identified *Leishmania* CDK12 as a strong candidate for drug discovery. Structural studies combined to resistance mechanism analysis confirmed CDK12 as a specific target for the molecule [99]. With satisfactory specificity as well as pharmacokinetic and toxicological properties, the compound has been declared a preclinical candidate, suggesting cheminformatics can indeed boost systematic approaches to discover new drugs against trypanosomatids [99].

5. Quiescence, a Double-Edged Sword in the Quest of New Trypanocidal Drugs

Dormancy or persister cell formation is an evolutionarily conserved adaptive mechanism for stress tolerance for bacterial pathogens. Persister cell development is often associated with the development of a subset of a population that is metabolically quiescent and hence cannot be intervened by drug treatment [100]. Such an adaptation enables the parasite to survive under immunological stress and drug exposure, reverting to normal proliferative mode once the stresses disappear. Such conditions are well exemplified by the latent infection of *Mycobacterium tuberculosis* which can persist for the entire lifespan in a metabolically dormant state [101]. Similar metabolic diversions from proliferative to dormant state are observed in eukaryotic pathogens including fungal and parasitic protozoan infections [102]. The hypnozoite liver stages of *Plasmodium*, often associated with relapse of infection even years after successful therapeutic clearance, is one such persister-like stage for *Plasmodium vivax* [103]. For trypanosomatids, semi-quiescence to quiescence have been detected for intracellular forms of several species of *Leishmania* and in *T. cruzi* [102]. Persister formation is particularly relevant clinically for *Leishmania*, as relapsing conditions like post-kala-azar dermal leishmaniasis (PKDL) occurring several years after treatment for visceral leishmaniasis and leishmaniasis recidivans occurring after the treatment of cutaneous leishmaniasis emerge from possible metabolically distinct parasites that circumvent drug treatment due to dormancy without acquiring resistance by signature genetic alterations [104]. Despite its clinical significance, there has been a lack of concerted effort to study persister development in trypanosomatids due to technical constraints including the labelling of quiescent cells to distinguish them from the normally proliferating population. In 2015, a detailed identification and characterization of the semi-quiescent physiological state was reported in *L. mexicana* intracellular amastigotes in infected BALB/c non-healing lesions with a prolific increase in doubling time to ~12 days compared to ~4 days in ex-vivo macrophage infections [105]. The semi-quiescent metabolic state was also characterized by low rates of transcription and protein turnover that is distinct from stationary phase or metacyclic promastigotes, and is possibly a response to complex growth restriction in the intracellular microenvironment in granulomas. They identified two distinct macrophage populations, one with ~100 cells and the other with an average of ~400 intracellular amastigotes, suggesting the existence of two distinct metabolic amastigote varieties. *L. mexicana* amastigotes are intrinsically more resistant to nitric oxide and build up large communal phagolysosomes, while *L. major* infection is eventually controlled by an adaptive Th1 immune response requiring inducible NOS (iNOS) [105]. Mandell et al. (2015) identified a definite fraction of amastigotes with barely detectable replication in a C57BL/6J mouse model of cutaneous *L. major* infection. This population was observed to harbor in less-infected macrophages and constituted almost 39% of amastigotes under the persistent infection condition, while a second subset of amastigotes retained the ability to replicate with a doubling time of around 60 h [106]. *L. major* lacking the Golgi GDP-mannose transporter required for lipophosphoglycan synthesis encoded by LPG2 (lpg2-) persist in the absence of pathology, and in mouse infections this knocked-out line attained a persister-like feature immediately after infection [106]. *L. braziliensis* amastigotes (both axenic and intracellular) bear characteristic features of quiescence, with a radical reduction of (i) the kDNA mini-circle abundance, (ii) the intracellular ATP level, (iii) the ribosomal components, and (iv) total RNA and protein levels [107]. The untargeted metabolomic profile revealed the significant depletion of amino acids, polyamines, and trypanothione, with increases in ergosterol and cholesterol biosynthesis. Dormancy attains further relevance for trypanosomatid infection, as regimens including short-term therapy of even 60 days for *T. cruzi* infection is not related to resistance development, and the parasite possibly alleviates drug-mediated clearance by adopting quiescence. In fact, in *T. cruzi*, non-proliferating amastigotes develop both in vitro and in vivo models of infection. *T. cruzi* amastigotes regularly and spontaneously cease replication and become non-responsive to effective trypanocidal drugs like benznidazole and nifurtimox [108]. One or two such dormant parasites are detectable in each infected cell after treatment. Such dormant parasites reinitiate proliferation after drug withdrawal. Exploring the intricacies of the alteration of physiological status for intracellular amastigotes in infected tissues by proteomic or transcriptomic approaches is impaired by the paucity of enrichment protocols. Each of these studies adopted various strategies to characterize and label persister cells. One such strategy exploited ^{2}H$_2$O labelling for determining DNA, RNA, protein, and membrane lipids. The in vitro deuterium labelling of deoxyribose could be achieved for promastigotes by maintaining 5% ^{2}H$_2$O in medium, and for the in vivo labelling of amastigotes, 5% ^{2}H$_2$O in the body water was established by providing mice with a bolus of 100% ^{2}H$_2$O followed by inclusion of 9% ^{2}H$_2$O in the drinking water for up to several months [105]. Differential labelling for replicative and non-replicative amastigotes is achieved with CellTrace Violet or CellTracker Red. After a brief pulse, the stain is either diluted out during cell division (for replicative form) or remains at the initial pulse level (for non-replicating forms). This approach can be combined with a fluorescent (tdTomato) or luciferase expression system to track viable parasites [108]. The incorporation of thymidine analogues 5-ethynyl-2′-deoxyuridine and 5-bromo-2′-deoxyuridine has been implemented to differentiate replicative and non-replicative cells in *Leishmania* spp. and *T. cruzi* [108,109]. Each of these approaches has been effective in tracing persister cells. Active translation or ribosomal action utilizes 70% of the total ATP generated in a viable cell, and in quiescent cells translational activity is highly compromised, with a concomitant decrease in the number of active ribosomes (~5-fold reduction in dormant compared to normal metabolic state). Hence, the reduced transcription of rDNA loci serves as a marker for quiescence and rDNA loci are part of a rare genomic landscape in trypanosomatids, which is regulated by a definite transcription factor [110]. In this context, the expression of the GFP gene under the 18S ribosomal DNA locus has been implemented as a biosensor for quiescence in laboratory and clinical strains of *L. braziliensis* and *L. mexicana*, and reduction of GFP expression was compatible with BrdU uptake analysis in vitro. With this approach, a superior FACS quantitative approach for persisters could be devised for recording quiescence development in mice (BALB/c) or hamsters (LVG Golden Syrian Hamster) models [109]. The study provided a clearer idea about metabolic diversity in amastigotes with the coexistence of shallow and deep quiescent stages. Quiescence is crucial for subclinical infections with its potential role in drug tolerance, and quiescent cells serve as reservoirs for transmission and elicit a protective response against subsequent infections in trypanosomatids, which warrants additional exploration [106]. The development of novel assay methods combined with identification of strategies to combat dormancy or exploit it in developing immunization strategies might expedite the success of elimination programs against trypanosomatid parasites.

6. Cytology-Driven MoA Profiling

In the last few years, we have witnessed an increase in the number of scientific reports on new potential drug candidates to treat leishmaniases and trypanosomiases. However, the vast majority lack insights or detailed mechanism of action evidence supporting further drug development and clinical trials. In this scenario, cell-based assays offer the contextualized relevance and complexity of living cells to track drug discovery approaches, especially when considering unicellular parasites. Kinetoplastids are classified in this category due to the presence of a kinetoplast—a dense structure made by DNA (kDNA) within their unique mitochondria. Therefore, mitochondrial function monitoring can be applied in order to provide hints on the MoA of drug candidates in the drug discovery pipeline. Cellular bioenergetics analysis based on extracellular flux can phenotypically characterize mitochondrial function and define the energetic status of aerobic and glycolytic metabolism, defining a range from quiescent to energetic profiling [111,112]. This approach was used to monitor oxygen consumption (mitochondrial respiration) vs. medium acidification rate (glycolysis) in *L. infantum* to metabolically characterize SbR mutants and evaluate the oxidative role of gold(I) complexes as metallodrug candidates to treat leishmaniasis [44]. This approach was also considered using host cells experimentally infected with *T. cruzi* intracellular amastigotes, monitoring not only the parasite's metabolism, but mimicking the natural conditions considering the context of endogenous conditions of infected cells [113]. These assays were performed on a Seahorse Extracellular Flux Analyzer, XF series (Agilent), and were initially used to monitor basal mitochondrial metabolism in *T. cruzi*, which is useful for drug screening purposes [114–117]. Microscopic imaging using cell-permeant mitochondrion-selective dyes such as MitoTracker or cell permeant acidotropic fluorophores like LysoTracker can be used to highlight ultrastructural alterations in essential organelles to make inferences about drug action and target elucidation by functional approaches [118]. These dyes can be used in high-content analysis approaches that have been shown as an alternative to monitor not only anti-parasitic drug action but also concomitant host toxicity analysis in the same assay for drug screening purposes [119]. Despite the above-mentioned fluorescent gene reporters, kDNA can be labelled to monitor cell replication for indirect drug activity measurement. The terminal deoxynucleotidyl transferase dUTP nick end labelling (TUNEL) technique allows the specific tagging of blunt DNA ends—a common feature in programmed cell death in mammalian cells. Conventional programmed cell death is not biochemically the same in trypanosomatids, and TUNEL signals are undetectable in trypanosome nuclei (genomic DNA). However, 25% of control (wild type, untreated) cells were reported to have TUNEL-positive kDNA. Treatments with eflornithine, nifurtimox, or melarsoprol did not change TUNEL signal, but pentamidine or suramin exposition reduced it, as an evidence of loss of kDNA following the latter treatments in a cell-cycle-dependent manner [120,121]. Trypanosomatids present closed mitosis (chromosomal condensation and segregation is maintained inside the nucleus during division), and the segregation of their single mitochondrial genome (kinetoplast) can be easily monitored by fluorescent microscopy during cell division in the presence of 4′,6′-diamidino-2-phenylindole (DAPI, a DNA-intercalating dye. This feature can be tracked under drug treatment to make inferences about mitosis or cytokinesis impairment. For example, non-treated *T. brucei* presented ~80% of cells with 1 nucleus and 1 kDNA pattern (1n1k), equivalent to G1 and S phase; ~15% were 1n2k (primarily G2 phase) and 5% were 2n2k (post mitosis). Suramin treatment switched profiling and 79% of the cells accumulated in >2n, indicating the blocking of cytokinesis in *T. brucei* [121]. A similar approach can be afforded using propidium iodide followed by flow cytometry analysis. Melarsoprol-treated *T. brucei* led to the accumulation of G2/M phase from 51% to 83%, indicating increasing replication but unsegregated nuclear genome, as an evidence of mitosis inhibition [121]. Genomic plasticity is a key factor in trypanosomatids, and plays an important role that must be taken into account when developing or testing new anti-trypanosomal drugs. In this context, DNA repair mechanisms are always being recruited, especially under stressful microenvironments like drug pressure. The enzyme uracil DNA glycosylase (UNG) participates in the DNA base excision repair (BER) pathway, and was found upregulated in *L. donovani* exposed to amphotericin B or sodium antimony gluconate. Curiously, drug-resistant clinical isolates of *L. donovani* from VL patients presented higher UNG expression [122], suggesting that LdUNG plays a key role in BER, conferring moderate resistance to oxidants; this opens new avenues as a potential target for combination therapy against leishmaniasis. The adoption of drug discovery strategies against trypanosomatids must consider drug-resistance studies and the evolutionary role of DNA repair in this context. Antibodies can be used to track specific markers of DNA damage in eukaryotes such as the phosphorylation of threonine 130 at the C terminus of histone γH2A in *T. brucei*, which is associated with a delay in S and G2 phases of the cell cycle [123].

7. Genome-Wide Approaches in Target and Resistance (Resistomics)

Functional genomics approaches are useful for identifying or validating a given drug target. This relies on strategies or tools that can be combined together with studies on drug resistance mechanisms to find clues for drug discovery. For example, the in vitro selection of drug-resistant parasites, followed by whole-genome or transcriptomic sequencing could unveil targets or signatures associated with the drug used for resistance selection. This was the case of compound 7, DDD853651/GSK3186899, selected from a chemical series of pyrazolopyrimidine scaffolds, active against *T. brucei* and used to select resistant *L. donovani* mutants as a strategy to understand the MoA and to prospect potential pathways or drug targets [99]. Whole-genome sequencing of these drug-resistant parasites revealed a single homozygous non-synonymous mutation in CRK12 (cyclin-dependent kinase 12 or cdc-2-related kinase 12), leading to a Gly 572 to Asp in the predicted catalytic domain of the enzyme, impairing electrostatic interactions and causing resistance to the pyrazolopyrimidine [99]. In this case, the resistance mechanism identification was useful to pinpoint the drug target involved in drug action. Among trypanosomatids, *T. cruzi* and *Leishmania* species (the latter belonging to the *L. (Leishmania)* subgenus) lack one or more components of the RNA interference (RNAi) machinery. However, knockdown by RNAi manipulations can be performed in *T. brucei* and *L. (Viannia)* subgenus spp., a very useful functional genomic tool to validate and identify new drug targets [124]. Inspired by these biological features, Alsford et al. (2011) described a new technique called RIT-Seq (RNAi target sequencing), where *T. brucei* were transfected with a library of interfering RNAs able to silence >99% of the mRNA in the parasite. This was followed by culturing in the presence of drug pressure in which the recovered parasites had their enriched plasmids sequenced [125]. This functional cloning technique allowed a genome-scale knockdown profiling in which the decrease of a given gene product is selected as a phenotypical marker for surviving under a stressful condition. In this way, the mechanisms underlying selective drug action and resistance can be screened in a high-throughput genome-scale RNAi panel [126]. A phenotyping genome-scale RNAi screen revealed, for example, the involvement of aquaglyceroporin 2 (AQP2) in melarsoprol and pentamidine susceptibility in African trypanosomes [127,128]. Melarsoprol is an arsenic-based drug, and similar to antimony-based compounds against *Leishmania* parasites, is taken up through aquaglyceroporin 1, which was associated with antimony resistance by using a dominant negative functional cloning strategy using a cosmid library [129]. Cosmid libraries can also be applied to select gain-of-function genes associated with a given phenotype, where the screening is based on overexpressing libraries. This approach was used to confirm previous and pinpoint new drug resistance markers in *Leishmania* parasites—a technique called Cos-seq or cosmid-based functional screening coupled to next-generation sequencing [130,131]. The most recent brother of the X-Seq family is a technique called Mut-Seq, or chemical mutagenesis coupled to next-generation sequencing. In this case, "Darwinian hands play dice" leading to stochastic mutations that could be kept when important for parasite survival under stressful pressure. This was elegantly applied to study miltefosine and paromomycin resistance mechanisms in *Leishmania* parasites. After using Mut-Seq to identify new targets and validate the essential role of kinase CDPK1 on paromomycin resistance in *Leishmania* using CRISPR-Cas9, Bhattacharya et al. (2019) suggested that

Mut-Seq screening is powerful tool to explore networks of drug resistance since CDPK1 was also involved in antimony resistance in the parasite [132]. Genome-wide approaches are very useful for capturing the main picture, and thus for choosing the most prominent biochemical pathway involved in drug action/resistance. This is also true when applying the revolutionary technique of genome editing: Clustered Regularly Interspaced Short Palindromic Repeats (CRISPR), CRISPR-associated gene 9 (Cas9)—CRISPR-Cas9. Beneke et al. (2017) developed a CRISPR-Cas9-based toolkit for the high-throughput genome editing of kinetoplastids that was further validated in single or multiple targets [133–135]. We are however currently revisiting concepts and moving from genome-wide approaches in parasite populations (or clones) to single-cell-based strategies to better understand the plasticity of *Leishmania* parasites that harbor mosaic aneuploidy—a feature that has impairments in the way the parasite will respond or not to a given drug. Using a single-cell genomic sequencing method, Negreira et al. (2020) identified 128 different karyotypes in 1560 *L. donovani* promastigotes [136]. They highlight the fact that some karyotypes presented pre-existing adaptations to antimony-based drugs, supporting a hypothesis raised even before this hint [137,138]. This reveals how complex it is to predict or open new avenues on MoA studies in trypanosomatids, and reinforces the evolutionary adaptions that guaranteed the establishment of trypanosomatids since the early Cretaceous [139]. Finally, and despite recent advances in genomic methods, there is still a relative paucity of functional annotations for a large number of gene products for trypanosomes, especially when compared to mammalian systems. In fact, this could explain why target-based methods lag behind phenotypic approaches in drug development for these parasites.

8. Metabolomics in Drug Screening

Like in a crime scene, studying the past is also a feasible alternative to tracking drug action and target identification. Metabolomics refers to the measurement of small metabolite molecules to investigate metabolic pathways, here in the context of drug discovery or target identification. Metabolite profiles are useful fingerprints offering clues on therapeutics targets in trypanosomatids, and can also be performed in the host to select signatures or markers associated with the dynamics of host–parasite interaction [140–145]. Metabolomics can also be applied to the rational development of defined minimal culture medium for in vitro drug screening purposes against trypanosomatids. In this regard, untargeted semi-quantitative or targeted quantitative metabolomics was used to decipher the major nutritional requirements of *T. brucei* and define all needs, removing unnecessary nutrients and improving drug sensitivity in activity studies [146]. Drug MoA can also be indirectly investigated through metabolomics, even without clear evidence on parasite alterations. Benznidazole is a 2-nitroimidazole prodrug that needs to be reduced in order to exert anti-trypanosomal activity against *T. cruzi*. Although benznidazole-treated parasites were minimally altered compared to untreated counterparts, metabolites concerning benznidazole linked to thiols such as trypanothione, glutathione, and cysteine indicates the thiol binding capacity of benznidazole on acting by disturbing redox homeostasis, leading to parasite death [147]. The cell redox system has also classically been related to antimony and resistance in *Leishmania* parasites. Combining untargeted metabolomics for initial screening coupled to ^{13}C traceability assays, Rojo et al. (2015) confirmed and compiled multi-target metabolic alterations not only in redox, but also in detoxification, biosynthetic processes and amino acid metabolism in *L. infantum*. Antimony-resistant parasites presented incremented proline and glutamate, supporting previous reports on high levels of glycolytic markers in resistant *Leishmania* as revealed by proteomics [148,149]. In summary, metabolomics approaches helped to identify MoA or resistance of several anti-trypanosomal drugs such as eflornithine or halogenated pyrimidines against *T. brucei*; miltefosine and antimony against *Leishmania* parasites [150]. Drug targets can also be mined in trypanosomatids by metabolomics pathway analysis using in silico approaches, as a predictive way based on pathway annotation and searching for analogous or specific enzymes [151].

9. Theranostic Approaches

The term theranostic, derived from the fusion of the words therapeutic and diagnostic, is here used to define strategies designed for diagnostic purposes that also act as therapeutic agents. Dual-function molecules or smart probes can be adapted for both parasite detection/identification and anti-trypanosomatid activity. This combination of diagnosis and therapeutics is still a growing field and there are very few studies on trypanosomatids. A group headed by professors Eduardo Coelho and Luiz Ricardo Goulart in Brazil proposed the use of phage display—a high-throughput proteomic technology to generate and screen peptides and antibodies—for the serodiagnosis and prevention of leishmaniasis as a theranostic approach [152]. Using this approach, the team identified a β-tubulin from *L. infantum* that was highly antigenic and immunogenic, presenting good performance on diagnostic efficacy and eliciting Th1 response in vitro with high IFN-γ and low IL-10 levels [153]. Recently, Singh et al. (2019) reviewed the literature on nanomedicine-based approaches to circumvent leishmaniasis and concluded that much progress was made in the field reaching considerable milestones on VL nanomedicine, but translational research is needed for the coming decade for developing effective theranostic solutions [154]. Thus, many current alternatives such as liposomes, nanoemulsions, niosomes, nanodiscs, solid lipids nanoparticles, quantum dots, nanotubes, polymer conjugates, and inorganic compounds could be applied to clinical settings.

10. Case Study: Proteasomal Inhibitors against *Leishmania*

Proteasome targeted inhibitor developments by Khare et al. (2016) and Wyllie et al. (2018) are among the few major break-throughs in the quest of safe, easily deliverable, and selective drugs against trypanosomatids in recent times [155,156]. Both studies targeted the identification of a common target for intervention for *Leishmania* spp., *T. cruzi*, and *T. brucei* spp. Khare et al. began their screen with a library of 3 million compounds against the three pathogens, and identified an azabenzoxazole (GNF5343) that was effective against the three [155]. A number of substitutions leading to a less-toxic version GNF6702 further optimized the compound. In mouse model of VL and CL, with oral delivery of 10 mg kg^{-1} for eight days, GNF6702 caused significant amelioration of liver parasitic burden. Similarly, it displayed prolific attenuation of parasite load in mouse models of Chagas disease and HAT. For leishmaniasis, Chagas disease, and HAT, the activities are comparable to the approved drugs miltefosine, benznidazole, and diminazene aceturate, respectively. In fact, for HAT it performed better than the in-use diminazene aceturate in terms of diminishing parasitic infection in brain. The primary mechanism of parasite growth inhibition by the compound series was the selective inhibition of the proteasome chymotrypsin-like activity. For analyzing resistance against the drug, they raised mutants against an early version of the drug, which showed 40-fold lower susceptibility to the drug. The phenotype was attributed to a homozygous mutation in the proteasome β4 subunit (PSMB4I29M/I29M) and a heterozygous mutation (PSMB4wt/F24L). These mutations led to reduced susceptibility to inhibition by the drug. Interestingly, the chymotrypsic catalytic center is hosted by a β5 subunit and a β4 subunit in close contact with a β5 subunit forming a plausible binding pocket for the drug. The study suggested proteasomal subunits as a selective target for the development of a common chemical scaffold against trypanosomatids. In concordance, an independent screen by Wylie et al. identified and studied a second candidate GSK3494245/DDD01305143/compound 8 [156]. The precursor of the compound was developed by scaffold hopping and substitutions from a basic component identified by a phenotypic screen of around 16,000 molecules against *T. cruzi*, and demonstrated efficacy against intra-macrophage amastigotes of *L. donovani*. The compound showed good in vitro metabolic stability (CLint = 0.8 mL min^{-1} g^{-1}) and selectivity over mammalian cells. They further addressed the compound in terms of duration of treatment by rate-of-kill assay that showed that induction of cell death is achievable within 72 h at nanomolar concentration range. Pharmacokinetic profiling for bioavailability and distribution revealed that it can be orally dosed to reach efficacious levels in a range of preclinical species, including mouse, rat, and dog. Moreover, virtually no significant safety or tolerability liabilities were detected by Ames test and in mouse lymphoma cells. For identifying the mechanism of action

for the drug, the authors preliminarily adopted RIT-seq technology [125]. The study suggested that knock-down of nonessential genes of ubiquitination pathway rendered reduced sensitivity to the drug, pinpointing proteasome as the possible point of intervention for the drug. The generation of resistant mutants led to the identification of independent mutations in the β5 subunit (G197C and G197S). The mutants were cross-resistant to GNF6702. Both mutations affected proteasomal activity, as determined in vitro by UbiQ-018 label (a fluorescent label for proteasomal subunits), and the mutations resulted in insensitivity to GSK3494245 (compound 8). The proteasomal inhibitors caused cytological changes in *Leishmania* promastigotes with accumulation of vesicular structures and induced cell cycle arrest in G2/M phase. CryoEM of *L. tarentolae* proteasome in combination with compound 8 identified a number of residues from β4 and β5 subunits. Additionally, the selectivity of the drug for kinetoplastid proteasome over human proteasome could be attributed to a lack of hydrophobic interaction, as F24 in *L. tarentolae* corresponds to S23 in human and π-stacking interaction. Both works identified a suitable target for developing a common anti-trypanosomatid drug development and developed human-trial-ready molecules that precisely target chymotrypsin-like protease action of kinetoplastid proteasome without affecting the human orthologues.

11. Perspectives and Concluding Remarks

At present, drugs for treating trypanosomatid diseases are far from ideal due to host toxicity, elevated cost, limited access, and increasing rates of drug resistance. Therefore, new oral, safe, short-course drugs are urgently needed. Moreover, these new drugs have to be safe and effective enough to treat patients who are asymptomatic, as well as patients who develop secondary conditions such as post-kala-azar dermal leishmaniasis [14].

In the vast majority of cases, trypanocidal agents are out of the scope of interest of the pharmaceutical industry, mainly because it is unclear how to make a profit by selling them. This situation is also becoming more frequent in the case of the discovery and development of antibiotics [157]. For this reason, drug-discovery research of novel trypanocidal compounds has been traditionally fueled by non-profit and governmental organizations. However, in the last decade, some pharmaceutical companies have become more engaged and have joined forces with academia as well as governmental and non-profit organizations to tackle NTDs. This is the case of The Drugs for Neglected Diseases initiative (DND*i*), a nonprofit research and development organization founded by Médecins sans Frontières (MSF), among other public–private partners, which has campaigned for change since 2004 to raise awareness of the trypanosomatids crisis among key policy- and decision-makers [158]. DND*i* performs high-throughput untargeted screenings of novel-drugs libraries for trypanosomatids in addition to identifying new drug candidates using targeted compounds from repurposing libraries. Since its creation, DND*i* has already provided seven treatments: ASAQ and ASMQ (two fixed-dose antimalarials), nifurtimox-eflornithine combination therapy for late-stage sleeping sickness, sodium stibogluconate and paromomycin (SSG + PM) combination therapy for VL in Africa, a set of combination therapies for VL in Asia, and a pediatric dosage form of benznidazole for Chagas disease. While combination therapies will improve the efficacy of the treatment and reduce the emergence of drug-resistant strains, currently we do not have enough effective molecules to guarantee durable therapeutic strategies. Consequently, more efforts should be deployed to discover and exploit novel families of trypanocidal drugs (with different modes of action), which could be rapidly integrated in combinatory treatments, or kept as drugs of last resort when current combinations fail.

An important bottleneck in the discovery and development of new trypanocidal drugs is the lack of well-validated molecular targets, which has traditionally hindered the use of classic target-based approaches (usually applied to the discovery of antibiotics) in the drug-discovery cascade. While it is true that this has fostered the development and implementation of sophisticated phenotypic in vitro assays, these assays encompass major challenges specific to each parasite (e.g., drugs must be active in the phagolysosome milieu when treating patients infected with *Leishmania*, drugs for HAT have to cross the blood–brain barrier, etc.). Moreover, once a hit has been identified in a phenotypic screen, different approaches (e.g., genomics and proteomics) should be deployed to identify the specific target(s), mode-of-action of the compound, and to predict any potential mechanism of drug resistance deployed by the parasite. This information is crucial to guarantee a rational and successful optimization of the hit, and serves to develop novel target-based drug discovery cascades.

Another major challenge in drug discovery for trypanosomatids is the lack of well-defined standards/criteria (e.g., strain, culture media, incubation times, etc.) for the selection and validation of hit compounds, which sometimes leads to opposing results between different research teams. Among these criteria, one of the critical ones is the selection of the most relevant animal model that is able to mimic the pathophysiological features and immunological responses observed in human hosts (e.g., BALB/c mice vs. Syrian golden hamsters as models for *L. donovani* and *L. infantum*; acute vs. chronic models for Chagas disease, etc.).

Moreover, in order to guarantee the success of drug discovery/repositioning in the fight against trypanosomatids, we have to generate high-quality data in many endemic countries (including field strains, drug-resistant strains, etc.), and to do so, we have to effectively increase the engagement of endemic countries in the R&D process [159].

New powerful and robust in vitro, in vivo, and in silico technologies have emerged in the last ten years. Moreover, we now have a more refined knowledge of the biology of these parasites, as well as the unprecedented ability to surgically manipulate trypanosomatids genome. The optimal use of these tools and knowledge will undoubtedly accelerate current drug discovery cascades, leading to the delivery of satisfactory treatment options for neglected patients with trypanosomatid infections.

Author Contributions: Conceptualization, C.F.-P., A.B. and R.L.d.M.-N.; writing—original draft preparation, C.F.-P., A.B., R.L.d.M.-N. and A.C.; writing—review and editing, C.F.-P., A.B. and R.L.d.M.-N.; project administration, C.F.-P.; funding acquisition, C.F.-P. All authors have read and agreed to the published version of the manuscript.

Funding: This work was supported by a Natural Sciences and Engineering Research Council of Canada Discovery Grant RGPIN-2017-04480 and by the Canada foundation for Innovation (www.innovation.ca), grant number 37324; both awarded to CFP. AC is supported by Fonds de Recherche du Québec-Nature et Technologies (FRQNT) scholarship program. RMN is a CNPq (Brazilian National Council for Scientific and Technological Development) Research Fellow (#310640/2017-2).

Acknowledgments: The authors want to thank Aida Mínguez-Menéndez for her help with the conception of the figures.

Conflicts of Interest: The authors declare no conflicts of interest.

References

1. Fernandez-Prada, C.; Minguez-Menendez, A.; Pena, J.; Tunes, L.G.; Pires, D.E.V.; Monte-Neto, R. Repurposed molecules: A new hope in tackling neglected infectious diseases. In *Silico Drug Design 1st Edition: Repurposing Techniques and Methodologies*; Roy, K., Ed.; Elsevier: Amsterdam, The Netherlands, 2019; pp. 119–160.

2. Arce, A.; Estirado, A.; Ordobas, M.; Sevilla, S.; Garcia, N.; Moratilla, L.; de la Fuente, S.; Martinez, A.M.; Perez, A.M.; Aranguez, E.; et al. Re-emergence of leishmaniasis in Spain: Community outbreak in Madrid, Spain, 2009 to 2012. *Eurosurveillance* **2013**, *18*, 20546. [CrossRef]

3. Uranw, S.; Hasker, E.; Roy, L.; Meheus, F.; Das, M.L.; Bhattarai, N.R.; Rijal, S.; Boelaert, M. An outbreak investigation of visceral leishmaniasis among residents of Dharan town, eastern Nepal, evidence for urban transmission of *Leishmania donovani*. *BMC Infect. Dis.* **2013**, *13*, 21. [CrossRef] [PubMed]

4. Abubakar, A.; Ruiz-Postigo, J.A.; Pita, J.; Lado, M.; Ben-Ismail, R.; Argaw, D.; Alvar, J. Visceral leishmaniasis outbreak in South Sudan 2009-2012: Epidemiological assessment and impact of a multisectoral response. *PLoS Negl. Trop Dis.* **2014**, *8*, e2720. [CrossRef] [PubMed]

5. Babuadze, G.; Alvar, J.; Argaw, D.; de Koning, H.P.; Iosava, M.; Kekelidze, M.; Tsertsvadze, N.; Tsereteli, D.; Chakhunashvili, G.; Mamatsashvili, T.; et al. Epidemiology of visceral leishmaniasis in Georgia. *PLoS Negl. Trop Dis.* **2014**, *8*, e2725. [CrossRef] [PubMed]

6. Lidani, K.C.F.; Andrade, F.A.; Bavia, L.; Damasceno, F.S.; Beltrame, M.H.; Messias-Reason, I.J.; Sandri, T.L. Chagas disease: From discovery to a worldwide health problem. *Front Pub. Health* **2019**, *7*, 166. [CrossRef]

7. Lee, B.Y.; Bacon, K.M.; Bottazzi, M.E.; Hotez, P.J. Global economic burden of Chagas disease: A computational simulation model. *Lancet Infect. Dis.* **2013**, *13*, 342–348. [CrossRef]

8. Ribeiro, V.; Dias, N.; Paiva, T.; Hagstrom-Bex, L.; Nitz, N.; Pratesi, R.; Hecht, M. Current trends in the pharmacological management of Chagas disease. *Int. J. Parasitol. Drugs Drug Resist.* **2019**, *12*, 7–17. [CrossRef]

9. Bern, C.; Montgomery, S.P.; Herwaldt, B.L.; Rassi, A.; Marin-Neto, J.A.; Dantas, R.O.; Maguire, J.H.; Acquatella, H.; Morillo, C.; Kirchhoff, L.V.; et al. Evaluation and treatment of Chagas disease in the united states a systematic review. *JAMA* **2007**, *298*, 2171–2181. [CrossRef]

10. Meymandi, S.; Hernandez, S.; Park, S.; Sanchez, D.R.; Forsyth, C. Treatment of Chagas disease in the United States. *Curr. Treat Options Infect. Dis.* **2018**, *10*, 373–388. [CrossRef]

11. Viotti, R.; Alarcon de Noya, B.; Araujo-Jorge, T.; Grijalva, M.J.; Guhl, F.; Lopez, M.C.; Ramsey, J.M.; Ribeiro, I.; Schijman, A.G.; Sosa-Estani, S.; et al. Latin American network for Chagas disease, NHEPACHA. Towards a paradigm shift in the treatment of chronic Chagas disease. *Antimicrob. Agents Chemother.* **2014**, *58*, 635–639. [CrossRef]

12. Alsford, S.; Kelly, J.M.; Baker, N.; Horn, D. Genetic dissection of drug resistance in trypanosomes. *Parasitology* **2013**, *140*, 1478–1491. [CrossRef] [PubMed]

13. Wilkinson, S.R.; Kelly, J.M. Trypanocidal drugs: Mechanisms, resistance and new targets. *Expert Rev. Mol. Med.* **2009**, *11*, e31. [CrossRef] [PubMed]

14. Burza, S.; Croft, S.L.; Boelaert, M. Leishmaniasis. *Lancet* **2018**, *392*, 951–970. [CrossRef]

15. Hotez, P.J. Global urbanization and the neglected tropical diseases. *PLoS Negl. Trop Dis.* **2017**, *11*, e0005308. [CrossRef] [PubMed]

16. Booth, M. Climate change and the neglected tropical diseases. *Adv. Parasitol.* **2018**, *100*, 39–126.

17. Robertson, S.A.; Renslo, A.R. Drug discovery for neglected tropical diseases at the Sandler Center. *Future Med. Chem.* **2011**, *3*, 1279–1288. [CrossRef]

18. Berenstein, A.J.; Magarinos, M.P.; Chernomoretz, A.; Aguero, F. A multilayer network approach for guiding drug repositioning in neglected diseases. *PLoS Negl. Trop Dis.* **2016**, *10*, e0004300. [CrossRef]

19. DiMasi, J.A.; Hansen, R.W.; Grabowski, H.G. The price of innovation: New estimates of drug development costs. *J. Health Econ.* **2003**, *22*, 151–185. [CrossRef]

20. Chong, C.R.; Sullivan, D.J.J. New uses for old drugs. *Nature* **2007**, *448*, 645–646. [CrossRef]

21. Novac, N. Challenges and opportunities of drug repositioning. *Trends Pharm. Sci.* **2013**, *34*, 267–272. [CrossRef]

22. Zheng, W.; Sun, W.; Simeonov, A. Drug repurposing screens and synergistic drug-combinations for infectious diseases. *Br. J. Pharm.* **2018**, *175*, 181–191. [CrossRef] [PubMed]

23. Kaufer, A.; Ellis, J.; Stark, D.; Barratt, J. The evolution of trypanosomatid taxonomy. *Parasit. Vectors* **2017**, *10*, 287. [CrossRef] [PubMed]

24. Field, M.C.; Horn, D.; Fairlamb, A.H.; Ferguson, M.A.; Gray, D.W.; Read, K.D.; De Rycker, M.; Torrie, L.S.; Wyatt, P.G.; Wyllie, S.; et al. Anti-trypanosomatid drug discovery: An ongoing challenge and a continuing need. *Nat. Rev. Microbiol.* **2017**, *15*, 217–231. [CrossRef] [PubMed]

25. Lukes, J.; Butenko, A.; Hashimi, H.; Maslov, D.A.; Votypka, J.; Yurchenko, V. Trypanosomatids are much more than just trypanosomes: Clues from the expanded family tree. *Trends Parasitol.* **2018**, *34*, 466–480. [CrossRef] [PubMed]

26. Gupta, N.; Goyal, N.; Rastogi, A.K. *In vitro* cultivation and characterization of axenic amastigotes of *Leishmania*. *Trends Parasitol.* **2001**, *17*, 150–153. [CrossRef]

27. Moreira, E.S.; Soares, R.M.; Petrillo-Peixoto Mde, L. Glucantime susceptibility of *Leishmania* promastigotes under variable growth conditions. *Parasitol. Res.* **1995**, *81*, 291–295.

28. De Rycker, M.; Hallyburton, I.; Thomas, J.; Campbell, L.; Wyllie, S.; Joshi, D.; Cameron, S.; Gilbert, I.H.; Wyatt, P.G.; Frearson, J.A.; et al. Comparison of a high-throughput high-content intracellular *Leishmania donovani* assay with an axenic amastigote assay. *Antimicrob. Agents Chemother.* **2013**, *57*, 2913–2922. [CrossRef]

29. Vermeersch, M.; da Luz, R.I.; Tote, K.; Timmermans, J.P.; Cos, P.; Maes, L. In vitro susceptibilities of *Leishmania donovani* promastigote and amastigote stages to antileishmanial reference drugs: Practical relevance of stage-specific differences. *Antimicrob. Agents Chemother.* **2009**, *53*, 3855–3859. [CrossRef]

30. Dorlo, T.P.; Balasegaram, M.; Beijnen, J.H.; de Vries, P.J. Miltefosine: A review of its pharmacology and therapeutic efficacy in the treatment of leishmaniasis. *J. Antimicrob. Chemother.* **2012**, *67*, 2576–2597. [CrossRef]

31. Brochu, C.; Wang, J.; Roy, G.; Messier, N.; Wang, X.Y.; Saravia, N.G.; Ouellette, M. Antimony uptake systems in the protozoan parasite *Leishmania* and accumulation differences in antimony-resistant parasites. *Antimicrob. Agents Chemother.* **2003**, *47*, 3073–3079. [CrossRef]

32. Fernandez-Prada, C.; Vincent, I.M.; Brotherton, M.C.; Roberts, M.; Roy, G.; Rivas, L.; Leprohon, P.; Smith, T.K.; Ouellette, M. Different mutations in a P-type ATPase transporter in *Leishmania* parasites are associated with cross-resistance to two leading drugs by distinct mechanisms. *PLoS Negl. Trop Dis.* **2016**, *10*, e0005171. [CrossRef] [PubMed]

33. Li, Q.; Zhao, Y.; Ni, B.; Yao, C.; Zhou, Y.; Xu, W.; Wang, Z.; Qiao, Z. Comparison of the expression profiles of promastigotes and axenic amastigotes in *Leishmania donovani* using serial analysis of gene expression. *Parasitol. Res.* **2008**, *103*, 821–828. [CrossRef] [PubMed]

34. Shadab, M.; Das, S.; Banerjee, A.; Sinha, R.; Asad, M.; Kamran, M.; Maji, M.; Jha, B.; Deepthi, M.; Kumar, M.; et al. RNA-Seq Revealed expression of many novel genes associated with *Leishmania donovani* persistence and clearance in the host macrophage. *Front Cell Infect. Microbiol.* **2019**, *9*, 17. [CrossRef] [PubMed]

35. Calvo-Alvarez, E.; Alvarez-Velilla, R.; Fernandez-Prada, C.; Balana-Fouce, R.; Reguera, R.M. Trypanosomatids see the light: Recent advances in bioimaging research. *Drug Discov. Today* **2015**, *20*, 114–121. [CrossRef] [PubMed]

36. Tegazzini, D.; Diaz, R.; Aguilar, F.; Pena, I.; Presa, J.L.; Yardley, V.; Martin, J.J.; Coteron, J.M.; Croft, S.L.; Cantizani, J. A replicative *in vitro* assay for drug discovery against *Leishmania donovani*. *Antimicrob. Agents Chemother.* **2016**, *60*, 3524–3532. [CrossRef] [PubMed]

37. Prada, C.F.; Alvarez-Velilla, R.; Balana-Fouce, R.; Prieto, C.; Calvo-Alvarez, E.; Escudero-Martinez, J.M.; Requena, J.M.; Ordonez, C.; Desideri, A.; Perez-Pertejo, Y.; et al. Gimatecan and other camptothecin derivatives poison *Leishmania* DNA-topoisomerase IB leading to a strong leishmanicidal effect. *Biochem. Pharm.* **2013**, *85*, 1433–1440. [CrossRef]

38. Balana-Fouce, R.; Prada, C.F.; Requena, J.M.; Cushman, M.; Pommier, Y.; Alvarez-Velilla, R.; Escudero-Martinez, J.M.; Calvo-Alvarez, E.; Perez-Pertejo, Y.; Reguera, R.M. Indotecan (LMP400) and AM13-55: Two novel indenoisoquinolines show potential for treating visceral leishmaniasis. *Antimicrob. Agents Chemother.* **2012**, *56*, 5264–5270. [CrossRef]

39. Seifert, K.; Escobar, P.; Croft, S.L. In vitro activity of anti-leishmanial drugs against *Leishmania donovani* is host cell dependent. *J. Antimicrob. Chemother.* **2010**, *65*, 508–511. [CrossRef]

40. Ginouves, M.; Simon, S.; Nacher, M.; Demar, M.; Carme, B.; Couppie, P.; Prevot, G. In vitro sensitivity of cutaneous *Leishmania* promastigote isolates circulating in French Guiana to a set of drugs. *Am. J. Trop. Med. Hyg.* **2017**, *96*, 1143–1150. [CrossRef] [PubMed]

41. Deep, D.K.; Singh, R.; Bhandari, V.; Verma, A.; Sharma, V.; Wajid, S.; Sundar, S.; Ramesh, V.; Dujardin, J.C.; Salotra, P. Increased miltefosine tolerance in clinical isolates of *Leishmania donovani* is associated with reduced drug accumulation, increased infectivity and resistance to oxidative stress. *PLoS Negl. Trop Dis.* **2017**, *11*, e0005641. [CrossRef] [PubMed]

42. Mowbray, C.E.; Braillard, S.; Speed, W.; Glossop, P.A.; Whitlock, G.A.; Gibson, K.R.; Mills, J.E.; Brown, A.D.; Gardner, J.M.; Cao, Y.; et al. Novel amino-pyrazole ureas with potent in vitro and in vivo antileishmanial activity. *J. Med. Chem.* **2015**, *58*, 9615–9624. [CrossRef] [PubMed]

43. Van den Kerkhof, M.; Mabille, D.; Chatelain, E.; Mowbray, C.E.; Braillard, S.; Hendrickx, S.; Maes, L.; Caljon, G. In vitro and in vivo pharmacodynamics of three novel antileishmanial lead series. *Int. J. Parasitol. Drugs Drug Resist.* **2018**, *8*, 81–86. [CrossRef]

44. Tunes, L.G.; Morato, R.E.; Garcia, A.; Schmitz, V.; Steindel, M.; Correa-Junior, J.D.; Dos Santos, H.F.; Frezard, F.; de Almeida, M.V.; Silva, H.; et al. Preclinical gold complexes as oral drug candidates to treat leishmaniasis are potent trypanothione reductase inhibitors. *ACS Infect. Dis.* **2020**, *6*, 1121–1139. [CrossRef] [PubMed]

45. Osorio, Y.; Travi, B.L.; Renslo, A.R.; Peniche, A.G.; Melby, P.C. Identification of small molecule lead compounds for visceral leishmaniasis using a novel *ex vivo* splenic explant model system. *PLoS Negl. Trop Dis.* **2011**, *5*, e962. [CrossRef] [PubMed]

46. Osorio, Y.E.; Travi, B.L.; Melby, P.C. An ex vivo splenic explant model system for the identification of small molecule therapeutics for visceral leishmaniasis. *FASEB J.* **2008**, *22*, 1122–1136.

47. Hefnawy, A.; Cantizani, J.; Pena, I.; Manzano, P.; Rijal, S.; Dujardin, J.C.; De Muylder, G.; Martin, J. Importance of secondary screening with clinical isolates for anti-leishmania drug discovery. *Sci. Rep.* **2018**, *8*, 11765. [CrossRef]

48. Smith, T.K.; Bringaud, F.; Nolan, D.P.; Figueiredo, L.M. Metabolic reprogramming during the *Trypanosoma brucei* life cycle. *F1000Research* **2017**, *6*, 683. [CrossRef]

49. MacGregor, P.; Szoor, B.; Savill, N.J.; Matthews, K.R. Trypanosomal immune evasion, chronicity and transmission: An elegant balancing act. *Nat. Rev. Microbiol.* **2012**, *10*, 431–438. [CrossRef]

50. Mackey, Z.B.; Baca, A.M.; Mallari, J.P.; Apsel, B.; Shelat, A.; Hansell, E.J.; Chiang, P.K.; Wolff, B.; Guy, K.R.; Williams, J.; et al. Discovery of trypanocidal compounds by whole cell HTS of Trypanosoma brucei. *Chem. Biol. Drug Des.* **2006**, *67*, 355–363. [CrossRef]

51. Sykes, M.L.; Avery, V.M. A luciferase based viability assay for ATP detection in 384-well format for high throughput whole cell screening of *Trypanosoma brucei brucei* bloodstream form strain 427. *Parasit. Vectors* **2009**, *2*, 54. [CrossRef]

52. Sykes, M.L.; Baell, J.B.; Kaiser, M.; Chatelain, E.; Moawad, S.R.; Ganame, D.; Ioset, J.R.; Avery, V.M. Identification of compounds with anti-proliferative activity against *Trypanosoma brucei brucei* strain 427 by a whole cell viability based HTS campaign. *PLoS Negl. Trop Dis.* **2012**, *6*, e1896. [CrossRef] [PubMed]

53. Faria, J.; Moraes, C.B.; Song, R.; Pascoalino, B.S.; Lee, N.; Siqueira-Neto, J.L.; Cruz, D.J.; Parkinson, T.; Ioset, J.R.; Cordeiro-da-Silva, A.; et al. Drug discovery for human African trypanosomiasis: Identification of novel scaffolds by the newly developed HTS SYBR Green assay for *Trypanosoma brucei*. *J. Biomol. Screen* **2015**, *20*, 70–81. [CrossRef] [PubMed]

54. Berninger, M.; Erk, C.; Fuss, A.; Skaf, J.; Al-Momani, E.; Israel, I.; Raschig, M.; Guntzel, P.; Samnick, S.; Holzgrabe, U. Fluorine walk: The impact of fluorine in quinolone amides on their activity against African sleeping sickness. *Eur. J. Med. Chem.* **2018**, *152*, 377–391. [CrossRef] [PubMed]

55. Torreele, E.; Bourdin Trunz, B.; Tweats, D.; Kaiser, M.; Brun, R.; Mazue, G.; Bray, M.A.; Pecoul, B. Fexinidazole: A new oral nitroimidazole drug candidate entering clinical development for the treatment of sleeping sickness. *PLoS Negl. Trop Dis.* **2010**, *4*, e923. [CrossRef] [PubMed]

56. Jacobs, R.T.; Nare, B.; Wring, S.A.; Orr, M.D.; Chen, D.; Sligar, J.M.; Jenks, M.X.; Noe, R.A.; Bowling, T.S.; Mercer, L.T.; et al. SCYX-7158, an orally-active benzoxaborole for the treatment of stage 2 human African trypanosomiasis. *PLoS Negl. Trop Dis.* **2011**, *5*, e1151. [CrossRef]

57. Villalta, F.; Rachakonda, G. Advances in preclinical approaches to Chagas disease drug discovery. *Expert Opin. Drug Discov.* **2019**, *14*, 1161–1174. [CrossRef]

58. de Souza, W.; de Carvalho, T.M.; Barrias, E.S. Review on *Trypanosoma cruzi*: Host cell interaction. *Int. J. Cell Biol.* **2010**, *2010*, 295394. [CrossRef]

59. Reyes, P.A.; Vallejo, M. Trypanocidal drugs for late stage, symptomatic Chagas disease (*Trypanosoma cruzi* infection). *Cochrane Database Syst. Rev.* **2005**. [CrossRef]

60. Chatelain, E.; Ioset, J.R. Phenotypic screening approaches for Chagas disease drug discovery. *Expert Opin. Drug Discov.* **2018**, *13*, 141–153. [CrossRef]

61. Bot, C.; Hall, B.S.; Bashir, N.; Taylor, M.C.; Helsby, N.A.; Wilkinson, S.R. Trypanocidal activity of aziridinyl nitrobenzamide prodrugs. *Antimicrob. Agents Chemother.* **2010**, *54*, 4246–4252. [CrossRef]

62. Canavaci, A.M.; Bustamante, J.M.; Padilla, A.M.; Perez Brandan, C.M.; Simpson, L.J.; Xu, D.; Boehlke, C.L.; Tarleton, R.L. In vitro and in vivo high-throughput assays for the testing of anti-*Trypanosoma cruzi* compounds. *PLoS Negl. Trop Dis.* **2010**, *4*, e740. [CrossRef]

63. Engel, J.C.; Ang, K.K.; Chen, S.; Arkin, M.R.; McKerrow, J.H.; Doyle, P.S. Image-based high-throughput drug screening targeting the intracellular stage of *Trypanosoma cruzi*, the agent of Chagas' disease. *Antimicrob. Agents Chemother.* **2010**, *54*, 3326–3334. [CrossRef] [PubMed]

64. Pena, I.; Pilar Manzano, M.; Cantizani, J.; Kessler, A.; Alonso-Padilla, J.; Bardera, A.I.; Alvarez, E.; Colmenarejo, G.; Cotillo, I.; Roquero, I.; et al. New compound sets identified from high throughput phenotypic screening against three kinetoplastid parasites: An open resource. *Sci. Rep.* **2015**, *5*, 8771. [CrossRef] [PubMed]

65. Alonso-Padilla, J.; Cotillo, I.; Presa, J.L.; Cantizani, J.; Pena, I.; Bardera, A.I.; Martin, J.J.; Rodriguez, A. Automated high-content assay for compounds selectively toxic to *Trypanosoma cruzi* in a myoblastic cell line. *PLoS Negl. Trop Dis.* **2015**, *9*, e0003493. [CrossRef]

66. De Rycker, M.; Thomas, J.; Riley, J.; Brough, S.J.; Miles, T.J.; Gray, D.W. Identification of trypanocidal activity for known clinical compounds using a new *Trypanosoma cruzi* hit-discovery screening cascade. *PLoS Negl. Trop Dis.* **2016**, *10*, e0004584. [CrossRef] [PubMed]

67. MacLean, L.M.; Thomas, J.; Lewis, M.D.; Cotillo, I.; Gray, D.W.; De Rycker, M. Development of *Trypanosoma cruzi in vitro* assays to identify compounds suitable for progression in Chagas' disease drug discovery. *PLoS Negl. Trop Dis.* **2018**, *12*, e0006612. [CrossRef]

68. Bernatchez, J.A.; Chen, E.; Hull, M.V.; McNamara, C.W.; McKerrow, J.H.; Siqueira-Neto, J.L. High-throughput screening of the ReFRAME library identifies potential drug repurposing candidates for *Trypanosoma cruzi*. *Microorganisms* **2020**, *8*, 472. [CrossRef]

69. Gupta, S.N. Visceral leishmaniasis: Experimental models for drug discovery. *Indian J. Med. Res.* **2011**, *133*, 27–39.

70. Loria-Cervera, E.N.; Andrade-Narvaez, F.J. Animal models for the study of leishmaniasis immunology. *Rev. Inst. Med. Trop. Sao Paulo* **2014**, *56*, 1–11. [CrossRef]

71. Sacks, D.L.; Melby, P.C. Animal models for the analysis of immune responses to leishmaniasis. *Curr. Protoc. Immunol.* **2015**, *108*, 19.2.1–19.2.24. [CrossRef]

72. Loeuillet, C.; Banuls, A.L.; Hide, M. Study of *Leishmania* pathogenesis in mice: Experimental considerations. *Parasit. Vectors* **2016**, *9*, 144. [CrossRef] [PubMed]

73. Mears, E.R.; Modabber, F.; Don, R.; Johnson, G.E. A review: The current in vivo models for the discovery and utility of new anti-leishmanial drugs targeting cutaneous leishmaniasis. *PLoS Negl. Trop Dis.* **2015**, *9*, e0003889. [CrossRef] [PubMed]

74. de Oliveira, C.I.; Brodskyn, C.I. The immunobiology of *Leishmania braziliensis* infection. *Front Immunol.* **2012**, *3*, 145. [CrossRef] [PubMed]

75. Blackwell, J.M. Genetic susceptibility to leishmanial infections: Studies in mice and man. *Parasitology* **1996**, *112*, S67–S74. [CrossRef]

76. Leon, C.M.; Montilla, M.; Vanegas, R.; Castillo, M.; Parra, E.; Ramirez, J.D. Murine models susceptibility to distinct *Trypanosoma cruzi* I genotypes infection. *Parasitology* **2017**, *144*, 512–519. [CrossRef]

77. Torres-Vargas, J.; Jimenez-Coello, M.; Guzman-Marin, E.; Acosta-Viana, K.Y.; Yadon, Z.E.; Gutierrez-Blanco, E.; Guillermo-Cordero, J.L.; Garg, N.J.; Ortega-Pacheco, A. Quantitative and histological assessment of maternal-fetal transmission of *Trypanosoma cruzi* in guinea pigs: An experimental model of congenital Chagas disease. *PLoS Negl. Trop Dis.* **2018**, *12*, e0006222. [CrossRef]

78. Becvar, T.; Siriyasatien, P.; Bates, P.; Volf, P.; Sadlova, J. Development of *Leishmania* (*Mundinia*) in guinea pigs. *Parasit. Vectors* **2020**, *13*, 181. [CrossRef]

79. Paranaiba, L.F.; Pinheiro, L.J.; Macedo, D.H.; Menezes-Neto, A.; Torrecilhas, A.C.; Tafuri, W.L.; Soares, R.P. An overview on *Leishmania* (*Mundinia*) *enriettii*: Biology, immunopathology, LRV and extracellular vesicles during the host-parasite interaction. *Parasitology* **2018**, *145*, 1265–1273. [CrossRef]

80. McCarroll, C.S.; Rossor, C.L.; Morrison, L.R.; Morrison, L.J.; Loughrey, C.M. A Pre-clinical animal model of *Trypanosoma brucei* infection demonstrating cardiac dysfunction. *PLoS Negl. Trop Dis.* **2015**, *9*, e0003811. [CrossRef]

81. Fulton, J.D.; Joyner, L.P.; Chandler, R.J. Studies on protozoa. Part II: The golden hamster (*Cricetus auratus*) and cotton rat (*Sigmodon hispidus*) as experimental hosts for *Leishmania donovani*. *Trans R. Soc. Trop. Med. Hyg.* **1950**, *44*, 105–112. [CrossRef]

82. Gomes-Silva, A.; Valverde, J.G.; Ribeiro-Romao, R.P.; Placido-Pereira, R.M.; Da-Cruz, A.M. Golden hamster (*Mesocricetus auratus*) as an experimental model for *Leishmania* (*Viannia*) *braziliensis* infection. *Parasitology* **2013**, *140*, 771–779. [CrossRef] [PubMed]

83. Cruz-Chan, J.V.; Aguilar-Cetina Adel, C.; Villanueva-Lizama, L.E.; Martinez-Vega, P.P.; Ramirez-Sierra, M.J.; Rosado-Vallado, M.E.; Guillermo-Cordero, J.L.; Dumonteil, E. A canine model of experimental infection with *Leishmania* (*L.*) *mexicana*. *Parasit. Vectors* **2014**, *7*, 361. [CrossRef] [PubMed]

84. Probst, R.J.; Wellde, B.T.; Lawyer, P.G.; Stiteler, J.S.; Rowton, E.D. Rhesus monkey model for *Leishmania major* transmitted by *Phlebotomus papatasi* sandfly bites. *Med. Vet. Entomol.* **2001**, *15*, 12–21. [CrossRef] [PubMed]

85. Freidag, B.L.; Mendez, S.; Cheever, A.W.; Kenney, R.T.; Flynn, B.; Sacks, D.L.; Seder, R.A. Immunological and pathological evaluation of rhesus macaques infected with *Leishmania major*. *Exp. Parasitol.* **2003**, *103*, 160–168. [CrossRef]

86. Souza-Lemos, C.; de-Campos, S.N.; Teva, A.; Porrozzi, R.; Grimaldi, G.J. In situ characterization of the granulomatous immune response with time in nonhealing lesional skin of *Leishmania braziliensis*-infected rhesus macaques (*Macaca mulatta*). *Vet. Immunol. Immunopathol.* **2011**, *142*, 147–155. [CrossRef]

87. Sathler-Avelar, R.; Vitelli-Avelar, D.M.; Mattoso-Barbosa, A.M.; Perdigao-de-Oliveira, M.; Costa, R.P.; Eloi-Santos, S.M.; Gomes Mde, S.; Amaral, L.R.; Teixeira-Carvalho, A.; Martins-Filho, O.A.; et al. Phenotypic features of circulating leukocytes from non-human primates naturally infected with *Trypanosoma cruzi* resemble the major immunological findings observed in human Chagas disease. *PLoS Negl. Trop Dis.* **2016**, *10*, e0004302. [CrossRef]

88. Folmer, R.H. Integrating biophysics with HTS-driven drug discovery projects. *Drug Discov. Today* **2016**, *21*, 491–498. [CrossRef]

89. Liu, R.; Li, X.; Lam, K.S. Combinatorial chemistry in drug discovery. *Curr. Opin. Chem. Biol.* **2017**, *38*, 117–126. [CrossRef]

90. Dos Santos, R.N.; Ferreira, L.G.; Andricopulo, A.D. Practices in molecular docking and structure-based virtual screening. *Methods Mol. Biol.* **2018**, *1762*, 31–50.

91. Chen, W.; Gong, L.; Guo, Z.; Wang, W.; Zhang, H.; Liu, X.; Yu, S.; Xiong, L.; Luo, J. A novel integrated method for large-scale detection, identification, and quantification of widely targeted metabolites: Application in the study of rice metabolomics. *Mol. Plant* **2013**, *6*, 1769–1780. [CrossRef]

92. Saeed, Y.; Bahram, H. Chemometrics tools in QSAR/QSPR studies: A historical perspective. *Chemom. Intell. Lab. Syst.* **2015**, *149*, 177–204.

93. Gourley, D.G.; Luba, J.; Hardy, L.W.; Beverley, S.M.; Hunter, W.N. Crystallization of recombinant Leishmania major pteridine reductase 1 (PTR1). *Acta Cryst. D Biol. Cryst.* **1999**, *55*, 1608–1610. [CrossRef] [PubMed]

94. Rashid, U.; Sultana, R.; Shaheen, N.; Hassan, S.F.; Yaqoob, F.; Ahmad, M.J.; Iftikhar, F.; Sultana, N.; Asghar, S.; Yasinzai, M.; et al. Structure based medicinal chemistry-driven strategy to design substituted dihydropyrimidines as potential antileishmanial agents. *Eur. J. Med. Chem.* **2016**, *115*, 230–244. [CrossRef] [PubMed]

95. Stevanovic, S.; Perdih, A.; Sencanski, M.; Glisic, S.; Duarte, M.; Tomas, A.M.; Sena, F.V.; Sousa, F.M.; Pereira, M.M.; Solmajer, T. In silico discovery of a substituted 6-methoxy-quinalidine with leishmanicidal activity in *Leishmania infantum*. *Molecules* **2018**, *23*, 772. [CrossRef]

96. De Luca, L.; Ferro, S.; Buemi, M.R.; Monforte, A.M.; Gitto, R.; Schirmeister, T.; Maes, L.; Rescifina, A.; Micale, N. Discovery of benzimidazole-based *Leishmania mexicana* cysteine protease CPB2.8DeltaCTE inhibitors as potential therapeutics for leishmaniasis. *Chem. Biol. Drug Des.* **2018**, *92*, 1585–1596. [CrossRef]

97. Romero-Estudillo, I.; Viveros-Ceballos, J.L.; Cazares-Carreno, O.; Gonzalez-Morales, A.; de Jesus, B.F.; Lopez-Castillo, M.; Razo-Hernandez, R.S.; Castaneda-Corral, G.; Ordonez, M. Synthesis of new alpha-aminophosphonates: Evaluation as anti-inflammatory agents and QSAR studies. *Bioorg. Med. Chem.* **2019**, *27*, 2376–23861. [CrossRef]

98. Purohit, P.; Pandey, A.K.; Singh, D.; Chouhan, P.S.; Ramalingam, K.; Shukla, M.; Goyal, N.; Lal, J.; Chauhan, P.M.S. An insight into tetrahydro-beta-carboline-tetrazole hybrids: Synthesis and bioevaluation as potent antileishmanial agents. *Medchemcomm* **2017**, *8*, 1824–1834. [CrossRef]

99. Wyllie, S.; Thomas, M.; Patterson, S.; Crouch, S.; De Rycker, M.; Lowe, R.; Gresham, S.; Urbaniak, M.D.; Otto, T.D.; Stojanovski, L.; et al. Cyclin-dependent kinase 12 is a drug target for visceral leishmaniasis. *Nature* **2018**, *560*, 192–197. [CrossRef]

100. Fisher, R.A.; Gollan, B.; Helaine, S. Persistent bacterial infections and persister cells. *Nat. Rev. Microbiol.* **2017**, *15*, 453–464. [CrossRef]

101. Mandal, S.; Njikan, S.; Kumar, A.; Early, J.V.; Parish, T. The relevance of persisters in tuberculosis drug discovery. *Microbiology* **2019**, *165*, 492–499. [CrossRef]

102. Barrett, M.P.; Kyle, D.E.; Sibley, L.D.; Radke, J.B.; Tarleton, R.L. Protozoan persister-like cells and drug treatment failure. *Nat. Rev. Microbiol.* **2019**, *17*, 607–620. [CrossRef] [PubMed]

103. Markus, M.B. Malaria Eradication and the Hidden Parasite Reservoir. *Trends Parasitol.* **2017**, *33*, 492–495. [CrossRef] [PubMed]

104. Le Rutte, E.A.; Zijlstra, E.E.; de Vlas, S.J. Post-Kala-Azar dermal leishmaniasis as a reservoir for visceral leishmaniasis transmission. *Trends Parasitol.* **2019**, *35*, 590–592. [CrossRef]

105. Kloehn, J.; Saunders, E.C.; O'Callaghan, S.; Dagley, M.J.; McConville, M.J. Characterization of metabolically quiescent *Leishmania* parasites in murine lesions using heavy water labeling. *PLoS Pathog.* **2015**, *11*, e1004683. [CrossRef]

106. Mandell, M.A.; Beverley, S.M. Continual renewal and replication of persistent *Leishmania major* parasites in concomitantly immune hosts. *Proc. Natl. Acad. Sci. USA* **2017**, *114*, E801–E810. [CrossRef] [PubMed]

107. Jara, M.; Berg, M.; Caljon, G.; de Muylder, G.; Cuypers, B.; Castillo, D.; Maes, I.; Orozco, M.D.C.; Vanaerschot, M.; Dujardin, J.C.; et al. Macromolecular biosynthetic parameters and metabolic profile in different life stages of *Leishmania braziliensis*: Amastigotes as a functionally less active stage. *PLoS ONE* **2017**, *12*, e0180532. [CrossRef]

108. Sanchez-Valdez, F.J.; Padilla, A.; Wang, W.; Orr, D.; Tarleton, R.L. Spontaneous dormancy protects *Trypanosoma cruzi* during extended drug exposure. *Elife* **2018**, *7*, e34039. [CrossRef]

109. Jara, M.; Maes, I.; Imamura, H.; Domagalska, M.A.; Dujardin, J.C.; Arevalo, J. Tracking of quiescence in *Leishmania* by quantifying the expression of GFP in the ribosomal DNA locus. *Sci. Rep.* **2019**, *9*, 18951. [CrossRef]

110. Narayanan, M.S.; Rudenko, G. TDP1 is an HMG chromatin protein facilitating RNA polymerase I transcription in African trypanosomes. *Nucleic. Acids. Res.* **2013**, *41*, 2981–2992. [CrossRef]

111. Ferrick, D.A.; Neilson, A.; Beeson, C. Advances in measuring cellular bioenergetics using extracellular flux. *Drug Discov. Today* **2008**, *13*, 268–274. [CrossRef]

112. Divakaruni, A.S.; Rogers, G.W.; Murphy, A.N. Measuring mitochondrial function in permeabilized cells using the Seahorse XF Analyzer or a Clark-Type oxygen electrode. *Curr. Protoc. Toxicol.* **2014**, *60*, 25.2.1–25.2.16. [CrossRef] [PubMed]

113. Gonzalez-Ortiz, L.M.; Sanchez-Villamil, J.P.; Celis-Rodriguez, M.A.; Lineros, G.; Sanabria-Barrera, S.; Serrano, N.C.; Rincon, M.Y.; Bautista-Nino, P.K. Measuring mitochondrial respiration in adherent cells infected with *Trypanosoma cruzi* Chagas, 1909 using Seahorse extracellular flux analyser. *Folia Parasitol.* **2019**, *66*, 16. [CrossRef]

114. Furtado, C.; Kunrath-Lima, M.; Rajao, M.A.; Mendes, I.C.; de Moura, M.B.; Campos, P.C.; Macedo, A.M.; Franco, G.R.; Pena, S.D.; Teixeira, S.M.; et al. Functional characterization of 8-oxoguanine DNA glycosylase of *Trypanosoma cruzi*. *PLoS ONE* **2012**, *7*, e42484. [CrossRef] [PubMed]

115. Li, Y.; Shah-Simpson, S.; Okrah, K.; Belew, A.T.; Choi, J.; Caradonna, K.L.; Padmanabhan, P.; Ndegwa, D.M.; Temanni, M.R.; Corrada Bravo, H.; et al. Transcriptome remodeling in *Trypanosoma cruzi* and human cells during intracellular infection. *PLoS Pathog.* **2016**, *12*, e1005511. [CrossRef] [PubMed]

116. Shah-Simpson, S.; Pereira, C.F.; Dumoulin, P.C.; Caradonna, K.L.; Burleigh, B.A. Bioenergetic profiling of *Trypanosoma cruzi* life stages using Seahorse extracellular flux technology. *Mol. Biochem. Parasitol.* **2016**, *208*, 91–95. [CrossRef] [PubMed]

117. Rufener, R.; Dick, L.; D'Ascoli, L.; Ritler, D.; Hizem, A.; Wells, T.N.C.; Hemphill, A.; Lundstrom-Stadelmann, B. Repurposing of an old drug: In vitro and in vivo efficacies of buparvaquone against *Echinococcus multilocularis*. *Int. J. Parasitol. Drugs Drug Resist.* **2018**, *8*, 440–450. [CrossRef]

118. Field, M.C.; Allen, C.L.; Dhir, V.; Goulding, D.; Hall, B.S.; Morgan, G.W.; Veazey, P.; Engstler, M. New approaches to the microscopic imaging of *Trypanosoma brucei*. *Microsc. Microanal.* **2004**, *10*, 621–636. [CrossRef]

119. Aulner, N.; Danckaert, A.; Rouault-Hardoin, E.; Desrivot, J.; Helynck, O.; Commere, P.H.; Munier-Lehmann, H.; Spath, G.F.; Shorte, S.L.; Milon, G.; et al. High content analysis of primary macrophages hosting proliferating *Leishmania* amastigotes: Application to anti-leishmanial drug discovery. *PLoS Negl. Trop Dis.* **2013**, *7*, e2154. [CrossRef]

120. Proto, W.R.; Coombs, G.H.; Mottram, J.C. Cell death in parasitic protozoa: Regulated or incidental? *Nat. Rev. Microbiol.* **2013**, *11*, 58–66. [CrossRef]

121. Thomas, J.A.; Baker, N.; Hutchinson, S.; Dominicus, C.; Trenaman, A.; Glover, L.; Alsford, S.; Horn, D. Insights into antitrypanosomal drug mode-of-action from cytology-based profiling. *PLoS Negl. Trop Dis.* **2018**, *12*, e0006980. [CrossRef]

122. Mishra, A.; Khan, M.I.; Jha, P.K.; Kumar, A.; Das, S.; Das, P.; Das, P.; Sinha, K.K. Oxidative stress-mediated overexpression of uracil DNA glycosylase in *Leishmania donovani* confers tolerance against antileishmanial drugs. *Oxid. Med. Cell Longev.* **2018**, *2018*, 4074357. [CrossRef] [PubMed]

123. Glover, L.; Horn, D. Trypanosomal histone gammaH2A and the DNA damage response. *Mol. Biochem. Parasitol.* **2012**, *183*, 78–83. [CrossRef] [PubMed]

124. Bellofatto, V.; Palenchar, J.B. RNA interference as a genetic tool in trypanosomes. *Methods Mol. Biol.* **2008**, *442*, 83–94. [PubMed]

125. Alsford, S.; Turner, D.J.; Obado, S.O.; Sanchez-Flores, A.; Glover, L.; Berriman, M.; Hertz-Fowler, C.; Horn, D. High-throughput phenotyping using parallel sequencing of RNA interference targets in the African trypanosome. *Genome Res.* **2011**, *21*, 915–924. [CrossRef] [PubMed]

126. Alsford, S.; Eckert, S.; Baker, N.; Glover, L.; Sanchez-Flores, A.; Leung, K.F.; Turner, D.J.; Field, M.C.; Berriman, M.; Horn, D. High-throughput decoding of antitrypanosomal drug efficacy and resistance. *Nature* **2012**, *482*, 232–236. [CrossRef]

127. Baker, N.; Glover, L.; Munday, J.C.; Aguinaga Andres, D.; Barrett, M.P.; de Koning, H.P.; Horn, D. Aquaglyceroporin 2 controls susceptibility to melarsoprol and pentamidine in African trypanosomes. *Proc. Natl. Acad. Sci. USA* **2012**, *109*, 10996–11001. [CrossRef]

128. Glover, L.; Alsford, S.; Baker, N.; Turner, D.J.; Sanchez-Flores, A.; Hutchinson, S.; Hertz-Fowler, C.; Berriman, M.; Horn, D. Genome-scale RNAi screens for high-throughput phenotyping in bloodstream-form African trypanosomes. *Nat. Protoc.* **2015**, *10*, 106–133. [CrossRef]

129. Marquis, N.; Gourbal, B.; Rosen, B.P.; Mukhopadhyay, R.; Ouellette, M. Modulation in aquaglyceroporin AQP1 gene transcript levels in drug-resistant *Leishmania*. *Mol. Microbiol.* **2005**, *57*, 1690–1699. [CrossRef]

130. Gazanion, E.; Fernandez-Prada, C.; Papadopoulou, B.; Leprohon, P.; Ouellette, M. Cos-Seq for high-throughput identification of drug target and resistance mechanisms in the protozoan parasite *Leishmania*. *Proc. Natl. Acad. Sci. USA* **2016**, *113*, E3012–E3021. [CrossRef] [PubMed]

131. Fernandez-Prada, C.; Sharma, M.; Plourde, M.; Bresson, E.; Roy, G.; Leprohon, P.; Ouellette, M. High-throughput Cos-Seq screen with intracellular *Leishmania infantum* for the discovery of novel drug-resistance mechanisms. *Int. J. Parasitol. Drugs Drug Resist.* **2018**, *8*, 165–173. [CrossRef]

132. Bhattacharya, A.; Leprohon, P.; Bigot, S.; Padmanabhan, P.K.; Mukherjee, A.; Roy, G.; Gingras, H.; Mestdagh, A.; Papadopoulou, B.; Ouellette, M. Coupling chemical mutagenesis to next generation sequencing for the identification of drug resistance mutations in *Leishmania*. *Nat. Commun.* **2019**, *10*, 5627. [CrossRef] [PubMed]

133. Beneke, T.; Madden, R.; Makin, L.; Valli, J.; Sunter, J.; Gluenz, E. A CRISPR Cas9 high-throughput genome editing toolkit for kinetoplastids. *R. Soc. Open Sci.* **2017**, *4*, 170095. [CrossRef] [PubMed]

134. Martel, D.; Beneke, T.; Gluenz, E.; Spath, G.F.; Rachidi, N. Characterisation of casein kinase 1.1 in *Leishmania donovani* using the CRISPR Cas9 toolkit. *Biomed. Res. Int.* **2017**, *2017*, 4635605. [CrossRef]

135. Beneke, T.; Demay, F.; Hookway, E.; Ashman, N.; Jeffery, H.; Smith, J.; Valli, J.; Becvar, T.; Myskova, J.; Lestinova, T.; et al. Genetic dissection of a *Leishmania* flagellar proteome demonstrates requirement for directional motility in sand fly infections. *PLoS Pathog.* **2019**, *15*, e1007828. [CrossRef] [PubMed]

136. Negreira, G.H.; Monsieurs, P.; Imamura, H.; Maes, I.; Kuk, N.; Yagoubat, A.; den Broeck, F.V.; Sterkers, Y.; Dujardin, J.-C.; Domagalska, M.A. Exploring the evolution and adaptive role of mosaic aneuploidy in a clonal *Leishmania donovani* population using high throughput single cell genome sequencing. *bioRxiv* **2020**, *2020*, 976233v1. Available online: www.biorxiv.org/content/10.1101/2020.03.05.976233v1 (accessed on 15 May 2020).

137. Ubeda, J.M.; Raymond, F.; Mukherjee, A.; Plourde, M.; Gingras, H.; Roy, G.; Lapointe, A.; Leprohon, P.; Papadopoulou, B.; Corbeil, J.; et al. Genome-wide stochastic adaptive DNA amplification at direct and inverted DNA repeats in the parasite *Leishmania*. *PLoS Biol.* **2014**, *12*, e1001868. [CrossRef]

138. Dumetz, F.; Cuypers, B.; Imamura, H.; Zander, D.; D'Haenens, E.; Maes, I.; Domagalska, M.A.; Clos, J.; Dujardin, J.C.; De Muylder, G. Molecular preadaptation to antimony resistance in *Leishmania donovani* on the Indian Subcontinent. *MSphere* **2018**, *3*, e00548-17. [CrossRef]

139. Poinar, G.J.; Poinar, R. Fossil evidence of insect pathogens. *J. Invertebr. Pathol.* **2005**, *89*, 243–250. [CrossRef]

140. Vargas, D.A.; Prieto, M.D.; Martinez-Valencia, A.J.; Cossio, A.; Burgess, K.E.V.; Burchmore, R.J.S.; Gomez, M.A. Pharmacometabolomics of meglumine antimoniate in patients with cutaneous leishmaniasis. *Front Pharm.* **2019**, *10*, 657. [CrossRef] [PubMed]

141. Hennig, K.; Abi-Ghanem, J.; Bunescu, A.; Meniche, X.; Biliaut, E.; Ouattara, A.D.; Lewis, M.D.; Kelly, J.M.; Braillard, S.; Courtemanche, G.; et al. Metabolomics, lipidomics and proteomics profiling of myoblasts infected with *Trypanosoma cruzi* after treatment with different drugs against Chagas disease. *Metabolomics* **2019**, *15*, 117. [CrossRef]

142. Ty, M.C.; Loke, P.; Alberola, J.; Rodriguez, A.; Rodriguez-Cortes, A. Immuno-metabolic profile of human macrophages after *Leishmania* and *Trypanosoma cruzi* infection. *PLoS ONE* **2019**, *14*, e0225588. [CrossRef]

143. McCall, L.I.; Tripathi, A.; Vargas, F.; Knight, R.; Dorrestein, P.C.; Siqueira-Neto, J.L. Experimental Chagas disease-induced perturbations of the fecal microbiome and metabolome. *PLoS Negl. Trop Dis.* **2018**, *12*, e0006344. [CrossRef] [PubMed]

144. McCall, L.I.; Morton, J.T.; Bernatchez, J.A.; de Siqueira-Neto, J.L.; Knight, R.; Dorrestein, P.C.; McKerrow, J.H. Mass spectrometry-based chemical cartography of a cardiac parasitic infection. *Anal. Chem.* **2017**, *89*, 10414–10421. [CrossRef] [PubMed]

145. Gazos-Lopes, F.; Martin, J.L.; Dumoulin, P.C.; Burleigh, B.A. Host triacylglycerols shape the lipidome of intracellular trypanosomes and modulate their growth. *PLoS Pathog.* **2017**, *13*, e1006800. [CrossRef]

146. Creek, D.J.; Nijagal, B.; Kim, D.H.; Rojas, F.; Matthews, K.R.; Barrett, M.P. Metabolomics guides rational development of a simplified cell culture medium for drug screening against *Trypanosoma brucei*. *Antimicrob. Agents Chemother.* **2013**, *57*, 2768–2779. [CrossRef] [PubMed]

147. Trochine, A.; Creek, D.J.; Faral-Tello, P.; Barrett, M.P.; Robello, C. Benznidazole biotransformation and multiple targets in *Trypanosoma cruzi* revealed by metabolomics. *PLoS Negl. Trop Dis.* **2014**, *8*, e2844. [CrossRef]

148. Rojo, D.; Canuto, G.A.; Castilho-Martins, E.A.; Tavares, M.F.; Barbas, C.; Lopez-Gonzalvez, A.; Rivas, L. A multiplatform metabolomic approach to the basis of antimonial action and resistance in *Leishmania infantum*. *PLoS ONE* **2015**, *10*, e0130675. [CrossRef] [PubMed]

149. Brotherton, M.C.; Bourassa, S.; Leprohon, P.; Legare, D.; Poirier, G.G.; Droit, A.; Ouellette, M. Proteomic and genomic analyses of antimony resistant *Leishmania infantum* mutant. *PLoS ONE* **2013**, *8*, e81899. [CrossRef]

150. Vincent, I.M.; Barrett, M.P. Metabolomic-based strategies for anti-parasite drug discovery. *J. Biomol. Screen* **2015**, *20*, 44–55. [CrossRef]

151. Alves-Ferreira, M.; Guimaraes, A.C.; Capriles, P.V.; Dardenne, L.E.; Degrave, W.M. A new approach for potential drug target discovery through in silico metabolic pathway analysis using *Trypanosoma cruzi* genome information. *Mem. Inst. Oswaldo Cruz.* **2009**, *104*, 1100–1110. [CrossRef]

152. Coelho, E.A.; Chavez-Fumagalli, M.A.; Costa, L.E.; Tavares, C.A.; Soto, M.; Goulart, L.R. Theranostic applications of phage display to control leishmaniasis: Selection of biomarkers for serodiagnostics, vaccination, and immunotherapy. *Rev. Soc. Bras. Med. Trop.* **2015**, *48*, 370–379. [CrossRef] [PubMed]

153. Costa, L.E.; Alves, P.T.; Carneiro, A.P.; Dias, A.C.S.; Fujimura, P.T.; Araujo, G.R.; Tavares, G.S.V.; Ramos, F.F.; Duarte, M.C.; Menezes-Souza, D.; et al. *Leishmania infantum* beta-tubulin identified by reverse engineering technology through phage display applied as theranostic marker for human visceral leishmaniasis. *Int. J. Mol. Sci.* **2019**, *20*, 1812. [CrossRef] [PubMed]

154. Singh, O.P.; Gedda, M.R.; Mudavath, S.L.; Srivastava, O.N.; Sundar, S. Envisioning the innovations in nanomedicine to combat visceral leishmaniasis: For future theranostic application. *Nanomedicine* **2019**, *14*, 1911–1927. [CrossRef] [PubMed]

155. Khare, S.; Nagle, A.S.; Biggart, A.; Lai, Y.H.; Liang, F.; Davis, L.C.; Barnes, S.W.; Mathison, C.J.; Myburgh, E.; Gao, M.Y.; et al. Proteasome inhibition for treatment of leishmaniasis, Chagas disease and sleeping sickness. *Nature* **2016**, *537*, 229–233. [CrossRef] [PubMed]

156. Wyllie, S.; Brand, S.; Thomas, M.; De Rycker, M.; Chung, C.W.; Pena, I.; Bingham, R.P.; Bueren-Calabuig, J.A.; Cantizani, J.; Cebrian, D.; et al. Preclinical candidate for the treatment of visceral leishmaniasis that acts through proteasome inhibition. *Proc. Natl. Acad. Sci. USA* **2019**, *116*, 9318–9323. [CrossRef]

157. Ardal, C.; Balasegaram, M.; Laxminarayan, R.; McAdams, D.; Outterson, K.; Rex, J.H.; Sumpradit, N. Antibiotic development–economic, regulatory and societal challenges. *Nat. Rev. Microbiol.* **2020**, *18*, 267–274. [CrossRef]

158. Pecoul, B. New drugs for neglected diseases: From pipeline to patients. *PLoS Med.* **2004**, *1*, e6. [CrossRef]

159. Croft, S.L. Neglected tropical diseases in the genomics era: Re-evaluating the impact of new drugs and mass drug administration. *Genome Biol.* **2016**, *17*, 46. [CrossRef]

Article

Genomic and Transcriptomic Analysis for Identification of Genes and Interlinked Pathways Mediating Artemisinin Resistance in *Leishmania donovani*

Sushmita Ghosh [1,2], **Aditya Verma** [1], **Vinay Kumar** [1], **Dibyabhaba Pradhan** [3], **Angamuthu Selvapandiyan** [2], **Poonam Salotra** [1] **and Ruchi Singh** [1,*]

1   ICMR-National Institute of Pathology, Safdarjung Hospital Campus, New Delhi 110029, India; sushmitaghosh1990@gmail.com (S.G.); adityaicmr@gmail.com (A.V.); vinaykr016@gmail.com (V.K.); poonamsalotra@hotmail.com (P.S.)
2   JH-Institute of Molecular Medicine, Jamia Hamdard, New Delhi 110062, India; selvapandiyan@jamiahamdard.ac.in
3   ICMR-AIIMS Computational Genomics Centre, Indian Council of Medical Research, New Delhi 110029, India; dbpinfo@gmail.com
*   Correspondence: ruchisp@gmail.com or ruchisingh.nip@gov.in

Received: 24 August 2020; Accepted: 14 September 2020; Published: 17 November 2020

Abstract: Current therapy for visceral leishmaniasis (VL), compromised by drug resistance, toxicity, and high cost, demands for more effective, safer, and low-cost drugs. Artemisinin has been found to be an effectual drug alternative in experimental models of leishmaniasis. Comparative genome and transcriptome analysis of in vitro-adapted artesunate-resistant (K133AS-R) and -sensitive wild-type (K133WT) *Leishmania donovani* parasites was carried out using next-generation sequencing and single-color DNA microarray technology, respectively, to identify genes and interlinked pathways contributing to drug resistance. Whole-genome sequence analysis of K133WT vs. K133AS-R parasites revealed substantial variation among the two and identified 240 single nucleotide polymorphisms (SNPs), 237 insertion deletions (InDels), 616 copy number variations (CNVs) (377 deletions and 239 duplications), and trisomy of chromosome 12 in K133AS-R parasites. Transcriptome analysis revealed differential expression of 208 genes (fold change ≥ 2) in K133AS-R parasites. Functional categorization and analysis of modulated genes of interlinked pathways pointed out plausible adaptations in K133AS-R parasites, such as (i) a dependency on lipid and amino acid metabolism for generating energy, (ii) reduced DNA and protein synthesis leading to parasites in the quiescence state, and (iii) active drug efflux. The upregulated expression of cathepsin-L like protease, amastin-like surface protein, and amino acid transporter and downregulated expression of the gene encoding ABCG2, pteridine receptor, adenylatecyclase-type receptor, phosphoaceylglucosamine mutase, and certain hypothetical proteins are concordant with genomic alterations suggesting their potential role in drug resistance. The study provided an understanding of the molecular basis linked to artemisinin resistance in *Leishmania* parasites, which may be advantageous for safeguarding this drug for future use.

Keywords: *Leishmania donovani*; whole-genome sequencing (WGS); transcriptome; artemisinin drug resistance

---

1. Introduction

Leishmaniasis is a major public health problem affecting the poor population of the world, mainly in the developing countries. The disease is endemic in 97 countries with 70,0000 to one million new cases per year [1,2]. Visceral leishmaniasis (VL), caused by the protozoan *Leishmania donovani*, is the most severe type, with frequent outbreaks and a greater mortality potential. In 2018, more than 95% of new cases reported to World Health Organization (WHO) occurred in 10 countries, including India [1,3]. Due to the lack of a vaccine and effective vector control, management of VL relies exclusively on a handful of chemotherapeutic agents, but most of the therapeutics, including pentavalent antimonials, miltefosine, and liposomal amphotericin B, are associated with serious drawbacks, such as being toxic and expensive, with a declining efficacy pertaining to an increase in the occurrence of resistance [4–6]. Therefore, there is a need to explore new safe, effective, and affordable treatment options for VL.

The antimalarial drug artemisinin and its derivatives have been found to also be effective against non-malarial parasites, such as *Leishmania*. There are several in vitro and in vivo studies demonstrating the antileishmanial activity of artemisinin and its derivatives with a high safety index [7–9]. As far as the mechanism of action is concerned, artemisinin and its derivatives have been reported to cause programmed cell death in *Leishmania* promastigotes by a loss of mitochondrial membrane potential, enabling externalization of phosphatidylserine, DNA fragmentation, and cell cycle arrest at the sub-G0/G1 phase [10]. The drug also works by the restoration of normal nitric oxide (NO) production by infected macrophages, initially impaired due to infection with *Leishmania* parasites [11–13]. Further, studies in mice suggest that administration of artemisinin results in the generation of iron-artemisinin adducts, which causes clearance of intracellular amastigotes [14]. However, antileishmanial activities and the possible mechanism of resistance to artemisinin in *Leishmania* parasites have been poorly explored. An understanding of the mechanisms of drug resistance in *Leishmania* is vital to protect existing drugs and for the development of new ones [15]. Drug-resistant parasites apply various strategies in order to survive under drug pressure, such as reduced drug uptake, active drug efflux, alteration of the drug targets, inactivation of drugs, etc. [16–22]. Various transcriptomic studies of drug-sensitive vs. -resistant parasites revealed that a number of genes have altered expression in drug-resistant parasites. Our group has previously shown by microarray analysis that approximately 3.9% and 2.9% of the total *Leishmania* genome representing various functional categories, such as metabolic pathways, transporters and cellular components among others, were differentially modulated (>2 fold) in experimentally selected miltefosine- and paromomycin-resistant lines, respectively [23,24].

Whole-genome sequence (WGS) analysis is another important tool used to detect mechanisms of drug resistance in *Leishmania*. It was earlier reported that in the absence of transcriptional control, *Leishmania* parasites have evolved mechanisms to alter mRNA levels by increased gene dosage through gene amplification, gene deletion, and aneuploidy in order to adapt to stress conditions, such as drug pressure [25–28]. The genome sequence of *Leishmania* field isolates from the Indian sub-continent revealed gene copy number variation (CNV) to be associated with susceptibility to sodium stibogluconate (SSG) [29]. Similarly, aneuploidy has been observed in the context of antimony, methotrexate, and nelfinavir resistance; however, the link between aneuploidy and drug resistance was circumstantial [25,26,30–32]. Additionally, single-nucleotide polymorphisms (SNPs) in drug targets or key enzymes constitute another strategy to survive under drug pressure. The acquisition of an inactivation mutation in the *L. donovani* miltefosine transporter gene (*Ld*MT) and/or its β-subunit (LdRos3) was reported to increase miltefosine resistance in both in vitro and in vivo studies as well as in clinical isolates [33–38].

Artemisinin resistance in malaria is associated with SNPs on chromosome 10, 13, and 14, and non-synonymous SNPs in the propeller domain of a kelch gene located on chromosome 13 [39,40]. Analysis of the transcriptome of *Plasmodium falciparum* isolates revealed a higher expression of unfolded protein response (UPR) in artemisinin resistance. Previously, we explored the mechanism of artesunate (a derivative of artemisinin) resistance in *Leishmania* parasites and showed that artesunate resistance in *Leishmania* is associated with parasite virulence, host immune modulation, and unfolded protein responses [41]. In the present study, the genome and transcriptome of artesunate-sensitive vs. -resistant *Leishmania* parasites were analyzed using next-generation sequencing (NGS) and single-color DNA microarray technology, respectively. Analysis of the genome structure and modulated gene expression identified several genes/pathways, which were further validated for their role in the selection of artesunate resistance in *Leishmania*. Expression analysis of Heat shock protein 70 (Hsp70) and Aquaglyceroporin 1 (AQP1) was validated in K133WT and K133AS-R cell lysate. In view of the important roles of ATP-binding cassette protein (ABC) transporters and the AQP1 gene in drug resistance in *Leishmania*, their roles were explored in artesunate resistance using respective inhibitors. Based on the analyses, a model was predicted for artesunate resistance in *Leishmania*.

2. Materials and Methods

2.1. Parasite and Culture Condition

L. donovani field isolate (K133WT), earlier derived from bone marrow aspirates of a VL patient and cryopreserved in a lab, was revived and propagated in medium M199 (Sigma-Aldrich, St. Louis, MO, USA) supplemented with 10% heat-inactivated fetal bovine serum (HI FBS, Gibco, Waltham, MA, USA), 100 IU/mL penicillin G, and 100 mg/mL streptomycin at 26 °C. The isolate was exposed to increasing concentrations (up to 50 μM) of artesunate drug (Sigma Aldrich, St. Louis, MO, USA) to obtain experimental artesunate-resistant parasites, which were designated as K133AS-R. The susceptibility of K133WT and K133AS-R parasites towards artesunate was determined, which showed that there was a 3.73-fold increase in the mean IC_{50} (50% inhibitory concentration) of K133AS-R parasites at the promastigote stage, with a value of 78.63 ± 9.17 μM vs. 21.08 ± 3.15 μM, and a >3-fold increase in the mean IC_{50} at the amastigote stage, with a value 73.09 ± 1.14 μM vs. 21.62 ± 3.24 μM for K133AS-R vs. K133WT isolates. This was reported in our previous study [41].

2.2. Genomic DNA Isolation from Parasite Culture

Genomic DNA (gDNA) from K133WT and K133AS-R promastigotes was isolated using a Wizard Genomic DNA purification kit (Promega, Madison, WI, USA) following the manufacturer's instructions. Quantification of the DNA was performed by optical density measurements in a Nanodrop and QubitFlex® 2.0 Fluorometer (Thermo Fisher Scientific, Waltham, MA, USA). The quality of the gDNA was checked on 1% agarose gel for the single intact band.

2.3. Genomic Library Preparation and Sequencing

Preparation of paired-end (PE) sequencing libraries of K133WT and K133AS-R was initiated with 200 ng of genomic DNA using a Truseq Nano DNA Library preparation kit (Illumina, Inc., SanDiego, CA, USA). The generated library was examined in a Bioanalyzer 2100 (Agilent Technologies, Santa Clara, CA, USA) using a high-sensitivity (HS) DNA chip and sequenced using the Illumina Hiseq 2000 platform according to the manufacturer's standard cluster generation and sequencing protocol [42]. Briefly, the mechanical shearing of gDNA by a Covaris instrument (Woburn, MA, USA) was done to generate fragments of 250–350 bp, after which fragmented ends were repaired and tailed with A at 3'. Thereafter, adapters were ligated, which was necessary for binding dual-barcoded libraries to the flow cell for sequencing. Finally, 314–355-bp libraries were generated and high-fidelity PCR amplification was done using HiFi PCR master reaction component mix to ensure maximum yield from limited amounts of starting material for sequencing on an Illumina Hiseq 2000 platform (Illumina Inc., San Diego, CA, USA) (2 × 150 bp chemistry). Whole-genome sequencing resulted in the generation of approximately 4 GB data per sample. The sequences of *L. donovani* K133WT and K133AS-R are available with NCBI GenBank as a BioProject with SRA accession no. PRJNA657979.

2.4. Whole-Genome Sequencing Data Analysis

Genomic data analysis was executed with minor modifications as described previously by Dumetz et al. 2017 [43]. The paired-end (PE) raw reads obtained from the sequencer were checked for the quality of the reads using FastQCv0.11.8 (Babraham Institute, Cambridge, UK) and were further trimmed to improve the quality of the reads using the Trimmomatic tool v0.38 (Usadellab. org, RWTH

Aachen University, Germany) [44]. The *L. donovani* strain LdBPK282A1 reference genome was indexed and high-quality pair-end reads were mapped using Burrows-Wheeler Aligner (BWA-MEM v0.7.5a algorithm, Broad Institute, Cambridge, MA, USA) [45]. The generated SAM file was converted into BAM format and duplicates were removed using Picard toolkit v1.119 (Broad Institute, Cambridge, MA, USA). Further, the BAM file was used for identifying SNPs and InDels using GATK Haplotype caller v3.4 (Broad Institute, Cambridge, MA, USA). Filtering of SNPs and InDels was performed using Bcf tools v0.1.18 (Sanger Institute, Cambridge shire, UK) from subdirectory of SAM tools (mapping quality cut off 25 and read depth of 15) and the variants were annotated with the SnpEff v4.3 tool (McGill University, Montreal, QC, Canada) [46,47].

Estimation of CNV along with chromosomal somy was done in accordance with the protocol designed by Downing et al. 2011 [29]. CNV estimation was done using CNVnator (https://github.com/abyzovlab/CNVnator) [48], and for somy assessment, the median read depth of each chromosome (d_i) was computed first followed by median depth estimation of 36 complete chromosomes (d_m). The somy state of an individual chromosome is determined as the ratio of (d_i/d_m) and the chromosome ploidy value is specified as $2 \times (d_i/d_m)$, considered often for diploid species [38]. The full cell-normalized chromosome somy (S)-value: S < 1.5, 1.5 < S < 2.5, and 2.5 < S < 3.5, was assigned to monosomy, disomy, and trisomy, respectively [43].

2.5. Functional Annotation and Classification of Unigenes

To identify all the unigenes present in K133WT and K133AS-R, a homology search was performed against the NCBI non redundant (NR) protein database in accordance with BLASTx program (NCBI, Bethesda, MD, USA) using a cutoff E-value of 10^{-05} and the maximal aligned results with the lowest E-value were chosen to annotate the unigenes [49,50]. The Gene Ontology (GO)-based annotation of the unigenes was carried out using Blast2GO version 3.0 (Biobam, Valencia, Spain) and Web Gene Ontology Annotation Plot (WEGO) was utilized to designate GO classification on the basis of the distribution of gene functions in different species [51–54]. The basis of the functional classification considered was biological processes, cellular components, and molecular functions.

2.6. Total RNA Isolation from Parasites

Early log-phase promastigotes (1×10^8) of both K133WT and K133AS-R were used to isolate total RNA using TRIzol reagent according to the manufacturer's instruction. Extracted RNA was cleaned up using a RNeasy Plus mini kit (Qiagen, Hilden, Germany). The absorbance of purified RNA was taken at 260 and 280 nm using a Nanodrop Spectrophotometer (Thermo Fisher Scientific, Waltham, MA, USA). The quality and integrity of RNA were assessed on an RNA 6000 Nano Assay Chips on Bioanalyzer 2100 (Agilent Technologies, Santa Clara, CA, USA). RNA of good quality based on the 260/280 values (Nanodrop, Thermo Scientific, Waltham, MA, USA), rRNA 28S/18S ratios, and RNA integrity number (RIN) was used for further analysis [24].

2.7. Oligonucleotide Array

Global mRNA expression profiling of K133WT and K133AS-R *L. donovani* was carried out usingsingle color microarray-based gene expression profiling. A high-density *Leishmania* multispecies 60-mer oligonucleotide array slide [8 × 15 K format] was used for the microarray experiment. The slide represented the entire genome of *L. infantum* and *L. major*. The microarray chip printed by Agilent Technologies (Santa Clara, CA, USA), contained a total of 9233 *Leishmania*-specific genes, including 540 control probes as described earlier [24,55,56].

2.8. RNA Labelling, Amplification, Hybridization, and Data Analysis

First, 200 ng of total RNA were converted to cDNA using oligodT primer tagged to T7 polymerase promoter at 40 °C. cDNA thus obtained was converted to cRNA using T7 RNA polymerase enzyme. The dye Cy3 was also incorporated during this step. Labeled cRNA was then cleaned using Qiagen

RNeasy Mini kit columns (Qiagen, Cat No: 74106, Hilden, Germany) and quality assessment was carried out using the Nanodrop ND-1000. Following this, Cy3-labeled cRNA was fragmented at 60 °C. Fragmented cRNA was hybridized on the array (AMADID: 027511) using the Gene Expression Hybridization kit (Agilent Technologies, Santa Clara, CA, USA) at 65 °C for 16 h in Sure hybridization Chambers. Hybridized slides were washed using Agilent Gene Expression wash buffers (Agilent Technologies, Santa Clara, CA, USA) and scanned on an Agilent Microarray Scanner (Agilent Technologies, Part Number G2600D). Images thus obtained were quantified using Agilent's Feature Extraction Software Version-10.7 (Santa Clara, CA, USA). Feature-extracted raw data were analyzed using the GeneSpring GX12.6.1 microarray data and pathway analysis tool (Santa Clara, CA, USA). Quartile (75th percentile) normalization was performed. Storey and bootstrapping analysis was performed for multiple testing corrections. The expression profile of K133AS-R parasites was extrapolated on a chromosome map of *Leishmania* parasites using custom R programs. The modulated expression of genes was identified using two criteria: (a) statistical and (b) biological. Statistical significance was determined by the t-test (unpaired) and a p value < 0.05 was considered as significant for both K133WT and K133AS-R parasites. The biological cutoff for up- or downregulation was 2-fold. Differentially regulated genes were analyzed for functional classification using the GeneDB, BLAST2GO, and AmiGO databases. The pathway analysis was carried out using the gene Spring GX12.6.7 (Santa Clara, CA, USA) and KEGG pathway analysis tool (Bethesda, MD, USA). Interacting partners of up- or downregulated genes in K133AS-R parasites were identified using the String 9.01 database [24].

2.9. Data Availability

The complete genome sequence was deposited in GenBank as BioProject number PRJNA657979: for *L. donovani* K133AS-R under the SRA accession number SRR12487478 and BioSample number SAMN15854505 and for *L. donovani* K133WT under the SRA accession number SRR12487479 and BioSample number SAMN15854504. The microarray data were deposited in the GEO NCBI database (http://www.ncbi.nlm.nih.gov/geo) in the MIAME format (GEO accession number GSE118460).

2.10. Quantitative Real-Time PCR (qPCR)

A total of 14 genes were selected from microarray data and validated for their differentially modulated expression by q-PCR (Supplementary Materials Table S1). First-strand cDNA was synthesized, from 5 µG of total RNA isolated from K133WT and K133AS-R promastigotes (early log phase), using the Superscript II RNAse H reverse transcriptase enzyme (Invitrogen, Carlsbad, CA, USA) and OligodT primers (Fermentas, Waltham, MA, USA). Equal amounts of cDNA were amplified in 25-µL reactions (in triplicate) containing 6 pmoL forward and reverse primers and 1 X Fast SYBR Green mastermix using a ABI 7500 Real-time PCR system (Applied Biosystems, Waltham, MA, USA). The relative amount of PCR products generated from each primer set was determined based on the threshold cycle (Ct) value and the amplification efficiencies. Gene expression levels were normalized using constitutively expressed genes encoding cystathionine-β-synthase (CBS) and glyceraldehyde-3-phosphate dehydrogenase (GAPDH). Quantification of the relative changes in the target gene expression was calculated using the $2^{-\Delta\Delta Ct}$ method. Primers for the targeted genes were designed using Primer express software version 3.0 (Applied Biosystems, Waltham, MA, USA) [57]. The list of genes, their functional relevance, and the primers used for real-time PCR are given in Supplementary Materials Table S1.

2.11. Western Blotting of Promastigote Cell Lysate

Preparation of the parasite lysate and Western blot analysis was performed following the method described earlier [58]. K133WT and K133AS-R cell lysates (100 µG) were separated by sodium dodecyl sulphate polyacrylamide gel electrophoresis (SDS-PAGE) on a 12% polyacrylamide gel and transferred to nitrocellulose membranes. The membrane strips were blocked and incubated sequentially with anti-AQP1 (1:1000), anti-HSP70 (1:500), or anti-tubulin (1:1000) (endogenous control) primary antibodies. Following this, the membrane was probed with Horse radish Peroxidase (HRP)-conjugated anti-rabbit IgG (1:80,000) produced in mice (Sigma Aldrich, St. Louis, MO, USA). Blot was developed using Western blot detection enhanced chemiluminescence (ECL) detection reagent (Merck, Burlington, MA, USA). The image was scanned with ChemiDoc (Bio-Rad, Hercules, CA, USA) and analyzed using Image Lab™ 5.1 software (Bio-Rad, Hercules, CA, USA) [58].

2.12. Cytotoxicity Assay

The 3-(4,5-dimethylthiazol-2-yl)-2,5-diphenyltetrazolium bromide (MTT) assay was used to assess the cytotoxicity of the inhibitors towards host macrophages and was performed following the manufacturer's instructions. Primary peritoneal macrophages (PECs) extracted from Balb/c mice were incubated with AQP1 inhibitor (Tocris Biosciences, Bristol, UK) (12.5–400 µM) or ABC transporters modulator verapamil (Sigma Aldrich, St. Louis, MO, USA) (6.25–200 µM) in a 96-well plate for 48 h at 37 °C in 5% CO_2. Following this, 25 µL of (5 mg/mL in 1 × PBS) MTT were added to each well and the plate was re-incubated at 37 °C in the dark. Then, 4 h later, all media was removed and 150 µL of Dimethyl Sulfoxide (DMSO) were added to each well, mixed well by pipetting, and incubated for 15 min in the dark. Absorbance was taken at 540 nm on an Infinite M200 multimode reader (Tecan, Switzerland). A decrease in the absorbance at 540 nm indicated a decrease in cell viability.

2.13. Artesunate Susceptibility in the Presence of Inhibitors

The susceptibility of K133WT and K133AS-R parasites towards artesunate was determined in the presence of the AQP1 inhibitor (Tocris Biosciences, Bristol, UK) and ABC transporter modulator, verapamil (Sigma, St. Louis, MO, USA). At the promastigote stage, both K133WT and K133AS-R isolates (1×10^5) were seeded into a 96-well plate with various concentrations of artesunate drug (1–650 µM) alone or in the presence of 40 µM of AQP1 inhibitor or 8 µM of verapamil and incubated at 25 °C. After 72 h of incubation, 50 µL of Resazurin (Sigma Aldrich, St. Louis, MO, USA) (0.0125% (*w/v*) in Phosphate Buffered Saline (PBS)) were added to each well and the plates were further incubated for 18 h. Fluorescence was measured at an excitation wavelength of 550 nm and emission wavelength of 590 nm on an Infinite M200 multimode reader (Tecan, Switzerland) to determine cell viability. Sigmoidal regression analysis was used to calculate IC_{50} [24].

At the amastigote stage, the mice PECs were infected with late log-phase promastigotes of K133WT or K133AS-R at a ratio of 10 parasites: 1 macrophage, plated into 8-well chamber slides and incubated for 16 h at 37 °C in 5% CO_2. Non-internalized promastigotes were washed off and infected macrophages were further incubated with various dilutions of artesunate drug (13, 26, 52, 104, 208, and 260 µM) with or without AQP1 inhibitor (Tocris Biosciences, Bristol, UK) (40 µM) or verapamil (8 µM). The inhibitor/modulator alone at the tried concentration was not lethal to either K133WT/K133AS-R isolates or host macrophages. Then, 48 h later, the slides were fixed and stained using Diff-Quik solutions. Macrophages were then examined for intracellular amastigotes at 1000 × magnification. The number of *L. donovani* amastigotes per 100 macrophages was counted and the survival rate of parasites relative to untreated macrophages was calculated to determine the IC_{50} value [24].

Ethics Approval

The ethics approval was obtained from the Institute Animal Ethics Committee of the ICMR-National Institute of Pathology, Safdarjung Hospital campus, New Delhi, India (Project No. NIP/IAEC-1502). The procedures for the care, use, and euthanasia of experimental animals were carried out under the guidelines of the Committee for the Purpose of Control and Supervision of Experiments on Animals (CPCSEA, Indira Paryavaran Bhawan, Jor Bagh, New Delhi) Government of India.

3. Results

3.1. Whole-Genome Sequence Diversity Data of K133AS-R Compared to K133WT

Comparative WGS data analysis of both in vitro-generated artemisinin-resistant parasite (K133AS-R) and the wild-type field isolate (K133WT) was performed to decipher the mechanisms responsible for drug resistance. Detailed analysis of SNPs and insertion-deletion mutations (InDels) was performed for K133WT and K133AS-R isolates relative to the *L. donovani* reference using GATK's Haplotype Caller (HC) [59]. WGS data analysis of K133WT showed a higher number of upstream gene variants followed by intergenic region and missense gene variants. Out of a total of 341 gene variants, 191 SNPs and 150 InDels were observed (Figure 1A). The maximum number of SNPs was observed on chromosome number 34, while no SNP was observed on chromosome number 5, 9, 11, 14, 21, and 26 out of a total of 36 chromosomes in *Leishmania*. Amongst the total InDels, 114 nucleotide insertions and 36 nucleotide deletions were observed.

The artemisinin-resistant parasite generated in vitro under drug selection pressure (K133AS-R) showed a higher number of upstream gene variants followed by intergenic region gene variants. Out of a total of 477 gene variants, 240 SNPs and 237 InDels were observed (Figure 1A).

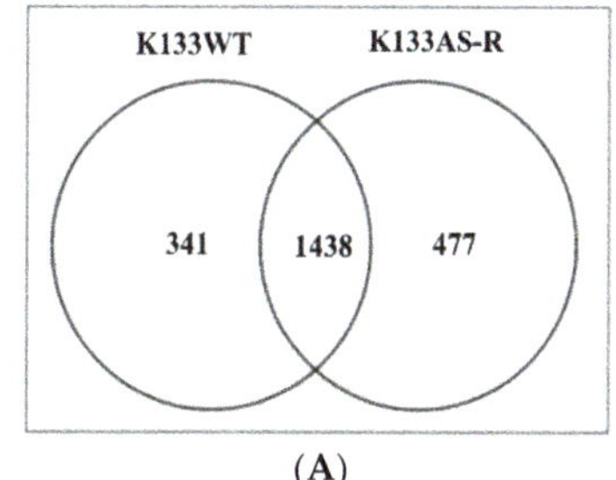

(**A**)

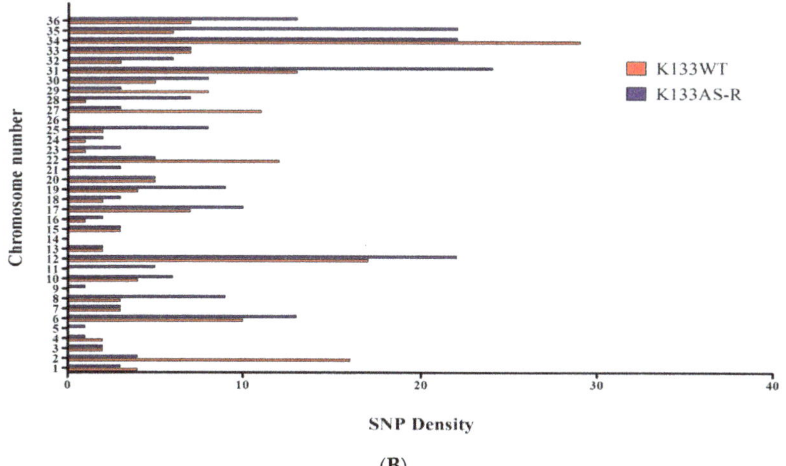

(**B**)

Figure 1. *Cont.*

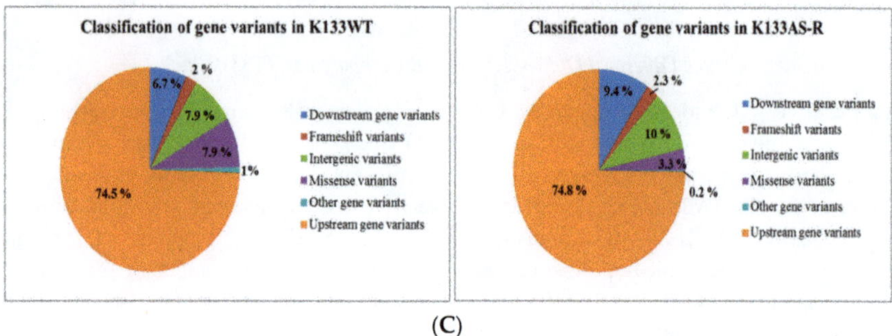

(**C**)

Figure 1. Comparative Analysis of Single Nucleotide Polymorphisms (SNPs) present in K133AS-R with K133WT. (**A**) Venn diagram showing the unique genes present in K133WT and K133AS-R. (**B**) Comparative SNP density analysis of K133WT vs. K133AS-R (**C**) Pie chart showing the percentage of different gene variants present in K133WT and K133AS-R.

The maximum number of SNPs was observed on chromosome number 31, while no SNPs were found on chromosome number 14 and 26, which is a common observation among K133WT and K133AS-R (Figure 1B). Among InDels, 173 nucleotide insertions and 64 nucleotide deletions were observed. Unique gene variants were also observed among K133WT and K133AS-R. In K133WT, the unique gene variants identified were upstream variant 254 (74.48%), downstream 23 (6.74%), missense 27 (7.9%), frameshift 7 (2.05%), and intergenic region gene variants 27 (7.9%), while other variations included disruptive / conservative in-frame insertions, one each and one conservative in-frame deletion, which is 0.3% of the total gene variation observed. In case of K133AS-R, the unique gene variants observed were 357 (74.84%) in upstream, 45 (9.43%) in downstream, 47 (9.85%) in intergenic region, 16 (3.35%) missense variant, 11 (2.05%) frameshift variants, and one stop-lost splice variant (0.21%) (Figure 1C). Insertions were observed to be the highest in the genome followed by transition, transversion, and deletion.

3.2. Differentially Expressed Genes in K133AS-R vs. K133WT

The data for K133AS-R revealed several differentially expressed genes, which are expected to contribute to drug resistance. Extensive variation in the expression of several genes like pteridine transporter and histone-encoding genes was observed in artemisinin-resistant isolates. Marked variation in the number of peptidases, such as metallopeptidase (LDBPK_330210), aminopeptidase P1 (LDBPK_020010), and lipases (LDBPK_341140), was also observed in K133AS-R. Enzymes involved in the lipid biochemical pathway, such as fatty acid elongation and fatty acid desaturation, were affected in artemisinin-resistant *Leishmania*, suggesting a decreased fluidity of the parasite membrane, which may be contributing towards drug resistance as observed in the case of miltefosine-resistant parasite [60]. Genes encoding phosphoglycan β-1,3 galactosyltransferase (involved in glycosylation of proteins), ATP binding cassette transporters (ABC transporters), ABCA2, ABCA7, and ABCA8 exhibited one missense and two frameshift mutations having moderate and high impact in K133AS-R parasites. Additionally, changes in folate/biopterin transporter (upstream gene variant, impact modifier), P-type H⁺-ATPase (frameshift mutation), and UDP-galactose transporter have been observed in the AS-R parasite. Cell surface protein-encoding genes viz. amastin-like proteins, and proteophosphoglycan (ppg3)-related protein displayed mutation in the K133AS-R isolate. Alterations in ceroidlipofuscinosis neuronal protein 3 (CLN3, LDBPK_061360) responsible for *Leishmania* virulence were also observed, showing six mutations, including missense mutation with moderate impact, which may have a direct effect on lysosomal function [61]. Moderate impacts on enzymes of the TCA cycle viz. citrate synthase, pyruvate kinase, and succinate dehydrogenase were noted in K133AS-R. In addition, specific genes present in K133AS-R that encode peptidase-like cysteine peptidase B, serine peptidase, heat shock proteins, upstream gene variants, and downstream gene variant with modifier impact are speculated to have direct or indirect role in pathogenesis. Interestingly, two novel gene mutations in K133AS-R including apical membrane antigen1 (AMA1, LDBPK_301480) (moderate impact, missense mutation) and cathepsin L-like protease were identified, whose role in pathogenesis has been reported previously [62].

3.3. Chromosomal Diversity in Artesunate-Resistant L. donovani

3.3.1. Chromosome Copy Number Variation (CNV)

Chromosome copy number analysis revealed large differences between K133WT and K133AS-R. The gene copy number variants' length observed was from 0.2 to 200 kilo base pair (kb) based on the size distribution of identified CNVs (Figure 2A). Of the total CNVs observed in K133AS-R, most were in the range of 1–5 kb size followed by 20 to 100 kb, accounting for 43.66% and 16.39%, respectively. A comparative analysis of local gene copy number variations between K133WT and K133AS-R was performed. In case of K133WT, a total of 586 CNVs were identified in which 365 deletions and 221 duplications were observed compared to K133AS-R, in which a total 616 CNVs were identified out of which 377 were deletions and 239 were duplications (Figure 2B).

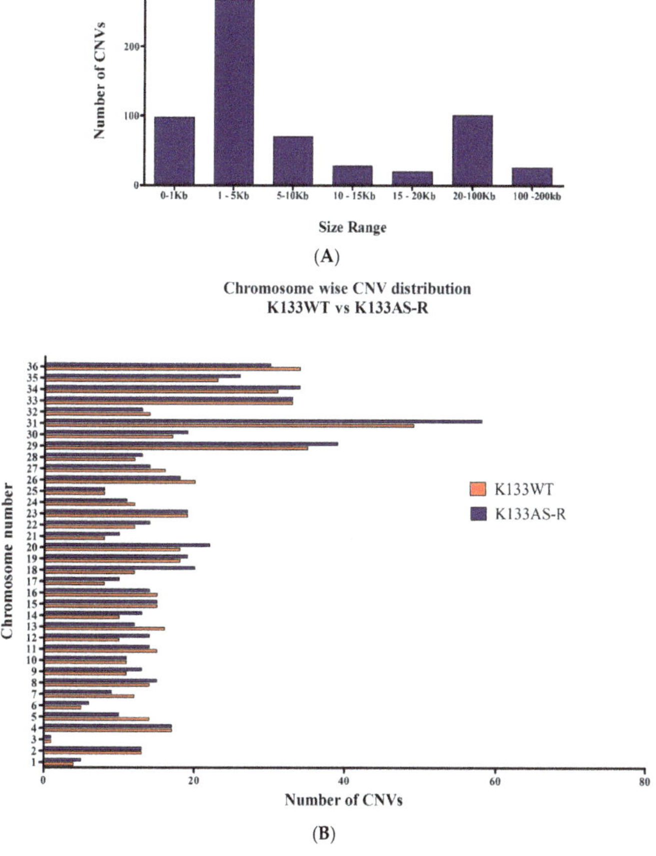

Figure 2. Analysis of CNV diversity in artesunate-resistant *L.donovani* (K133AS-R). (**A**) Size distribution of CNVs detected in the K133AS-R genome (**B**) Comparative CNV analysis of K133WT vs. K133AS-R.

3.3.2. Variance in Allelic Frequency Due to Change in Chromosomal Somy/Ploidy

In most of the cases, drug resistance in pathogenic microorganism correlates with gene expression changes, which somehow are concordant with the chromosomal ploidy changes. Normalized read depth data are generally used to assess copy number variation as somy estimation does not always show a result in integral values, since it depicts the average of a population of the cell that does not strictly show identical karyotypes. To determine somy, the two-loop method was used [38,63]. Chromosomal somy data analysis of K133WT and K133AS-R shows that most of the chromosomes were disomic. Deflection from this pattern was detected in chromosome number 14 and 32, which were monosomic, while chromosome 5, 8, 20, 23, and 31 displayed the trisomy condition in both K133WT and K133AS-R. Chromosome 12 only was found in the trisomy condition, a unique observation in K133AS-R (Figure 3).

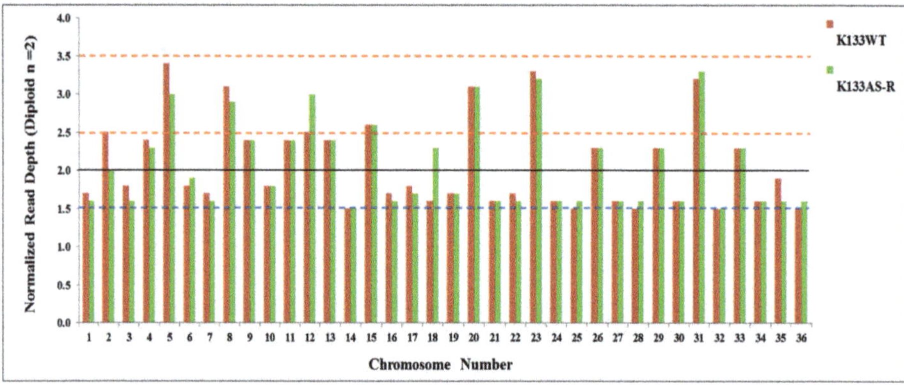

Figure 3. Chromosomy estimation in *L. donovani* parental line (K133WT) and artemisinin-resistant lines K133AS-R. The solid line represents median coverage and it was assigned a value of 2, considering that diploid is the principal ploidy state in *Leishmania*. The dotted line represents the calculated values for other somies (blue- monosomy; between two dotted red-trisomy).

3.4. Functional Annotation and Classification of K133WT and K133AS-R Unigenes

Additional validation, functional annotation, and classification of K133WT and K133AS-R unigenes derived from reference-based assembly data was performed as described in the methodology section. Out of 7671 genes retrieved in K133WT and 7792 genes in K133AS-R, a total of 7652 (99.75%) in K133WT and 7778 (99.82%) genes were found with BLAST hits. K133AS-R unigenes were annotated with at least one biological term from GO information, while the remaining 19 genes in K133WT and 14 in K133AS-R did not result in any BLAST hit. Species distribution analysis based on BLASTx results with BLAST hit sharing showed high sequence similarity with *L. donovani* and *L. infantum* sequences (Figure 4A,B). In K133WT, 2925 GO terms were allocated to biological processes, 2987 terms to cellular components, and 3152 GO terms to molecular functions, while in K133AS-R, 2942 GO terms were assigned to biological processes, 3005 terms to cellular components, and 3178 GO terms to molecular functions. Within the biological process category, cellular metabolism, cellular component organization, or biogenesis was most abundant. Within the cellular component category, GO terms corresponded to the cell and organelle part, membrane part, and protein-containing complexes (Figure 4C,D). Under molecular function category, GO terms mostly corresponded to different catalytic, binding, and transporter activity, which were abundant among unigenes [52].

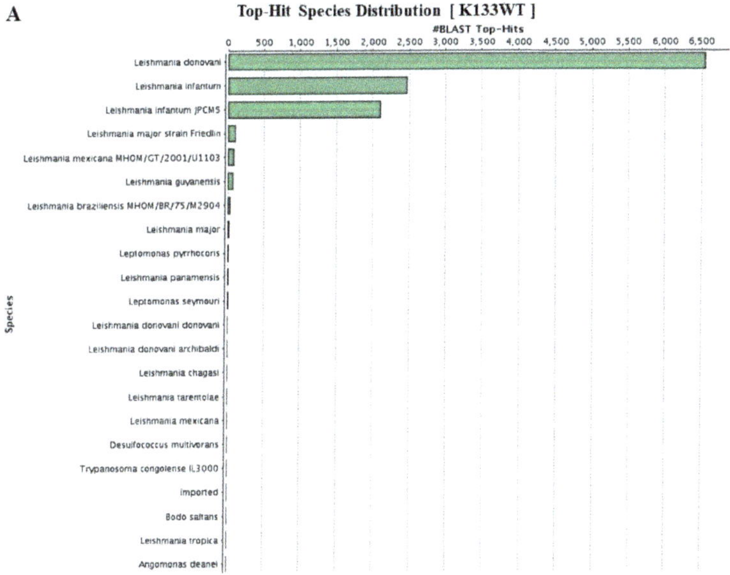

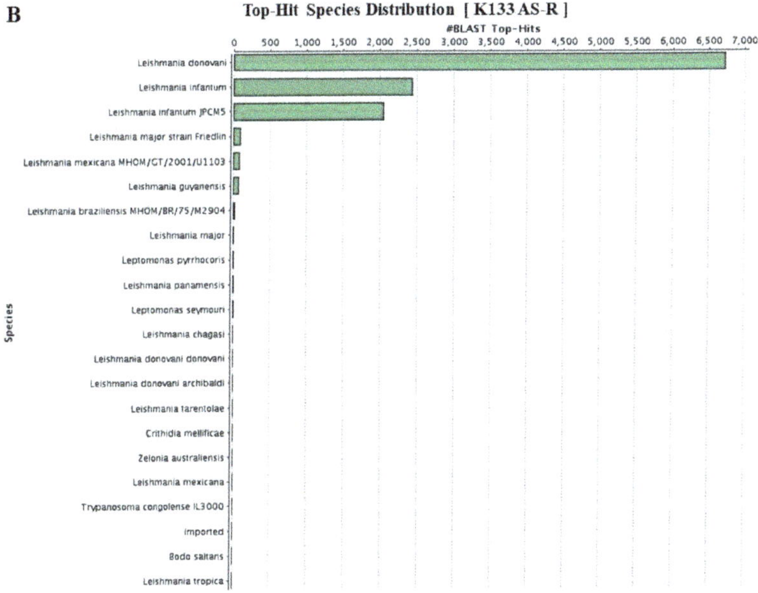

Figure 4. *Cont.*

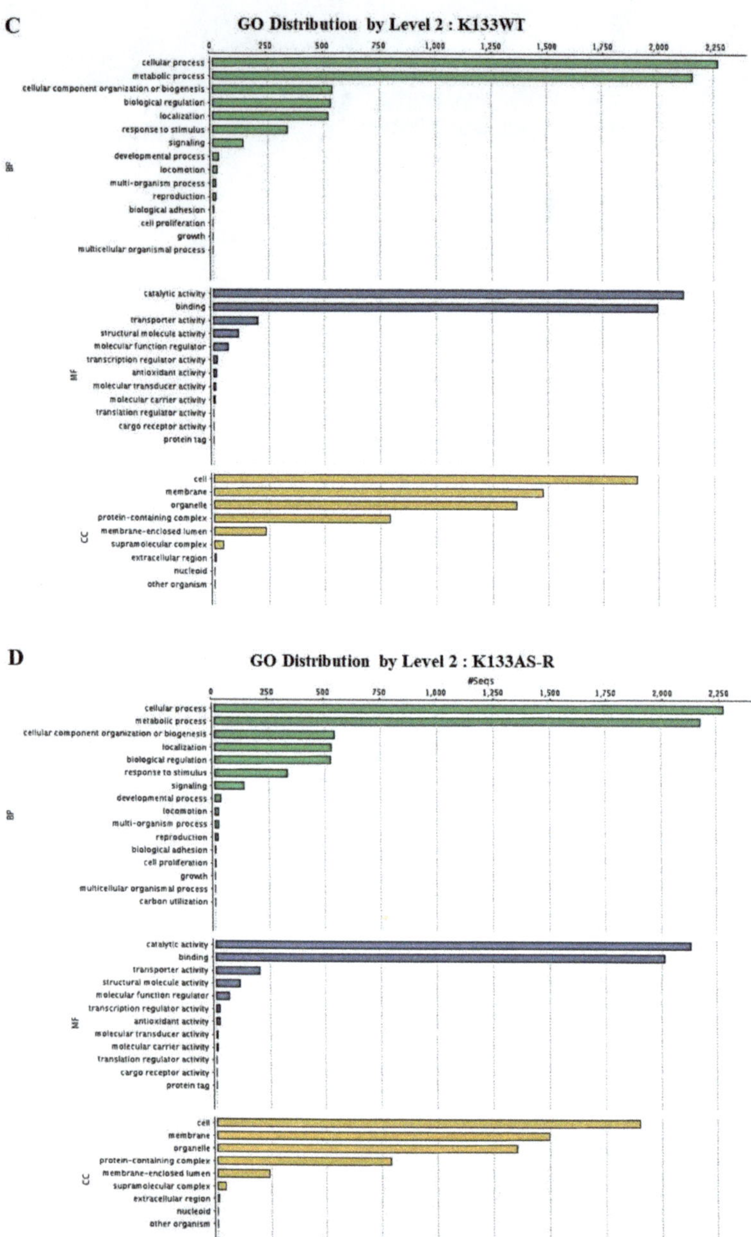

Figure 4. Characterization of K133WT and K133AS-R unigenes based on an NCBI non redundant (Nr) protein database search. (**A**) Species distribution of the top Blast hits for the K133WT assembled unigenes and (**B**) Species distribution of the top Blast hits for the K133AS-R assembled unigenes with a cutoff E-value of 10^{-05}. Gene Ontology (GO) annotation for all the assembled unigenes in K133WT (**C**) and K133AS-R (**D**) GO-terms were assigned to functionally annotate the genes based on BLAST search results using the Blast2GO program (Biobam BioInformatics, Valencia, Spain). The results were classified based in three functional categories, Green bar represents biological function (BF); Blue: molecular function (MF); and Yellow: cellular component (CC).

3.5. Comparative Transcriptome Analysis of K133WT vs. K133AS-R Parasites

Gene expression analysis using one-color DNA microarray experiment, of K133WT vs. K133AS-R isolate, revealed a modulated expression of 208 genes (approximately 2.26%) in drug-resistant parasites. The plot $\log_2$ transformed expression ratio of K133AS-R (red line) vs. K133WT (green line) as a function of the chromosomal location of microarray probes is shown in Supplementary Materials Figure S1. Out of 208 differentially modulated genes, 102 genes (1.11%) were upregulated and 106 genes (1.15%) were downregulated in K133AS-R parasites. The overall expression pattern of mRNA is shown in Supplementary Materials Table S2.

The gene expression level on the genomic scale was analyzed using a chromosome map (Figure 5A). The chromosome map showed that chromosome 18, 25, 31, and 33 contained higher numbers of upregulated genes while chromosome 33 and 36 contained higher numbers of downregulated genes in the K133AS-R isolates. Among the upregulated genes, the highest number were present on chromosome 31, which included AQP1 (LinJ.31.0030), amastin (LinJ.31.0460, LmjF.31.0450), and a few uncharacterized proteins. Upregulated proteins include autophagocytosis protein (LinJ.33.0320), protein having RNA ligase (LinJ.33.0580) activity and transaminase (LinJ.33.1410) activity on chromosome 33, protein involved in trpanothione biosynthesis process (LinJ.18.1660) on chromosome 18, Kinesin (LinJ.25.2150), and DNA-directed RNA polymerase II (LinJ.25.1350) on chromosome 25. The maximum number of downregulated genes were present on chromosome 33, among which more than 50% were hypothetical uncharacterized proteins. Other downregulated genes on chromosome 33 included translation initiation factor 2 (LinJ.33.2880), small nuclear ribonucleoprotein complex (LinJ.33.3340), H1 histone-like protein (LinJ.33.339/0), and metallocarboxipeptidase (LinJ.33.2670). Genes showing downregulated expression on chromosome 36 included isoleucyl-t-RNA synthetase (LinJ.36.5870), translation elongation factor 1-β (LinJ.36.1490), glucose transporters (LinJ.36.6550, LmjF.36.6290, LinJ.36.6560), phosphoglycerate mutase family member 5 (LinJ.36.4270), and ubiquitin protein ligase (LinJ.36.6600).

Genes showing differential expression (both up- and downregulated genes) were classified into various functional categories and a number of altered pathways in K133AS-R parasites were identified with the help of several databases and bioinformatics tools as mentioned above in Section 2.8. The percentage of genes exhibiting altered expression with genes remained unaltered in K133AS-R parasites is shown in Figure 5B. Among the 208 genes showing modulated expression in K133AS-R parasites, a total of 144 genes were categorized into function and distributed into eight different functional categories (Figure 5C). All the 144 genes with their functional categories are enlisted in Supplementary Materials Table S3.

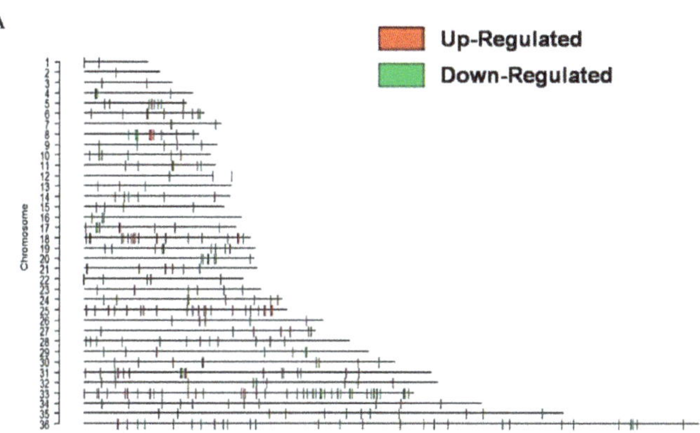

Figure 5. *Cont.*

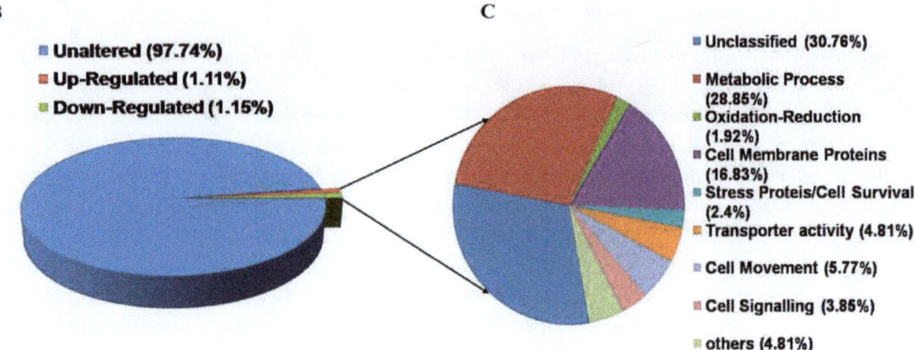

Figure 5. Comparative transcriptome profiling of K133WT and K133AS-R isolate. (**A**) Comparative gene expression of K133WT vs. K133AS-R parasites on the chromosome map. Chromosome map for differential gene expression was generated using Custom R program. Red lines indicate upregulated genes whereas green lines indicate downregulated genes in the K133AS-R parasite. (**B**) Percentage of differentially expressed genes in K133AS-R parasites. The percentage of modulated genes was calculated from the total 9170 genes obtained in Quality Control (QC) after filtering. Overall, 1.11% of genes were upregulated (red) whereas 1.15% of genes were downregulated (green); however, 97.74% of genes remained unaltered in K133AS-R parasites. (**C**) Categorization of genes showing differential expression in K133AS-R parasites according to GO functional categories. GO categories of differentially expressed genes in K133AS-R parasites suggested that genes belonging to various functional categories, such as metabolic processes, oxidation-reduction, cell membrane proteins, stress proteins, transporter activity, cell movement, and cell signaling, showed modulated expression. Unclassified proteins included hypothetical proteins with unknown function (that have not been characterized experimentally).

3.6. Validation of Modulated Gene Expression Using qPCR

Fourteen differentially expressed genes were selected for validation of expression analysis based on their role in various metabolic pathways and artesunate resistance. The selected 14 genes (9 upregulated and 5 downregulated) were validated for their expression in K133WT and K133AS-R parasites by qPCR. The fold change in the gene expression of K133AS-R/K133WT observed in q-PCR was compared with that observed in microarray experiments (Figure 6). The results obtained by qPCR for selected genes agreed with the transcriptome data derived by microarray experiments.

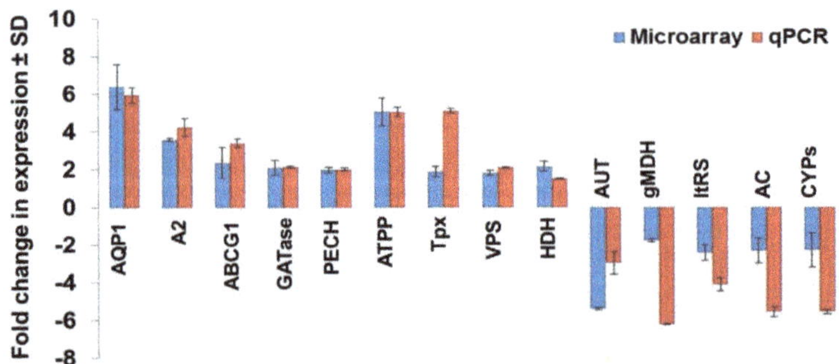

Figure 6. Validation of modulated expression of selected genes by qPCR. Selected 14 genes showing modulated expression in a microarray were validated for their altered expression by q-PCR in three independent RNA preparations. Fold changes in the gene expression of K133AS-R parasites with respect to K133WT parasites ± SD, obtained by q-PCR and microarray experiments, are represented here. The q-PCR data were normalized using two endogenous controls, glyceraldehyde 3-phosphate dehydrogenase (GAPDH) and cystathionine β-synthase (CBS).

3.7. Targeted Protein Profiling of AQP1 and HSP70 in K133WT and K133AS-R Leishmania Parasites by Western Blotting

Western blot analysis revealed that the expression of AQP1 was 1.6-fold higher in K133AS-R parasites whereas that of HSP70 was 5.46-fold lower in K133AS-R parasites as compared to K133WT parasites (Figure 7).

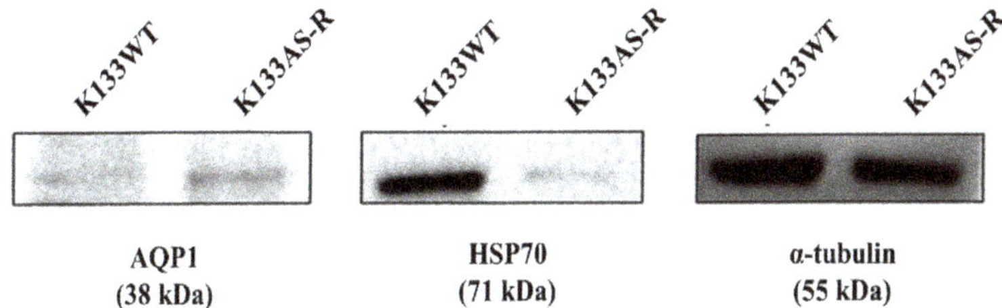

Figure 7. Expression analysis of AQP1 and HSP70 by Western blotting. Western blot analysis for the expression of AQP1 and HSP70 and α tubulin (endogenous control) protein was performed using 100 µgpromastigote cell lysates of K133WT and K133AS-R parasites. Proteins separated on a 12% SDS–PAGE gel, transferred to nitrocellulose membranes that were probed with anti-AQP1, anti-HSP70, or anti-α tubulin antibody followed by horseradish peroxidase (HRP)-conjugated antibody and developed using enhanced chemiluminescence (ECL).

3.8. Susceptibility of K133WT and K133AS-R Parasites in the Presence of the AQP1 Inhibitor and Modulator of ABC Transporters

The susceptibility of K133WT and K133AS-R parasites towards artesunate was determined in the presence of AQP1 inhibitor and modulator of ABC transporters, verapamil. The cytotoxicity of the AQP1 inhibitor or verapamil determined for host macrophages (mice PECs) by the MTT assay revealed that the cytotoxic concentration 50% (CC_{50}) of the AQP1 inhibitor was 233.47 ± 40.19 and that of verapamil was 111 ± 14.17 (Figure S2). The IC_{50} of K133AS-R parasites towards artesunate significantly decreased by 1.9-fold in the presence of the AQP1 inhibitor and 2.2-fold in the presence of verapamil at the promastigote stage (Figure 8A). Surprisingly, at the intracellular amastigote stage, K133WT parasites showed a significant increase of >4-fold in the IC_{50}; however, no significant alteration was observed in the IC_{50} of K133AS-R parasites towards artesunate in the presence of the AQP1 inhibitor. Further, in the presence of verapamil at the amastigote stage, IC_{50b} of artesunate for the K133AS-R parasites decreased by 2-fold (Figure 8B). However, there was no significant alteration in IC_{50} of K133WT parasites in the presence of the AQP1 inhibitor at the promastigote stage and in the presence of verapamil at the promastigote or amastigote stage (Figure 8A,B).

3.9. Analysis of Modulated Genes and Pathways in K133AS-R Parasites

Based on all the observations, a model depicting all the adaptations in K133AS-R parasites was proposed (Figure 9), which suggests the following genes/pathways are affected in K133AS-R parasites.

3.9.1. Autophagy, UPR, and Oxidative Stress

In K133AS-R, Atg8 (LinJ.19.0860) that plays an important role in formation of autophagosome was downregulated. Downregulated expression of HSP70 was also observed in K133AS-R parasites. On the other hand, an upregulated expression of lipoate protein ligase (LinJ.36.3230) involved in lipoic acid biosynthesis was observed in K133AS-R parasites, which eventually lead to upregulated expression of γ-glutamyl cysteine synthetase (GSH1).

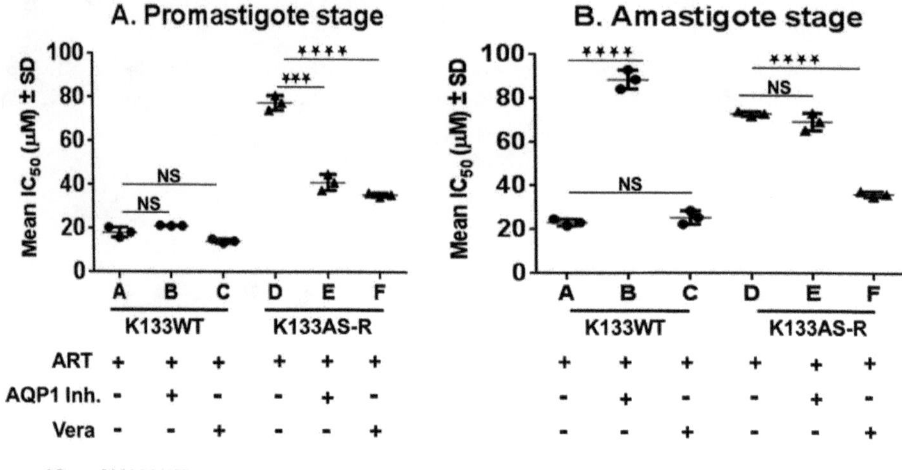

Figure 8. Susceptibility of K133WT/K133AS-R isolates in the presence of the AQP1 inhibitor or the modulator to ABC transporter, verapamil. In vitro susceptibility of the sensitive wild-type strain K133 WT/artemisinin-resistant strain K133AS-R isolates towards artesunate in the presence of the AQP1 inhibitor (AQP1 Inh.) and verapamil (Vera) at (**A**) the promastigote stage and (**B**) amastigote stage. $IC_{50} \pm SD$ of three independent experiments in duplicates is represented here. *** represents $p \leq 0.001$, **** represents $p \leq 0.0001$, NS represents not significant, Circle represents IC_{50} of K133WT, Triangle represents IC_{50} of K133AS-R.

3.9.2. Carbohydrate, Lipid, and Amino Acid Metabolism

K133AS-R parasites showed downregulated expression of gene phosphoacetylglucosamine mutase (LmjF07.0805) involved in the conversion of N-acetyl-α-D-glucosamine-1-phosphate to N-acetyl-D-glucosamine-6-phosphate, which later forms fructose-6-phosphate. Further, various glucose transporters, such as the glucose transporter, Imgt2 (LinJ.36.6550, LmjF36.6290), were also downregulated, suggesting downregulation in carbohydrate metabolism. On the other hand, genes involved in amino acid and lipid metabolism, such as methylmalonyl CoA mutase (LinJ.27.0310; involved in isoleucine, valine, and leucine metabolism), glutamine aminotransferases (LinJ.33.1410; involved in glutamine metabolism), myo-inositol-1-phosphate synthase (LinJ.14.1450), and a hypothetical protein having lipase activity (LinJ.13.0200) involved in lipid metabolism, showed upregulated expression.

3.9.3. DNA Synthesis and Translation Machinery

Genes responsible for DNA replication like nucleoside transporter 1 (LinJ.36.2040), H1 histone-like protein (LinJ.33.3390), and endonuclease/exonuclease activity (LinJ.28.1000), were downregulated. Genes involved in protein translation, such as translation initiation factor IF-2 (LinJ.33.2880), Isoleucyl-tRNA synthetase (LinJ.36.5870), and 28S ribosomal RNA (LmjF27.rRNA.32), were also downregulated. On the other hand, small RNA molecules that play an essential role in RNA biogenesis and guide chemical modifications of ribosomal RNAs (rRNAs) and other RNA genes (tRNA and snRNAs) U1snRNA, U2 snRNA, and U3 snRNA were upregulated. Further, genes involved in protein degradation, such as metallopeptidase (LinJ.11.0640) and carboxypeptidases (LinJ.33.2670), were downregulated.

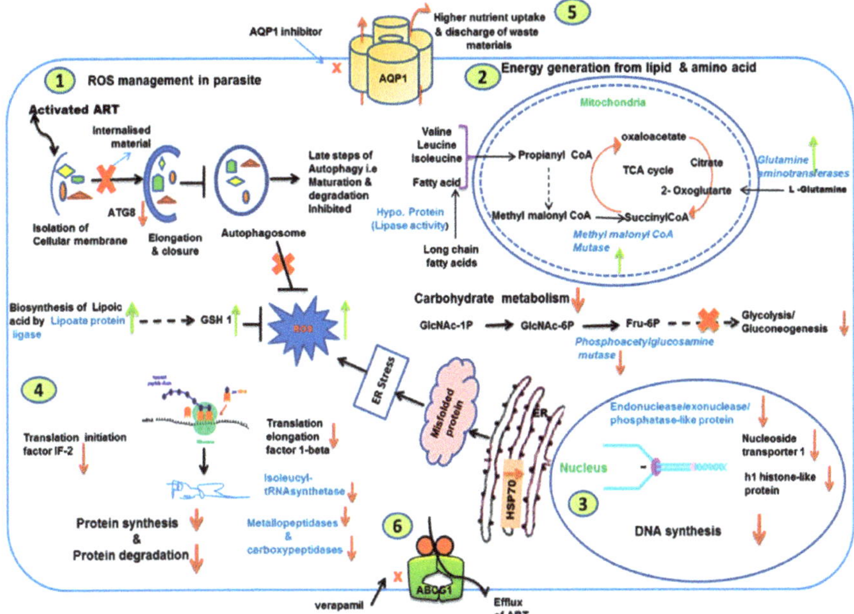

Figure 9. Transcriptome predicted adaptations contributing artesunate resistance in *L. donovani*: Genes altered in K133AS-R parasites are represented here. Genes marked with an up and down arrow represent, respectively, the upregulated genes and the downregulated genes in K133AS-R parasites. 1, 2, 3, 4, 5, and 6 are probable adaptations in K133AS-R parasites. (1.) Downregulation of Atg8 and HSP70 leads to increased ROS production, which was compensated by upregulation in the expression of GSH1, (2.) Upregulated expression of enzymes involved in amino acid and lipid metabolism and downregulated expression of the enzyme involved in carbohydrate metabolism, suggesting a dependency on these metabolites for energy generation, (3.) Reduced DNA synthesis that leads to the parasites in the quiescence state may be responsible for artesunate resistance in *Leishmania*, (4.) Reduced protein synthesis and reduced protein degradation, (5.) Upregulated expression of AQP1 leads to higher nutrient uptake and increased discharge of waste material and metabolic end product from the parasites and (6.) Upregulated expression of ABC transporter (ABCG1) and partial reversion or resistance in the presence of the ABC transporter modulator verapamil suggested probable involvement of the ABC transporter in the efflux of artesunate drug.

3.9.4. Modulated Expression of Transporters

Aquaglyceroporin (AQP1) (LinJ.31.0030), UDP-galactose transporter (LPG5B) (LinJ.18.0400), and ATP-binding cassette protein subfamily G, member 1, putative (ABCG1) (LmjF06.0080) were upregulated in K133 AS-R parasites. On the other hand, an ABC transporter family-like protein (LinJ.33.3410A), nuclear transport factor 2 (LinJ.10.0900), and amino acid transporters, AAT19 (LinJ.07.1340) and AAT22 (LinJ.22.0100), showed downregulated expression.

3.10. Correlation of Whole-Genome Sequencing Analysis with Transcriptomic Data

Advancement in genomics and transcriptomics technologies have conferred considerable enhancement of our knowledge related to the set of changes that occur within the parasite at the molecular level resulting in the evolution of drug resistance. Comparative analysis of the genome and transcriptome data of the two distinct strains of *L. donovani* (K133WT and K133AS-R) provided a correlation of the mechanism behind the development of artemisinin resistance. Eight upregulated and 10 downregulated genes in the transcriptome data matched with the NGS data. Out of eight upregulated genes, two amastin-like surface protein genes (LinJ08_V3.0700/LdBPK_080710) and (LinJ34_V3.0700/LdBPK_43111505) showed one and five mutations respectively; the remaining six genes had upstream gene variants. The types of mutations observed were insertion, deletion, and transition. Genes that correlated in both included the amino acid transporter ATP11 and cathepsin—L like cysteine protease. Out of 10 downregulated genes, ABCA2 (LinJ11_V3.1230/LdBPK_111210) and receptor-type adenylatecyclase b (fragment) (LinJ17_V3.0140/LdBPK_170120) displayed a frameshift mutation at two sites. The genes that were observed to be downregulated included phosphoacetylglucosamine mutase-like protein (LinJ07_V30930/LdBPK_070930) and pteridine transporter (LinJ06_V3.1320/LdBPK_061320). Cathepsin L-like protease, a type of lysosomal endopeptidases, is present in both the promastigote and amastimogote stage of *Leishmania* species and involved in crucial biological process of parasites, such as evasion of the host immune system [63–65].

4. Discussion

Sesquiterpene, artemisinin, a secondary metabolite extracted from *Artemisia annua*, is an important antimalarial drug that has shown antimicrobial and antiviral activities [66,67]. Several in vitro and in vivo studies suggested potential antileishmanial activity of this drug [8–10,68]. However, the possibility of the emergence of resistance following the use of artemisinin as antileishmanial treatment cannot be denied. In our previous study, we reported that in vitro-selected artesunate-resistant *Leishmania* parasites were more virulent, successfully modulating the host cell defense mechanism, and exhibited altered expression of genes involved in the unfolded protein response, as compared to sensitive parasites [41]. The present study aimed to explore the genome and transcriptome of artesunate-resistant *Leishmania* parasites in order to understand the mechanism of resistance and to safeguard this drug for future use. Next-generation sequencing (NGS) platforms have advanced to provide a precise and comprehensive means for the detection of molecular mutations. Genomic and transcriptomic analyses would help in the advancement of our understanding of the biology of *Leishmania*. This comparative analysis of whole-genome sequences attempted to explicate genetic factors responsible for drug resistance in *L. donovani*. Here, we demonstrated that the in vitro-selected artesunate-resistant (K133AS-R) parasite was quite distinct from the sensitive wild-type (K133WT) at the genome and transcriptome level.

Major findings of the study are summarized in three sections. Firstly, from the genomic landscape, we found a high number of SNPs and InDel, many of them having a pronounced influence (stop codon gained/lost and frame-shifts) on essential biological functions. Briefly, in K133AS-R, upstream gene variants were higher followed by intergenic region gene variants. Out of the total number of gene mutations, SNPs were high as compared to InDel. The highest number of SNPs was observed on chromosome number 12, 31, 34, and 35. Among InDel mutations, insertions were greater than deletions. Non-coding mutations, such as upstream gene variants and downstream gene variants, affect regulatory elements and lead a to loss of function that results in reduced gene expression, or a gain of function resulting in differential gene expression [69]. In this study, we analyzed that selective forces are majorly acting on non-coding regions of the genome. Secondly, the major changes observed were concerned with local copy number variations (CNVs). In K133AS-R, higher deletion occurred as compared to duplication and CNV lengths in the range of 1–5 kbp were either deleted or duplicated, depicting that changes occurred at small sequences rather than larger sequences. The highest number of CNVs were observed in K133AS-R on chromosome no 31, 29, 20, and 18. In the absence of regulation of gene expression at the initiation site, duplication/deletion of specific genes in a genomic sequence modulates the transcript level and its products [69,70]. Complex chromosomal copy number variation is often observed in *Leishmania* parasites due to their asexual mode of replication [26]. Thirdly, second-generation sequencing data obtained with the Illumina analyzer sets out remarkable read depth coverage throughout the chromosomes of *Leishmania*, and both the K133AS-R and K133 WT exhibited a uniform read depth in all chromosomes that is disomic except chromosome number 12 in

K133 AS-R. A read depth greater than two-fold was observed in case of chromosome 12, suggesting that the chromosome is present in the trisomy condition. Chromosomy variation in *Leishmania* is a well-known adaptive strategy in response to experimental drug resistance selection [71]. Aneuploidy is mostly influenced by the environmental condition and is more prevalent in promastigotes under in vitro conditions than in amastigotes present inside the vertebrate host. It arises through unlicensed replication due to a lack of proper cell cycle regulation and/or mitotic non-disjunction [43]. Further, GO terms based functional annotation of genes lead to classification into different categories, including metabolic, cellular processes, and biological regulation, which include the response to stimulus, cell signaling, and growth. WGS data analysis gives ample information regarding genetic variation compared to other sequencing approaches that include SNPs, InDel, as well as structural variants *viz* CNVs, inversion, translocation, and ploidy variation in chromosomes [72].

Analysis of transcriptome data by microarray and further experimental validation of differentially expressed proteins resulted in several important findings. To maintain cellular homeostasis, the eukaryotic cells have developed specialized mechanisms, such as lysis of intracellular proteins and organelles, which regulate cellular functions like enzymatic activity, removal of toxic or misfolded proteins, and the production of free amino acids to ensure cell survival under stressful conditions. The eukaryotic cells are known to perform these functions by the process of autophagy, which is believed to have originated at a later point during evolution [73]. Artesunate causes high levels of ROS generation within the cell. Further, it has been reported in cancer cells that autophagy plays a cytoprotective role within cells by inhibiting ROS. In K133AS-R, downregulated expression of Atg8 suggested reduced inhibition of ROS. In addition, deceased expression of HSP70, both at transcript and protein levels, suggested an accumulation of a higher number of misfolded proteins, resulting in higher ER stress and finally higher ROS production in K133AS-R parasites. On the other hand, upregulated expression of lipoate protein ligase leads to upregulated expression of GSH and thus plays an important role in glutathione biosynthesis and response to oxidative stress. This may be a compensatory approach of K133AS-R parasites to survive under oxidative stress.

Downregulated expression of gene phosphoacetylglucosamine mutase, which is eventually involved in the formation of fructose-6-phosphate, an important component of glycolysis/ gluconeogenesis, suggested downregulation of these pathways in K133AS-R parasites. In addition, the downregulated expression of various glucose transporters suggested that K133AS-R parasites may not depend on carbohydrate metabolism for energy requirements. Hence, artesunate-resistant parasites may depend on amino acids and lipids for energy generation as inferred by the upregulated expression of methylmalonyl CoA mutase (involved in isoleucine, valine, and leucine metabolism) and glutamine aminotransferases (involved in glutamine metabolism) while myo-inositol-1-phosphate synthase and a hypothetical protein are involved in lipid metabolism.

Leucin-rich AMA1 protein secreted by many *Leishmania* species, including *L. donovani*, helps them to interact with cholesterol present in the host cell membrane and thereby assist the internalization of parasites [74–76]. Amastin, a transmembraneglycoprotein, encoded by a large gene family initially reported in the amastigote stage of trypanosomes and later observed as a surface protein expressed in *Leishmania* species (encoded by six copies of genes) plays an important role in visceralization [77]. The importance of amastin in the pathogenesis of *Leishmania* species is well documented in a previous study [55]. The data analysis showed that the parasites may undergo genomic alterations to express certain genes differentially to adapt to the drug-induced selection pressure.

K133AS-R parasites showed downregulated expression of several genes involved in DNA synthesis and translation machinery. Reduced DNA/protein synthesis leads to an arrest of parasites in a quiescent state, which may be responsible for drug resistance as reported in case of artemisinin resistance in malaria [78]. Further, there was a downregulated expression of metallo- and carboxy-peptidase involved in protein degradation, which may be an adaptive approach of K133AS-R parasites to overcome reduced protein synthesis.

In *Leishmania*, AQP1 plays an important role in providing nutrients from the host organism, mainly glucose, amino acids, and fatty acids. These may also be responsible for discarding waste and metabolic end-products, such as lactate, from the parasite's cytosol [79]. In the presence of AQP1 inhibitor, drug-resistant mutants showed a significant increase in susceptibility towards artesunate at the promastigote stage; however, no significant alteration in drug susceptibility was observed in drug-sensitive parasites, indicating an important role of AQP1 in the selection of artesunate resistance in *Leishmania*. AQP1 has been reported to be involved in the uptake of antimonial drugs and its downregulated expression has been found to be associated with drug resistance [80,81]. Interestingly, in artesunate resistance, higher expression of AQP1 both at the mRNA and protein levels was observed to be associated with drug resistance. Another interesting observation was the decrease in the susceptibility of drug-sensitive parasites towards artesunate at the intracellular amastigote stage, whereas no significant alteration in drug susceptibility was observed with drug-resistant parasites.

Higher expression of ABC transporters has been widely reported in drug resistance in *Leishmania* [24,81–83]. Upregulated expression of ABCG1 (ABCG subfamily) was observed in artesunate resistance. Further, the use of the ABC transporter verapamil resulted in a significant increase in the susceptibility of K133AS-R parasites towards artesunate both at the promastigote and amastigote stage, suggesting an important role of ABCG1 in the selection of drug resistance. However, functional characterization of ABCG1 needs to be carried out in order to establish its role in artesunate resistance. LPG5B (UDP-galactose transporter) plays diverse roles in parasite survival, like the control of parasite binding to the sand fly midgut wall, resistance to lysis by complement, protection from oxidative damage, and delayed fusion of phagolysosomes. Upregulated expression of LPG5B in K133AS-R may be helpful to parasites for their survival under drug pressure.

Comparative genome as well as transcriptome data analysis resulted in several major findings, such as upregulation of cathepsin L-like protease, amastin-like surface protein, and amino acid transporter at both the genome as well as the RNA level. Downregulated genes that were observed to be in sync with NGS data were ABCG2, Pteridine receptor, receptor-type adenylatecyclase, phosphoaceylglucosamine mutase-like protein, and certain hypothetical proteins.

Our data explicate a better insight in genomic and transcriptomics alteration that occurs during artemisinin stress under in vitro conditions and would act as a baseline for further studies involving the applicability of genomic changes encountered in the study of the clinically resistant and sensitive *L. donovani* isolated from patients of leishmaniasis. Overall, this study highlights genes and interlinked pathways contributing to artemisinin resistance using *Leishmania* as a model and highlights putative mechanisms that have applicability not only to malaria but also other diseases against which the drug is found to be effective.

Supplementary Materials: The following are available online at http://www.mdpi.com/2073-4425/11/11/1362/s1, Figure S1: Comparative transcriptional responses following ART adaptation in *L. donovani*, Figure S2: Cytotoxicity of (A) AQP1 inhibitor and (B) Verapamil to host macrophages (mice PECs). Percentage cell viability ± SD with the increasing drug concentration has been plotted here. The data was obtained from three independent experiments, Table S1: List of genes validated for their modulated expression by Quantitative real time-PCR, Table S2: Pattern of up-regulated and down-regulated gene expression in K133 AS-R parasite, Table S3: Genes differentially modulated with their functional categories in K133 AS-R *L. donovani* parasites.

Author Contributions: Conceptualization, R.S., P.S., A.V. and S.G.; Methodology, S.G. and A.V.; Software, R.S. and D.P.; Validation, S.G., A.V., V.K. and D.P.; Formal Analysis, A.V., S.G. and V.K.; Investigation, S.G. and A.V.; Resources, R.S., P.S., A.S. and D.P.; Writing—Original Draft Preparation, A.V., S.G., V.K. and R.S.; Writing—Review & Editing, R.S., P.S. and A.S.; Supervision, R.S., P.S. and A.S.; Project Administration, R.S.; Funding Acquisition, R.S. All authors have read and agreed to the published version of the manuscript.

Funding: This work was supported by the Indian Council of Medical Research (ICMR, India) grant no. 53/16/2012-CMB/BMS and 6/9-7(188)2018-ECDII to RS. SG and AV are grateful to ICMR for financial support.

Conflicts of Interest: The authors declare no conflict of interest.

References

1. Burza, S.; Croft, S.L.; Boelaert, M. Leishmaniasis [seminar]. *Lancet* **2018**, *392*, 951–970. [CrossRef]
2. Alvar, J.; Vélez, I.D.; Bern, C.; Herrero, M.; Desjeux, P.; Cano, J.; Jannin, J.; de Boer, M. Leishmaniasis worldwide and global estimates of its incidence. *PLoS ONE* **2012**, *7*, e35671. [CrossRef] [PubMed]
3. Torres-Guerrero, E.; Quintanilla-Cedillo, M.R.; Ruiz-Esmenjaud, J.; Arenas, R. Leishmaniasis: A review. *F1000Research* **2017**, *6*. [CrossRef] [PubMed]
4. Salih, N.A.; van Griensven, J.; Chappuis, F.; Antierens, A.; Mumina, A.; Hammam, O.; Boulle, P.; Alirol, E.; Alnour, M.; Elhag, M.S. Liposomal amphotericin B for complicated visceral leishmaniasis (kala-azar) in eastern Sudan: How effective is treatment for this neglected disease? *Trop. Med. Int. Health* **2014**, *19*, 146–152. [CrossRef] [PubMed]
5. Rijal, S.; Ostyn, B.; Uranw, S.; Rai, K.; Bhattarai, N.R.; Dorlo, T.P.C.; Beijnen, J.H.; Vanaerschot, M.; Decuypere, S.; Dhakal, S.S. Increasing failure of miltefosine in the treatment of Kala-azar in Nepal and the potential role of parasite drug resistance, reinfection, or noncompliance. *Clin. Infect. Dis.* **2013**, *56*, 1530–1538. [CrossRef] [PubMed]
6. Sundar, S.; Singh, A.; Rai, M.; Prajapati, V.K.; Singh, A.K.; Ostyn, B.; Boelaert, M.; Dujardin, J.-C.; Chakravarty, J. Efficacy of miltefosine in the treatment of visceral leishmaniasis in India after a decade of use. *Clin. Infect. Dis.* **2012**, *55*, 543–550. [CrossRef]
7. Islamuddin, M.; Chouhan, G.; Want, M.Y.; Tyagi, M.; Abdin, M.Z.; Sahal, D.; Afrin, F. Leishmanicidal activities of *Artemisia annua* leaf essential oil against visceral leishmaniasis. *Front. Microbiol.* **2014**, *5*, 626. [CrossRef]
8. Ghaffarifar, F.; Heydari, F.E.; Dalimi, A.; Hassan, Z.M.; Delavari, M.; Mikaeiloo, H. Evaluation of apoptotic and antileishmanial activities of Artemisinin on promastigotes and BALB/C mice infected with *Leishmania major*. *Iran. J. Parasitol.* **2015**, *10*, 258.
9. Want, M.Y.; Islamuddin, M.; Chouhan, G.; Ozbak, H.A.; Hemeg, H.A.; Dasgupta, A.K.; Chattopadhyay, A.P.; Afrin, F. Therapeutic efficacy of artemisinin-loaded nanoparticles in experimental visceral leishmaniasis. *Colloids Surf. B Biointerfaces* **2015**, *130*, 215–221. [CrossRef]
10. Sen, R.; Bandyopadhyay, S.; Dutta, A.; Mandal, G.; Ganguly, S.; Saha, P.; Chatterjee, M. Artemisinin triggers induction of cell-cycle arrest and apoptosis in *Leishmania donovani* promastigotes. *J. Med. Microbiol.* **2007**, *56*, 1213–1218. [CrossRef]
11. Ebrahimisadr, P.; Ghaffarifar, F.; Hassan, Z.M. In-vitro evaluation of antileishmanial activity and toxicity of artemether with focus on its apoptotic effect. *Iran. J. Pharm. Res. IJPR* **2013**, *12*, 903. [PubMed]
12. Nemati, S.; Nahrevanian, H.; Haniloo, A.; Farahmand, M. Investigation on nitric oxide and C-reactive protein involvement in antileishmanial effects of artemisinin and glucantim on cutaneous leishmaniasis. *Adv. Stud. Biol.* **2013**, *5*, 27–36. [CrossRef]
13. Rodrigues, I.A.; Mazotto, A.M.; Cardoso, V.; Alves, R.L.; Amaral, A.C.F.; Silva, J.R.D.A.; Pinheiro, A.S.; Vermelho, A.B. Natural products: Insights into leishmaniasis inflammatory response. *Mediat. Inflamm.* **2015**, *2015*. [CrossRef] [PubMed]
14. Sen, R.; Saha, P.; Sarkar, A.; Ganguly, S.; Chatterjee, M. Iron enhances generation of free radicals by Artemisinin causing a caspase-independent, apoptotic death in *Leishmania donovani* promastigotes. *Free Radic. Res.* **2010**, *44*, 1289–1295. [CrossRef]
15. Hefnawy, A.; Berg, M.; Dujardin, J.-C.; De Muylder, G. Exploiting knowledge on *Leishmania* drug resistance to support the quest for new drugs. *Trends Parasitol.* **2017**, *33*, 162–174. [CrossRef]
16. Ouellette, M.; Drummelsmith, J.; Papadopoulou, B. Leishmaniasis: Drugs in the clinic, resistance and new developments. *Drug Resist. Updat.* **2004**, *7*, 257–266. [CrossRef]
17. Pourshafie, M.; Morand, S.; Virion, A.; Rakotomanga, M.; Dupuy, C.; Loiseau, P.M. Cloning of S-adenosyl-l-methionine: C-24-Δ-sterol-methyltransferase (ERG6) from *Leishmania donovani* and characterization of mRNAs in wild-type and amphotericin B-resistant promastigotes. *Antimicrob. Agents Chemother.* **2004**, *48*, 2409–2414. [CrossRef]
18. Jhingran, A.; Chawla, B.; Saxena, S.; Barrett, M.P.; Madhubala, R. Paromomycin: Uptake and resistance in *Leishmania donovani*. *Mol. Biochem. Parasitol.* **2009**, *164*, 111–117. [CrossRef]

19. Castanys-Muñoz, E.; Alder-Baerens, N.; Pomorski, T.; Gamarro, F.; Castanys, S. A novel ATP-binding cassette transporter from *Leishmania* is involved in transport of phosphatidylcholine analogues and resistance to alkyl-phospholipids. *Mol. Microbiol.* **2007**, *64*, 1141–1153. [CrossRef]

20. Mandal, S.; Maharjan, M.; Singh, S.; Chatterjee, M.; Madhubala, R. Assessing aquaglyceroporin gene status and expression profile in antimony-susceptible and-resistant clinical isolates of *Leishmania donovani* from India. *J. Antimicrob. Chemother.* **2010**, *65*, 496–507. [CrossRef]

21. Maltezou, H.C. Drug resistance in visceral leishmaniasis. *J. Biomed. Biotechnol.* **2009**, *2010*. [CrossRef] [PubMed]

22. Deep, D.K.; Singh, R.; Bhandari, V.; Verma, A.; Sharma, V.; Wajid, S.; Sundar, S.; Ramesh, V.; Dujardin, J.C.; Salotra, P. Increased miltefosine tolerance in clinical isolates of *Leishmania donovani* is associated with reduced drug accumulation, increased infectivity and resistance to oxidative stress. *PLoS Negl. Trop. Dis.* **2017**, *11*, e0005641. [CrossRef] [PubMed]

23. Kulshrestha, A.; Sharma, V.; Singh, R.; Salotra, P. Comparative transcript expression analysis of miltefosine-sensitive and miltefosine-resistant *Leishmania donovani*. *Parasitol. Res.* **2014**, *113*, 1171–1184. [CrossRef]

24. Verma, A.; Bhandari, V.; Deep, D.K.; Sundar, S.; Dujardin, J.C.; Singh, R.; Salotra, P. Transcriptome profiling identifies genes/pathways associated with experimental resistance to paromomycin in *Leishmania donovani*. *Int. J. Parasitol. Drugs Drug Resist.* **2017**, *7*, 370–377. [CrossRef]

25. Ubeda, J.-M.; Légaré, D.; Raymond, F.; Ouameur, A.A.; Boisvert, S.; Rigault, P.; Corbeil, J.; Tremblay, M.J.; Olivier, M.; Papadopoulou, B. Modulation of gene expression in drug resistant *Leishmania* is associated with gene amplification, gene deletion and chromosome aneuploidy. *Genome Biol.* **2008**, *9*, R115. [CrossRef] [PubMed]

26. Leprohon, P.; Legare, D.; Raymond, F.; Madore, E.; Hardiman, G.; Corbeil, J.; Ouellette, M. Gene expression modulation is associated with gene amplification, supernumerary chromosomes and chromosome loss in antimony-resistant *Leishmania infantum*. *Nucleic Acids Res.* **2009**, *37*, 1387–1399. [CrossRef]

27. Mukherjee, A.; Langston, L.D.; Ouellette, M. Intrachromosomal tandem duplication and repeat expansion during attempts to inactivate the subtelomeric essential gene GSH1 in *Leishmania*. *Nucleic Acids Res.* **2011**, *39*, 7499–7511. [CrossRef]

28. Ouameur, A.A.; Girard, I.; Légaré, D.; Ouellette, M. Functional analysis and complex gene rearrangements of the folate/biopterin transporter (FBT) gene family in the protozoan parasite *Leishmania*. *Mol. Biochem. Parasitol.* **2008**, *162*, 155–164. [CrossRef]

29. Downing, T.; Imamura, H.; Decuypere, S.; Clark, T.G.; Coombs, G.H.; Cotton, J.A.; Hilley, J.D.; de Doncker, S.; Maes, I.; Mottram, J.C. Whole genome sequencing of multiple *Leishmania donovani* clinical isolates provides insights into population structure and mechanisms of drug resistance. *Genome Res.* **2011**, *21*, 2143–2156. [CrossRef]

30. Mukherjee, A.; Boisvert, S.; do Monte-Neto, R.L.; Coelho, A.C.; Raymond, F.; Mukhopadhyay, R.; Corbeil, J.; Ouellette, M. Telomeric gene deletion and intrachromosomal amplification in antimony-resistant *L eishmania*. *Mol. Microbiol.* **2013**, *88*, 189–202. [CrossRef]

31. Kumar, P.; Lodge, R.; Raymond, F.; Ritt, J.; Jalaguier, P.; Corbeil, J.; Ouellette, M.; Tremblay, M.J. Gene expression modulation and the molecular mechanisms involved in Nelfinavir resistance in *Leishmania donovani* axenic amastigotes. *Mol. Microbiol.* **2013**, *89*, 565–582. [CrossRef] [PubMed]

32. Leprohon, P.; Fernandez-Prada, C.; Gazanion, É.; Monte-Neto, R.; Ouellette, M. Drug resistance analysis by next generation sequencing in *Leishmania*. *Int. J. Parasitol. Drugs Drug Resist.* **2015**, *5*, 26–35. [CrossRef] [PubMed]

33. Pérez-Victoria, F.J.; Castanys, S.; Gamarro, F. *Leishmania donovani* resistance to miltefosine involves a defective inward translocation of the drug. *Antimicrob. Agents Chemother.* **2003**, *47*, 2397–2403. [CrossRef] [PubMed]

34. Pérez-Victoria, F.J.; Sánchez-Cañete, M.P.; Castanys, S.; Gamarro, F. Phospholipid translocation and miltefosine potency require both *L. donovani* miltefosine transporter and the new protein LdRos3 in *Leishmania* parasites. *J. Biol. Chem.* **2006**, *281*, 23766–23775. [CrossRef] [PubMed]

35. Seifert, K.; Pérez-Victoria, F.J.; Stettler, M.; Sánchez-Cañete, M.P.; Castanys, S.; Gamarro, F.; Croft, S.L. Inactivation of the miltefosine transporter, LdMT, causes miltefosine resistance that is conferred to the amastigote stage of *Leishmania donovani* and persists in vivo. *Int. J. Antimicrob. Agents* **2007**, *30*, 229–235. [CrossRef]

36. Srivastava, S.; Mishra, J.; Gupta, A.K.; Singh, A.; Shankar, P.; Singh, S. Laboratory confirmed miltefosine resistant cases of visceral leishmaniasis from India. *Parasites Vectors* **2017**, *10*, 1–11. [CrossRef]

37. Cojean, S.; Houzé, S.; Haouchine, D.; Huteau, F.; Lariven, S.; Hubert, V.; Michard, F.; Bories, C.; Pratlong, F.; Le Bras, J. *Leishmania* resistance to miltefosine associated with genetic marker. *Emerg. Infect. Dis.* **2012**, *18*, 704. [CrossRef]

38. Mondelaers, A.; Sanchez-Cañete, M.P.; Hendrickx, S.; Eberhardt, E.; Garcia-Hernandez, R.; Lachaud, L.; Cotton, J.; Sanders, M.; Cuypers, B.; Imamura, H. Genomic and molecular characterization of miltefosine resistance in *Leishmania infantum* strains with either natural or acquired resistance through experimental selection of intracellular amastigotes. *PLoS ONE* **2016**, *11*, e0154101. [CrossRef]

39. Takala-Harrison, S.; Clark, T.G.; Jacob, C.G.; Cummings, M.P.; Miotto, O.; Dondorp, A.M.; Fukuda, M.M.; Nosten, F.; Noedl, H.; Imwong, M. Genetic loci associated with delayed clearance of *Plasmodium falciparum* following artemisinin treatment in Southeast Asia. *Proc. Natl. Acad. Sci. USA* **2013**, *110*, 240–245. [CrossRef]

40. Ariey, F.; Witkowski, B.; Amaratunga, C.; Beghain, J.; Langlois, A.-C.; Khim, N.; Kim, S.; Duru, V.; Bouchier, C.; Ma, L. A molecular marker of artemisinin-resistant *Plasmodium falciparum* malaria. *Nature* **2014**, *505*, 50–55. [CrossRef]

41. Verma, A.; Ghosh, S.; Salotra, P.; Singh, R. Artemisinin-resistant *Leishmania* parasite modulates host cell defense mechanism and exhibits altered expression of unfolded protein response genes. *Parasitol. Res.* **2019**, *118*, 2705–2713. [CrossRef] [PubMed]

42. Bronner, I.F.; Quail, M.A.; Turner, D.J.; Swerdlow, H. Improved protocols for illumina sequencing. *Curr. Protoc. Hum. Genet.* **2013**, *79*, 12–18. [CrossRef]

43. Dumetz, F.; Imamura, H.; Sanders, M.; Seblova, V.; Myskova, J.; Pescher, P.; Vanaerschot, M.; Meehan, C.J.; Cuypers, B.; De Muylder, G. Modulation of aneuploidy in *Leishmania donovani* during adaptation to different in vitro and in vivo environments and its impact on gene expression. *MBio* **2017**, *8*. [CrossRef] [PubMed]

44. Bolger, A.M.; Lohse, M.; Usadel, B. Trimmomatic: A flexible trimmer for Illumina sequence data. *Bioinformatics* **2014**, *30*, 2114–2120. [CrossRef]

45. Li, H.; Durbin, R. Fast and accurate short read alignment with Burrows–Wheeler transform. *Bioinformatics* **2009**, *25*, 1754–1760. [CrossRef]

46. Cingolani, P.; Platts, A.; Wang, L.L.; Coon, M.; Nguyen, T.; Wang, L.; Land, S.J.; Lu, X.; Ruden, D.M. A program for annotating and predicting the effects of single nucleotide polymorphisms, SnpEff: SNPs in the genome of *Drosophila melanogaster* strain w1118; iso-2; iso-3. *Fly (Austin)* **2012**, *6*, 80–92. [CrossRef]

47. Li, H.; Wysoker, A. Durbin R; 1000 Genome project data processing subgroup. The sequence alignment/Map format and SAMtools. *Bioinformatics* **2009**, *25*, 2078–2079. [CrossRef]

48. Abyzov, A.; Urban, A.E.; Snyder, M.; Gerstein, M. CNVnator: An approach to discover, genotype, and characterize typical and atypical CNVs from family and population genome sequencing. *Genome Res.* **2011**, *21*, 974–984. [CrossRef]

49. Altschul, S.F.; Madden, T.L.; Schäffer, A.A.; Zhang, J.; Zhang, Z.; Miller, W.; Lipman, D.J. Gapped BLAST and PSI-BLAST: A new generation of protein database search programs. *Nucleic Acids Res.* **1997**, *25*, 3389–3402. [CrossRef]

50. Kent, W.J. BLAT—the BLAST-like alignment tool. *Genome Res.* **2002**, *12*, 656–664. [CrossRef]

51. Conesa, A.; Götz, S.; García-Gómez, J.M.; Terol, J.; Talón, M.; Robles, M. Blast2GO: A universal tool for annotation, visualization and analysis in functional genomics research. *Bioinformatics* **2005**, *21*, 3674–3676. [CrossRef] [PubMed]

52. Rai, A.; Yamazaki, M.; Takahashi, H.; Nakamura, M.; Kojoma, M.; Suzuki, H.; Saito, K. RNA-seq transcriptome analysis of *Panax japonicus*, and its comparison with other *Panax species* to identify potential genes involved in the saponins biosynthesis. *Front. Plant Sci.* **2016**, *7*, 481. [CrossRef] [PubMed]

53. Consortium, G.O. Creating the gene ontology resource: Design and implementation. *Genome Res.* **2001**, *11*, 1425–1433. [CrossRef] [PubMed]

54. Ye, J.; Fang, L.; Zheng, H.; Zhang, Y.; Chen, J.; Zhang, Z.; Wang, J.; Li, S.; Li, R.; Bolund, L. WEGO: A web tool for plotting GO annotations. *Nucleic Acids Res.* **2006**, *34*, W293–W297. [CrossRef] [PubMed]

55. Rochette, A.; McNicoll, F.; Girard, J.; Breton, M.; Leblanc, É.; Bergeron, M.G.; Papadopoulou, B. Characterization and developmental gene regulation of a large gene family encoding amastin surface proteins in *Leishmania* spp. *Mol. Biochem. Parasitol.* **2005**, *140*, 205–220. [CrossRef]

56. Leprohon, P.; Légaré, D.; Ouellette, M. Intracellular localization of the ABCC proteins of *Leishmania* and their role in resistance to antimonials. *Antimicrob. Agents Chemother.* **2009**, *53*, 2646–2649. [CrossRef]

57. Bhandari, V.; Sundar, S.; Dujardin, J.C.; Salotra, P. Elucidation of cellular mechanisms involved in experimental paromomycin resistance in *Leishmania donovani*. *Antimicrob. Agents Chemother.* **2014**, *58*, 2580–2585. [CrossRef]

58. Sharma, V.; Sharma, P.; Selvapandiyan, A.; Salotra, P. *Leishmania donovani*-specific U b-related modifier-1: An early endosome-associated ubiquitin-like conjugation in *Leishmania donovani*. *Mol. Microbiol.* **2016**, *99*, 597–610. [CrossRef]

59. Samarasinghe, S.R.; Samaranayake, N.; Kariyawasam, U.L.; Siriwardana, Y.D.; Imamura, H.; Karunaweera, N.D. Genomic insights into virulence mechanisms of *Leishmania donovani*: Evidence from an atypical strain. *BMC Genom.* **2018**, *19*, 843. [CrossRef]

60. Rakotomanga, M.; Saint-Pierre-Chazalet, M.; Loiseau, P.M. Alteration of fatty acid and sterol metabolism in miltefosine-resistant *Leishmania donovani* promastigotes and consequences for drug-membrane interactions. *Antimicrob. Agents Chemother.* **2005**, *49*, 2677–2686. [CrossRef]

61. Besteiro, S.; Tonn, D.; Tetley, L.; Coombs, G.H.; Mottram, J.C. The AP3 adaptor is involved in the transport of membrane proteins to acidocalcisomes of Leishmania. *J. Cell Sci.* **2008**, *121*, 561–570. [CrossRef] [PubMed]

62. Tonkin, M.L.; Roques, M.; Lamarque, M.H.; Pugnière, M.; Douguet, D.; Crawford, J.; Lebrun, M.; Boulanger, M.J. Host cell invasion by apicomplexan parasites: Insights from the co-structure of AMA1 with a RON2 peptide. *Science* **2011**, *333*, 463–467. [CrossRef] [PubMed]

63. Camacho, E.; González-De la Fuente, S.; Rastrojo, A.; Peiró-Pastor, R.; Solana, J.C.; Tabera, L.; Gamarro, F.; Carrasco-Ramiro, F.; Requena, J.M.; Aguado, B. Complete assembly of the *Leishmania donovani* (HU3 strain) genome and transcriptome annotation. *Sci. Rep.* **2019**, *9*, 1–15. [CrossRef]

64. Lustigman, S.; Zhang, J.; Liu, J.; Oksov, Y.; Hashmi, S. RNA interference targeting cathepsin L and Z-like cysteine proteases of *Onchocerca volvulus* confirmed their essential function during L3 molting. *Mol. Biochem. Parasitol.* **2004**, *138*, 165–170. [CrossRef] [PubMed]

65. Dalton, J.P.; Neill, S.O.; Stack, C.; Collins, P.; Walshe, A.; Sekiya, M.; Doyle, S.; Mulcahy, G.; Hoyle, D.; Khaznadji, E. *Fasciola hepatica* cathepsin L-like proteases: Biology, function, and potential in the development of first generation liver fluke vaccines. *Int. J. Parasitol.* **2003**, *33*, 1173–1181. [CrossRef]

66. Appalasamy, S.; Lo, K.Y.; Ch'ng, S.J.; Nornadia, K.; Othman, A.S.; Chan, L.-K. Antimicrobial activity of artemisinin and precursor derived from in vitro plantlets of *Artemisia annua* L. *Biomed. Res. Int.* **2014**, *2014*. [CrossRef]

67. Roy, S.; He, R.; Kapoor, A.; Forman, M.; Mazzone, J.R.; Posner, G.H.; Arav-Boger, R. Inhibition of human cytomegalovirus replication by artemisinins: Effects mediated through cell cycle modulation. *Antimicrob. Agents Chemother.* **2015**, *59*, 3870–3879. [CrossRef]

68. Want, M.Y.; Islammudin, M.; Chouhan, G.; Ozbak, H.A.; Hemeg, H.A.; Chattopadhyay, A.P.; Afrin, F. Nanoliposomal artemisinin for the treatment of murine visceral leishmaniasis. *Int. J. Nanomed.* **2017**, *12*, 2189. [CrossRef]

69. Spielmann, M.; Kakar, N.; Tayebi, N.; Leettola, C.; Nürnberg, G.; Sowada, N.; Lupiáñez, D.G.; Harabula, I.; Flöttmann, R.; Horn, D. Exome sequencing and CRISPR/Cas genome editing identify mutations of ZAK as a cause of limb defects in humans and mice. *Genome Res.* **2016**, *26*, 183–191. [CrossRef]

70. Günzl, A. RNA polymerases and transcription factors of trypanosomes. In *RNA Metabolism in Trypanosomes*; Springer: Berlin/Heidelberg, Germany, 2012; pp. 1–27.

71. Barja, P.P.; Pescher, P.; Bussotti, G.; Dumetz, F.; Imamura, H.; Kedra, D.; Domagalska, M.; Chaumeau, V.; Himmelbauer, H.; Pages, M. Haplotype selection as an adaptive mechanism in the protozoan pathogen *Leishmania donovani*. *Nat. Ecol. Evol.* **2017**, *1*, 1961–1969. [CrossRef]

72. Gilissen, C.; Hehir-Kwa, J.Y.; Thung, D.T.; van de Vorst, M.; van Bon, B.W.M.; Willemsen, M.H.; Kwint, M.; Janssen, I.M.; Hoischen, A.; Schenck, A. Genome sequencing identifies major causes of severe intellectual disability. *Nature* **2014**, *511*, 344–347. [CrossRef] [PubMed]

73. Levine, B.; Kroemer, G. Autophagy in the pathogenesis of disease. *Cell* **2008**, *132*, 27–42. [CrossRef] [PubMed]

74. Kedzierski, L.; Montgomery, J.; Bullen, D.; Curtis, J.; Gardiner, E.; Jimenez-Ruiz, A.; Handman, E. A leucine-rich repeat motif of *Leishmania* parasite surface antigen 2 binds to macrophages through the complement receptor 3. *J. Immunol.* **2004**, *172*, 4902–4906. [CrossRef] [PubMed]

75. Gokulasuriyan, R.K.; Ghosh, M. Comparative in-silico genome analysis of *Leishmania (Leishmania) donovani*: A step towards its species specificity. *Meta Gene* **2014**, *2*, 782–798.

76. Laha, B.; Verma, A.K.; Biswas, B.; Sengodan, S.K.; Rastogi, A.; Willard, B.; Ghosh, M. Detection and characterization of an albumin-like protein in *Leishmania donovani*. *Parasitol. Res.* **2019**, *118*, 1609–1623. [CrossRef]
77. Jackson, A.P. The evolution of amastin surface glycoproteins in trypanosomatid parasites. *Mol. Biol. Evol.* **2010**, *27*, 33–45. [CrossRef]
78. Bridgford, J.L.; Xie, S.C.; Cobbold, S.A.; Pasaje, C.F.A.; Herrmann, S.; Yang, T.; Gillett, D.L.; Dick, L.R.; Ralph, S.A.; Dogovski, C. Artemisinin kills malaria parasites by damaging proteins and inhibiting the proteasome. *Nat. Commun.* **2018**, *9*, 1–9. [CrossRef]
79. Beitz, E. Aquaporins from pathogenic protozoan parasites: Structure, function and potential for chemotherapy. *Biol. Cell* **2005**, *97*, 373–383. [CrossRef]
80. Marquis, N.; Gourbal, B.; Rosen, B.P.; Mukhopadhyay, R.; Ouellette, M. Modulation in aquaglyceroporin AQP1 gene transcript levels in drug-resistant *Leishmania*. *Mol. Microbiol.* **2005**, *57*, 1690–1699. [CrossRef]
81. Kumar, D.; Singh, R.; Bhandari, V.; Kulshrestha, A.; Negi, N.S.; Salotra, P. Biomarkers of antimony resistance: Need for expression analysis of multiple genes to distinguish resistance phenotype in clinical isolates of *Leishmania donovani*. *Parasitol. Res.* **2012**, *111*, 223–230. [CrossRef]
82. Leprohon, P.; Légaré, D.; Girard, I.; Papadopoulou, B.; Ouellette, M. Modulation of *Leishmania* ABC protein gene expression through life stages and among drug-resistant parasites. *Eukaryot. Cell* **2006**, *5*, 1713–1725. [CrossRef] [PubMed]
83. Purkait, B.; Kumar, A.; Nandi, N.; Sardar, A.H.; Das, S.; Kumar, S.; Pandey, K.; Ravidas, V.; Kumar, M.; De, T. Mechanism of amphotericin B resistance in clinical isolates of *Leishmania donovani*. *Antimicrob. Agents Chemother.* **2012**, *56*, 1031–1041. [CrossRef] [PubMed]

 *genes*

Review

Trypanosoma cruzi Genome: Organization, Multi-Gene Families, Transcription, and Biological Implications

Alfonso Herreros-Cabello [1], Francisco Callejas-Hernández [1], Núria Gironès [1,2,*] and Manuel Fresno [1,2,*]

1   Centro de Biología Molecular Severo Ochoa, Consejo Superior de Investigaciones Científicas, Universidad Autónoma de Madrid, Cantoblanco, 28049 Madrid, Spain; alfonso.herreros@uam.es (A.H.-C.); bio.fcallejas@gmail.com (F.C.-H.)
2   Instituto Sanitario de Investigación Princesa, 28006 Madrid, Spain
*   Correspondence: ngirones@cbm.csic.es (N.G.); mfresno@cbm.csic.es (M.F.)

Received: 7 September 2020; Accepted: 12 October 2020; Published: 14 October 2020

Abstract: Chagas disease caused by the parasite *Trypanosoma cruzi* affects millions of people. Although its first genome dates from 2005, its complexity hindered a complete assembly and annotation. However, the new sequencing methods have improved genome annotation of some strains elucidating the broad genetic diversity and complexity of this parasite. Here, we reviewed the genomic structure and regulation, the genetic diversity, and the analysis of the principal multi-gene families of the recent genomes for several strains. The telomeric and sub-telomeric regions are sites with high recombination events, the genome displays two different compartments, the core and the disruptive, and the genome plasticity seems to play a key role in the survival and the infection process. *Trypanosoma cruzi* (*T. cruzi*) genome is composed mainly of multi-gene families as the trans-sialidases, mucins, and mucin-associated surface proteins. Trans-sialidases are the most abundant genes in the genome and show an important role in the effectiveness of the infection and the parasite survival. Mucins and MASPs are also important glycosylated proteins of the surface of the parasite that play a major biological role in both insect and mammal-dwelling stages. Altogether, these studies confirm the complexity of *T. cruzi* genome revealing relevant concepts to better understand Chagas disease.

Keywords: *Trypanosoma cruzi* strain; sequencing methods; genome plasticity; gene expression; trans-sialidases; mucins

---

1. General Aspects of *T. cruzi* Biology

Trypanosomatidae family includes parasites of vertebrates, invertebrates, and plants. Due to their adaptation to different environmental conditions and high biological diversity, these protists produce a major impact on all biotic communities [1,2]. *Trypanosoma cruzi* (*T. cruzi*) is the parasite that causes the Chagas disease or American Trypanosomiasis, a chronic endemic illness of Central and South America, and a neglected tropical disease. Chagas disease is characterized by an acute phase with low mortality and symptomatology. Then, the patients can remain in an asymptomatic phase for life or, after many years without any sign of disease, develop a symptomatic chronic phase with cardiomyopathy, megavisceras, or both [3]. Moreover, these variations in the disease outcomes are related to the high genetic variability of the parasite [4–7].

T. cruzi presents a very complex life cycle that includes an invertebrate hematophagous triatomine vector and a broad range of mammalian hosts [8]. In both insect and mammalian hosts, four different major developmental stages were identified [9,10]. The non-infective epimastigotes are present in the midgut of triatomines where they differentiate into infective metacyclic trypomastigotes that after the infection of host cells are differentiated into the replicative amastigotes [11]. Finally, these amastigotes replicate by binary fission and lyse the cell differentiating to bloodstream trypomastigotes that can infect other cells of the host.

The mitochondrial DNA of *T. cruzi* is formed by a network of concatenated circular molecules of maxicircles and minicircles that is called the kinetoplast. This structure contains dozens of maxicircles (20–40 kb) and thousands of minicircles (0.5–10 kb) with varying sizes depending on species [12,13]. Maxicircles contain the characteristic mitochondrial genes of other eukaryotes and consist of two regions: the coding region and the divergent/variable region, very difficult to sequence due to its repetitive sequences [14]. Minicircles are exclusive to trypanosomatids and they are directly involved in U-insertion/deletion editing system as they encode guide RNAs (gRNAs) [15]. Moreover, it is suggested that both molecule populations are heterogeneous showing strain-specific variations [16,17].

T. cruzi reproduction is usually asexual by binary division, but there are evidences of natural hybridization, genetic exchange between strains and sexual reproduction [18–21]. Also, the population genetics of *T. cruzi* generated a significant interest, producing two opposing views. A clonal theory was proposed considering *T. cruzi* as the paradigm of the predominant clonal evolution (PCE) model of pathogens, displaying that this parasite shares many features with other parasitic protozoa, fungi and bacteria [22,23]. However, other researchers have demonstrated that *T. cruzi* could reproduce sexually by a mechanism consistent with classic meiosis, and have suggested that the PCE model in this parasite does not reflect the biological reality [21,24].

In mitosis the genome of *T. cruzi* does not condense to form chromosomes, preventing its visualization by conventional techniques [25,26]. Instead, parasite karyotype was determined by molecular biology techniques, such as pulsed-field gel electrophoresis (PFGE) in combination with Southern blot. These studies revealed a large molecular variability in size and number of chromosomes between strains and even among clones of the same strain [27,28]. The parasite is usually described as diploid, and the size of chromosomes varies from 0.45 Mb to 4 Mb and the number from 19 to 40. Experiments by flow cytometry have estimated the genome size between 80 and 150 Mb [29].

2. Classification of *T. cruzi* Strains

There are many genetically different strains of *T. cruzi* [30,31]. Therefore, field investigators have looked for methods to classify these strains mostly according to their biological and genomic differences. The first classification was established in 1999 in a Satellite Meeting held at Fiocruz [32]. An expert committee reviewed the available data establishing two principal subgroups named *T. cruzi* I and *T. cruzi* II (Figure 1A). This classification was proposed considering biological and biochemical characteristics and molecular approaches such as the mini-exon studies and the 24Sα ribosomal DNA sequence.

Ten years later knowledge of the molecular diversity of the parasite increased and multilocus genotyping analyses revealed six distinct Discrete Typing Units (DTUs) [30], which in turn classified in two major subdivisions called DTU I and DTU II. DTUs are defined as "sets of stocks that are genetically more related to each other than to any other stock and that are identifiable by common genetic, molecular or immunological markers" [33]. Furthermore, based on phylogenetic information from multilocus enzyme electrophoresis (MLEE) and random amplified polymorphic DNA (RAPD) markers the DTU II was split into five DTUs (IIa-e) [34,35], and DTUs I and IIb correspond, respectively, to the *T. cruzi* I and *T. cruzi* II groups recommended by the original committee in 1999 (Figure 1B). This new classification considered that DTUs I and IIb were the ancestral strains, DTUs IId and IIe were the products of a minimum of two hybridization events [36–38], and DTUs IIa and IIc as ancestral hybrids.

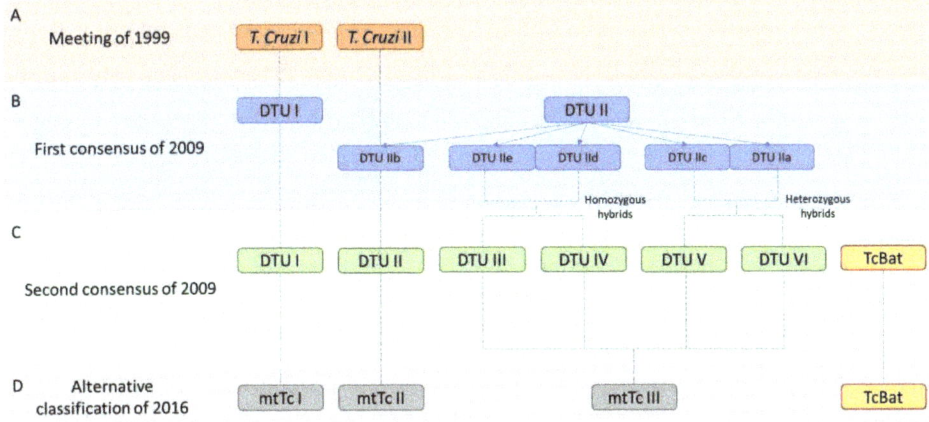

Figure 1. Different classifications of *Trypanosoma cruzi* since 1999. (**A**) Classification of the meeting of 1999. (**B**) First consensus classification of 2009. (**C**) Second consensus classification of 2009. (**D**) Alternative classification proposed in 2016.

However, a second revision that same year (2009) proposed a final classification in 6 DTUs [30]. DTUs I and II were the ancestral strains, DTUs III-IV those with at least one recombination event between DTUs I and II (homozygote hybrids), and DTUs V-VI were heterozygote hybrids of the DTUs II and III (Figure 1C). A new strain detected in bats was also included in the classification as TcBat [39] and with subsequent studies based on diverse molecular markers, it is considered to be the seventh DTU [40].

Finally, in 2016, Barnabé et al. [41] questioned the statistical validity of this classification. They performed a phylogenetic reconstruction by maximum likelihood trees based on the most common mitochondrial genes in databases. They proposed a new aggrupation considering the expression of three genes, two mitochondrial (*CytB* and *COII*) and one nuclear (*Gpi*). This new classification established three groups, the ancestral mtTcI and mtTcII, and the mtTcIII that grouped all the hybrid strains. They included the TcBat as an independent strain, although it was phylogenetically related to the mtTcI (Figure 1D).

3. The Genomes of *T. cruzi*: A New Update

The first version of a *T. cruzi* genome was published in 2005 [42] from the CL Brener strain. Interestingly, genomes for *Leishmania major* [43] and *Trypanosoma brucei* (*T. brucei*) [44] were simultaneously published in the same year.

The CL Brener strain was the most analyzed until then, with reproducible models in vitro, capable of producing an acute phase and being susceptible to Benznidazole [45]. In contrast to *Leishmania major* or *T. brucei* that had around 20–25% of repetitions in the genome, *T. cruzi* presented around 50%, making genome analysis and assembly more difficult [46]. Therefore, this first genome did not achieve the expected quality and remains incomplete, although it has been the principal reference for many researchers until today, despite the increasing availability of new and better genome sequences.

To date, there are several genomes of *T. cruzi* in the databases of the National Center for Biotechnology Information (NCBI) and TriTrypDB. This contributed to the study and understanding of the phenotypic, pathogenic, or complex variations among strains. Table 1 displays a summary of the recently available genomes in databases for the most studied strains. Some of these genomes were constructed from short-read sequencing methods (i.e., Illumina/Roche 454), such as Y [47], 231 [48], Sylvio X10/1 [49], G [50], or B7 strain of *T. cruzi marinkellei* [51]. Although these methods produce a high number of reads and have a low error rate, a relevant problem is the incapability to generate a complete chromosome reconstruction from short reads, causing very fragmented genomes in the case of complex genomes as trypanosomatids. This could lead to over-, under- or miss-representation of genes or complete chromosomic regions. In this regard, long-read sequencing methods (i.e., PacBio, Nanopore) could be a better choice for the trypanosomatids genomes [52], as the case of Bug2148 strain [53]. This technology allows the sequencing of long genetic fragments avoiding the complex and repetitive nature of the parasite. It could contribute to obtaining genomes with less redundant sequences and more completed, although the assembly size is still below the estimations made by DNA measurements (80–150 Mb) [29]. However, the error rate is bigger using long-read methods (and needs to be minimized by increasing the sequencing coverage) than in short-read methods. Therefore, in recent years, some laboratories chose the combination of both techniques to improve the assembly process, as the Berenice [54] or TCC and Dm28c [55] strains. In fact, the use of long-read sequencing methods generates contigs of more than 1 Mb, probably covering whole chromosomes. This allows the assembly of a genome in the smaller number of contigs, as happens with Berenice, Dm28c, TCC and Bug2148 strains (Table 1), obtaining the largest contig N50. Other researchers suggested that the copy number of conserved genes of *T. cruzi*, such as the monoglyceride lipase gene could be used as misassemble control [56].

Table 1. Data of the most recent genomes of the best-studied strains of *T. cruzi* and the B7 strain of *T. cruzi marinkellei*. BNEL: CL Brener Non-Esmeraldo-like; BEL: CL Brener Esmeraldo-like; PacBio: Pacific Biosciences. Contig N50: is a statistic median such that the 50% of the whole assembly is contained in contigs equal to or larger than this value.

Strain	DTU	Size (Mbp)	Contigs	Contig N50	%GC	Date of Version	Sequencing Method	References
G	I	25.17	1450	74,655	47.40	November 2018	Roche 454	[50]
Dm28c	I	53.27	636	317,638	51.60	May 2018	Illumina + PacBio	[55]
Sylvio X10/1	I	38.59	27,019	2307	51.20	October 2012	Roche 454 + Illumina	[49,57]
Berenice	II	40.80	934	148,957	51.20	June 2020	Illumina + Nanopore	[54]
Y	II	39.34	10,127	11,782	51.43	October 2017	Illumina	[47]
231	III	35.36	8469	14,202	48.60	January 2018	Illumina	[48]
Bug2148	V	55.22	934	196,760	51.63	October 2017	PacBio	[53]
CL	VI	65.00	7764	73,547	39.80	November 2018	Roche 454	[50]
TCC	VI	87.06	1236	264,196	51.70	May 2018	Illumina + PacBio	[55]
CL Brener	VI	89.94	32,746	14,669	51.70	July 2005	Sanger	[42]
BNEL	VI	32.53	41	870,934	43.94	December 2015	Sanger	[58]
BEL	VI	32.53	41	870,934	40.35	December 2015	Sanger	[58]
T. c. marinkellei B7 strain	—	38.65	23,154	2846	50.90	October 2012	Roche 454 + Illumina	[51]

Moreover, it was demonstrated that transcriptomic data may be useful to correct and re-annotate previous assembled genomes. Besides, in the case of Sylvio X10/1, RNAseq data was used to improve the previous genome annotation showing that 79.95% of the genome corresponds to the coding sequence, while the previous genomic analysis established only a 37.73% [57]. These results also suggested that the haploid genome for Sylvio X10/1 may be higher than previously reported (at least 51 Mb).

In the NCBI the reference genome is the hybrid CL Brener genome of 2005 [42,58] and presently many researchers rely on this information. CL Brener is a hybrid strain, where their homologous chromosomes presented different length and genetic content. Furthermore, this strain was separated in two haplotypes, named as Brener Esmeraldo-like and Brener Non-Esmeraldo-like, which genomes are also deposited in databases. Full length chromosome sequencing was performed with this hybrid strain, using a combined strategy based on bacterial artificial chromosome (BAC) ends sequencing and synteny maps with *T. brucei* [58], obtaining 41 virtual chromosomes (Table 1). Despite the continuous re-annotations of these genomes, they are far from being the best reference considering all the new and more completed genomes obtained with current techniques of long and short-read sequencing as Y [47], Bug2148 [53], Berenice [54] or Dm28c [55] strains. Therefore, we need to pose again which genome is appropriate as a reference for *T. cruzi* research and if the existence of just one genome reference is useful due to the high heterogeneity of the parasite. Moreover, and more importantly, some of the different DTUs of *T. cruzi* showed relevant differences in pathogenicity in mice [6]. This forces us to understand the differences at a genomic level and each strain would need a specific genomic analysis. Also, this high pathogenic, biological, and genetic diversity of the *T. cruzi* strains, even within DTUs, suggests that DTUs might not be a definitive form of classification, and it was hypothesized if *T. cruzi* could be a complex of species rather than a unique specie [59].

4. Genetic Diversity and Genome Structure of *T. cruzi*

4.1. Ploidy

Different studies confirmed the complexity of the *T. cruzi* genome, with different chromosome lengths between clones of the same strain, strains of distinct DTUs, or strains of the same DTU [26,28]. However, ploidy or chromosomal copy number variation (CCNV) analysis in this parasite could not be studied until the arrival of the Next Generation Sequencing (NGS) approaches.

Aneuploidy was studied in detail in *Leishmania*, whose "mosaic aneuploidies" are ploidy variations between isolates from the same strain and even between individual cells from the same population. These aneuploidies are related to drug resistance, gene expression regulation, or host adaptation [60–62]. Otherwise, in *T. brucei* a ploidy stability exists, including the subspecies *T. b. gambiense* and *T. b. rhodesiense* [63].

Regarding *T. cruzi*, the CCNV analysis depends on the quality of the assembled reference genome. Studies including strains of different DTUs revealed that as in *Leishmania*, the aneuploidy pattern varies among and within strains and DTUs [26]. However, the used reference genome was from CL Brener, which is not the most completed genome that we have in databases. Despite this limitation, it was concluded that the strains from DTU I seem to be more stable, while the strains from DTUs II and III present a high degree of aneuploidies as monosomies, trisomies, or tetrasomies [64].

These results suggest that the aneuploidies events could be used by *T. cruzi* to expand their genes and promote alterations in gene expression, something that may be critical for parasites that depend on post-transcriptional mechanisms to control gene expression. Although aneuploidies are mainly associated with debilitating phenotypes in many eukaryotes, they may be involved in species-specific adaptations during trypanosomatid evolution, affecting, for example, multi-gene families that are critical for the establishment of a productive infection in the mammalian hosts [65].

4.2. Genome Composition

Besides the different mechanisms to control gene expression such as polycistronic transcription, RNA editing, nuclear compartmentalization, or trans-splicing [66,67], *T. cruzi* presents genomic plasticity and an unusual gene organization among strains. Tandemly repeated sequences take up more than 50% of the *T. cruzi* genome and, although the parasite is considered a diploid organism, it presents variations in chromosome number and aneuploidy arrangements between strains and clones of the same strain [26,56,68].

The genome plasticity of *T. cruzi* is related to the genetic composition and a compartmentalization in two principal large regions of protein-coding genes was established. The first one is the core compartment, where we can find highly conserved genes with known function and genes without an assigned function typically annotated as hypothetical conserved genes that present synteny conservation with other species such as *Leishmania major* and *T. brucei*. The second one is the non-syntenic disruptive compartment, which is mainly composed by genes that evolve constantly, such as those that belong to surface multi-gene families (trans-sialidases, MASPs, or mucins). Both core and disruptive compartments show opposite G + C content and gene organization, with high differences in their regulatory sites [26,55].

T. cruzi genome is formed by three types of DNA. (1) Coding sequence of single-copy genes that are conserved between strains and species. (2) Coding sequence of multi-copy gene families, such as surface proteins or virulent factors. (3) Non-coding sequences and repetitive sequences, such as tandem repeats, retrotransposable elements and short repeat elements, which represent more than half of the genome affecting the methods of short-read sequencing above all as we explained before. Interestingly, around 50% of the genetic content of *T. cruzi* has unknown functions [47], which correlates with proteome studies of CL Brener, Dm28c, Y, and VFRA strains [69–72] in which around 40–50% of total proteins were of unknown function. This indicates how much we do not know yet about *T. cruzi* biology.

Regarding the single-copy genes, it was estimated that *T. cruzi* has more than 215 of these genes [54]. Although in the hybrid strains these genes might be underestimated according to previous results [47], due to the conservation of these genes and the apparition of new variations. Recent results in Y and Bug2148 strains confirmed this theory, with 183 and 400 detected single-copy genes, respectively [47]. The identification of these genes may help to understand the differential behaviors among strains as different pathogenicity, immune evasion, or life cycle.

4.3. Telomeric Regions

Telomeric and sub-telomeric regions in *T. cruzi* are sites of frequent DNA recombination that generate extensive genetic variations [73]. Therefore, they present a continuous evolutionary process. This concerns the relative abundance and organization of different genes, such as trans-sialidases, DGF-1 (dispersed gene family 1), RNA-helicases, RHS (retrotransposon hot spot genes), and N-acetyl-transferases [74,75]. In other protist parasites, as *T. brucei* or *Plasmodium falciparum*, sub-telomeric regions also present an important role in events of antigenic variation [76,77]. Trans-sialidase-like genes were located close to telomeric regions in *T. cruzi*, which generates new gene variations through non-homologous recombination. It was suggested that double-strand breaks produced in the sub-telomeric regions by retrotransposon nucleases are repaired by homologous recombination, but when the repair includes non-homologous chromatids there is a possibility to generate new gene variants [73]. This mechanism could contribute to the immune evasion of the parasite. Technically, it could also contribute to the collapsed assemblies of repetitive regions in sequencing. These sequences, the tandem repeats and/or other short repetitive genomic motifs, which correspond to telomeric and sub-telomeric regions, may produce an increment of fragmented genomes in *T. cruzi*, in comparison with other related species as *Leishmania* or *T. brucei*.

4.4. G + C Content

The %G + C is an indirect measure of the complexity of the genomes. Regarding the core and disruptive compartments of *T. cruzi*, they present a different content of G + C. While the core compartment has a 48% of G + C, the disruptive compartment has a 53%. In fact, it was hypothesized that genes with elevated recombination probability and constant evolution present high levels of guanines and cytosines [78,79]. We demonstrated in Y and Bug2148 strains that trans-sialidase-containing contigs (including pseudogenes) have a slightly higher %G + C content [47], suggesting that previous assemblies collapsed by repetitive sequences as those enriched in G + C [75,80]. These studies confirmed that variations in the %G + C were correlated with specific telomeric repeats described for *T. cruzi*, as the hexameric repeat TTAGGG and poly Ts structures [75,80]. Furthermore, in mammalian cells the %G + C content was correlated with mRNA expression, being the G + C-rich genes those with more efficient expression [81].

4.5. Replication Origin

Chromosomes of eukaryotic organisms are replicated from hundreds to thousands of DNA replication origins (ORIs), which are specified by the binding of the origin recognition complex (ORC). ORIs were mapped in *T. brucei* by marker frequency analysis sequencing (MFA-seq) coupled to ChIP analysis of the ORC [82]. These studies displayed that all mapped *T. brucei* ORIs are located at the boundaries of the transcription units. This was also detected in another specie as *Leishmania major*, where replication initiation sites are close to the genomic locations where the RNA pol II finishes, suggesting a strong correlation between the transcription kinetics and the replication initiation [83]. These studies also revealed more than 5000 potential sites of ORIs by SNS-seq techniques (Small nascent strand purification coupled with deep sequencing). However, another study detected by MFA-seq just one origin per chromosome in *Leishmania major* [84]. This happens because MFA-seq might detect mainly constitutive origins, while SNS-seq techniques may not reflect the frequency of origin activation, since these techniques might also identify flexible and/or dormant origins. Considering all those results, the complete replication of the genome in *T. brucei* and *Leishmania major* may require not merely constitutive ORIs that are fired in every cell cycle, but also further flexible and/or dormant ORIs, which cannot coincide with ORC binding and are fired stochastically [85].

Regarding *T. cruzi*, the ORIs of CL Brener strain were recently analyzed by MFA-seq [86], mapping 103 and 110 putative consensus ORIs in each haplotype of this hybrid strain. Moreover, the analysis displayed that some replication initiation sites map to the borders of the transcription units, as in *Leishmania major* and *T. brucei*. Interestingly, the majority of the putative predicted ORIs presented a great abundance within coding DNA sequences and showed a great G + C content enrichment (65% of average), while the genomic regions had a maximum of 54%. Also, another analysis with the same strain of *T. cruzi* by DNA combing, which can detect any replication initiation event (including constitutive, flexible and dormant origins, but without reference to genome location), displayed a median inter-origin distance of 1711 kb [87].

Considering the chromosomal location, while some ORIs of *T. cruzi* are located in non-transcribed regions as those seen in *T. brucei* and *Leishmania major*, many others are strategically localized at sub-telomeric regions (with a strong focus on DGF-1 genes), where they can produce genetic variability of multi-gene families [86]. The transcription orientation toward telomeres suggests that the abundance of putative ORIs in sub-telomeric regions produces head-on transcription-replication collisions since the replisomes go toward the centers of the chromosomes. These results suggest that collisions between DNA replication and transcription are recurrent in the *T. cruzi* genome and produce genetic variability, as suggested by the increase in SNP levels in the sub-telomeric regions and the DGF-1 genes containing putative ORIs [86].

5. Transcription of *T. cruzi*

Transcription in *T. cruzi* is polycistronic. Protein-coding genes are organized into non-overlapping clusters on the same DNA strand sometimes with unrelated predicted functions and separated by relatively short intergenic regions. Polycistronic transcripts are processed to produce mature mRNAs [88]. *T. cruzi* gene clusters can range from 30 to 500 kb separated by divergent or convergent strand-switch regions, or in a head-to-tail orientation whereby transcription terminates and then restarts from the same strand [57,89].

These strand-switch regions present a different nucleotide composition compared to the rest of the genome and a higher intrinsic curvature associated with transcriptional regulation [90]. In both *T. cruzi* and *T. brucei* canonical signals for RNA polymerase II promoters have not already been identified, except for the genes encoding the spliced leader (SL) [91]. In trypanosomatids, the transcription start sites and histone variants implicated in the transcription initiation process were described mainly at the divergent strand-switch regions [92,93]. Otherwise, the convergent strand-switch regions contain preferentially sites of transcription termination as well as RNA polymerase III transcribed tRNA genes [94].

Up to hundreds of genes are transcribed at the same time by the RNA pol II in large Polycistronic Transcription Units (PTUs). The final mRNA maturation occurs by trans-splicing and polyadenylation processes (Figure 2). The trans-splicing is a special form of RNA processing by which two mRNAs encoded in different genome locations react to constitute a unique transcript [95]. In *T. cruzi* it consists of the insertion of a sequence of 39 nucleotides in the 5′ of each transcript, known as mini-exon or SL. This SL is transcribed from a tandem array as a precursor of around 140 nucleotides and is the target for the capping modification. The insertion of this Cap-SL gives stability to the mRNA and causes the excision of each mRNA of the PTU allowing the final polyadenylation [88,96].

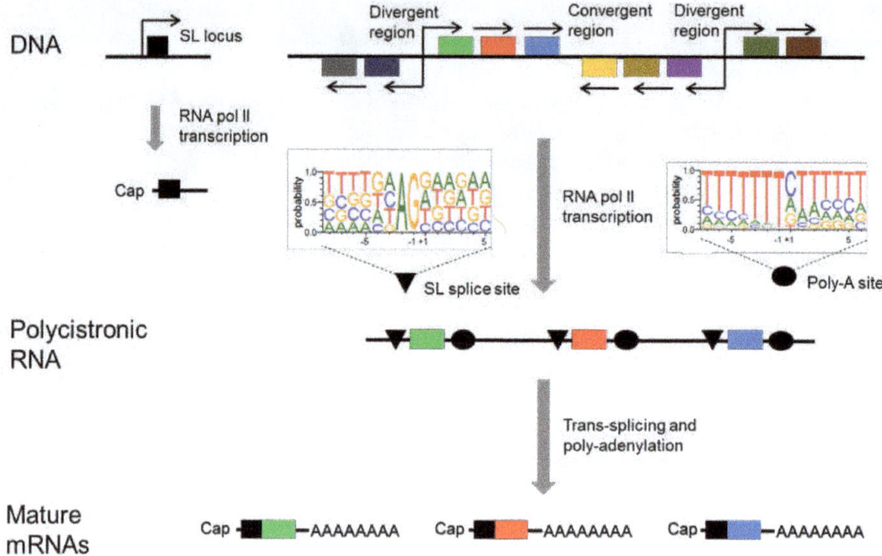

Figure 2. Transcription process of *T. cruzi*. RNA polymerase II produces polycistronic RNAs that are modified by trans-splicing and polyadenylation. The final mature mRNAs contain the Cap with the SL and the poly A tail. SL: spliced leader.

The AG dinucleotide was described as the consensus sequence for the SL trans-splicing in *T. cruzi* [57], Leishmania major [97] and *T. brucei* [98]. However, small differences were detected between all of them in the nucleotide composition surrounding the AG dinucleotide, suggesting that different specific mechanisms are involved in the mRNA maturation among these species. For example, considering the first residue before the AG dinucleotide, the most probable in *T. cruzi* is an adenine, as in *T. brucei*, while in Leishmania major is a cytosine. Also, at position -4 a guanine is the most probable nucleotide in *T. cruzi* and Leishmania major, in contrast to *T. brucei* where a poly T tract starts and continues up to 50 nucleotides upstream. Interestingly, this pyrimidine enrichment is one of the principal differences between these trypanosomatids. In *T. cruzi* and *T. brucei* this C-T pattern is conserved just in the upstream 5′ region, while in Leishmania major represent about the 70% of the nucleotides upstream and downstream the AG dinucleotide. Besides, whereas the downstream region in *T. cruzi* is composed of purine nucleotides (A–G) up to 60%, in *T. brucei* A-T dinucleotides are the most frequent bases, indicating that *T. cruzi* and Leishmania major transcripts present a more proportional nucleotide composition than *T. brucei*.

The AAUAAA polyadenylation signal of eukaryotes is not present in trypanosomatids. Recent studies published by our group demonstrated that *T. cruzi* shows a single nucleotide that seems to be the most probable signal of polyadenylation start, being cytosine the most frequent nucleotide (45.3%) and thymine the less frequent (6.79%) [57]. This differs from other trypanosomatids species, as *Leishmania major* and *T. brucei* that presents a AA dinucleotide [97,98] as the most probable signal for polyadenylation. Furthermore, the surrounding genomic regions are also different. Whereas *T. cruzi* displays an abundant thymine composition in the upstream region, and a higher T-A composition in the downstream, *Leishmania major* shows a more variable sequence composition in both upstream and downstream regions, and *T. brucei* a uniform pattern in both extremes composed by T-A nucleotides.

These results suggest that the mRNA maturation processes in *T. cruzi* may differ notably from *Leishmania major* and *T. brucei*.

Genes in trypanosomatids do not have promoter regions to regulate gene expression and their regulation is mainly at the post-transcriptional level, with a key role of the 3′ UTR regions. The principal mechanisms of regulation are the stability or instability of the transcripts, gene duplication, histone regulation, and translation efficiency [96,99]. Therefore, despite the genes of the same polycistron are transcribed in an equal proportion, differences in their expression were detected in distinct life cycle stages or growth conditions [97,98,100]. This could explain the selection of highly repetitive sequences in the parasite through evolution [46,101,102] by the aggregation of tandem repeats, retrotransposons, and repetitive short sequences in chromatin remodeling [103].

However, the concrete mechanisms involved in the regulation of the gene expression in *T. cruzi* are still unknown and were not further studied as in other species, such as *Leishmania* or *T. brucei* [104,105]. In this last specie, for example, it was demonstrated that in heat-shock conditions the genes close to the transcription initiation sites are down-regulated, while genes in a distal position increase their expression [106].

6. Principal Multi-Gene Families of *T. cruzi*

T. cruzi possesses several multi-gene families, some with hundreds of members, which contribute to the repetitive nature of the parasite's genome, such as the retrotransposons or the tandem repeats. Most of these multi-gene families code for surface proteins, which play different key roles in the *T. cruzi* life cycle, from the establishment of an effective host-cell interaction and invasion until the protection against the host immune system. Furthermore, these multi-gene families present a huge expansion and constant evolution that produces a great diversity among strains [107].

Therefore, many efforts to unravel the structure, distribution, and functions of these multi-gene families were made. Several groups identified in the disruptive compartment of the *T. cruzi* genome multi-gene families as trans-sialidases (TSs), mucins and MASPs, whereas RHS, GP63 and DGF-1 families were located in both disruptive and core compartments [55]. Copy numbers of these multi-gene families in the genomes of strains of *T. cruzi* and B7 strain of *T. cruzi marinkellei* are displayed in Figure 3. According to data, and considering all strains as whole, the most expanded multi-gene family is the TS family, following by MASPs, RHS, mucins and DGF-1, although this is not so for all strains with available genomes. There is a high variability among strains that may be related to a strain-specific genetic profile, the accuracy of the assembled genomes, and the genomic plasticity. This produces a great diversity that could explain the different infection kinetics, virulence and/or immune responses that were detected between *T. cruzi* strains [6,7,108].

Here, we focus on the principal multi-gene families in terms of diversity, abundance and function that belong to the disruptive compartment of the *T. cruzi* genome: TSs, mucins and MASPs.

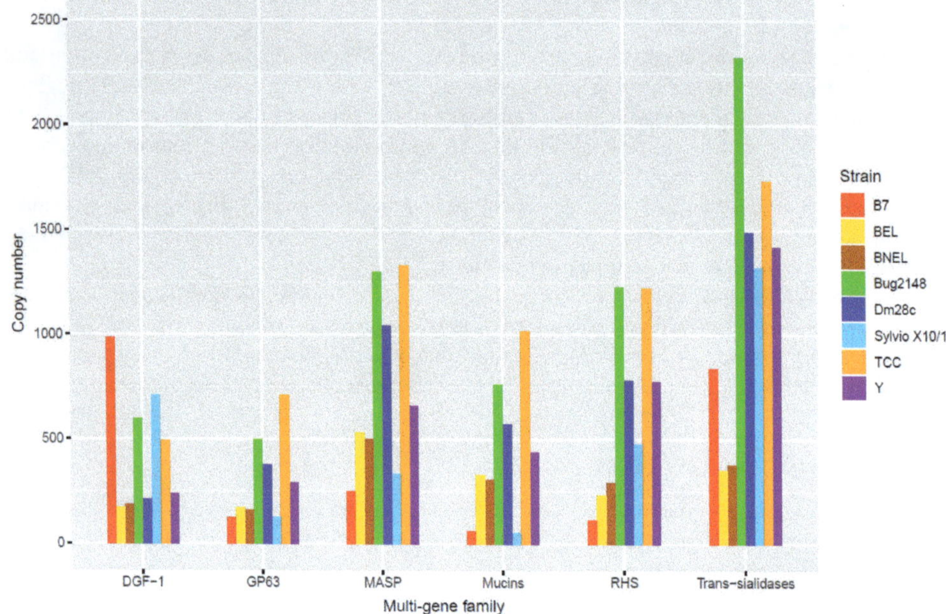

Figure 3. Genome copy number of the most abundant multi-gene families of *T. cruzi* and the B7 strain of *T. cruzi marinkellei*. BNEL: CL Brener Non-Esmeraldo-like; BEL: CL Brener Esmeraldo-like; DGF-1: Dispersed Gene Family 1; GP63: Glycoprotein 63; MASP: Mucin-Associated Surface Proteins; RHS: Retrotransposon Hot Spot genes.

6.1. Trans-Sialidase (TS) Family

The membrane of parasites as *T. cruzi*, *T. brucei*, or *Trypanosoma rangeli* (*T. rangeli*) is covered by many surface proteins, and most of them are TS or TS-like proteins that are critical for the interactions with the exogenous environment. The TS family is much smaller in *T. brucei* than in *T. cruzi*, and it is absent in *Leishmania major* [42,46,109]. In *T. cruzi*, TS members are localized on the membrane surface of metacyclic, bloodstream trypomastigotes, and intracellular amastigotes and are involved in host-parasite interaction processes [110–113]. They can present a glycosylphosphatidylinositol (GPI) anchor, although this can be removed by the action of a phosphatidylinositol phospholipase C, and then TSs can be released into the bloodstream. TSs are mainly distributed along the flagellum, cell body, and flagellar pocket of the parasite as mucins [114,115].

It is the largest family in *T. cruzi* considering all strains and all their members share the VTVxNVxLYNR motif [116], although some of them present a degeneration of this motif. The first estimates based on the CL Brener genome displayed that the TS family had around 1430 members and 639 pseudogenes [42,112,117] and subsequent studies with strains as Y, Dm28c, TCC or Sylvio X10/1 obtained similar numbers. However, Bug2148 strain displayed 2325 copies [47], almost the double, which could be caused by the hybrid origin of the strain, although the percentage of TS genes with respect to the total of the genetic content is very similar to other strains. Moreover, many TSs are found near the telomeric and sub-telomeric regions, which may cause collapsed assemblies and lead to under or over-representations of the genes. This implies that part of the TS expansion is due to their chromosomal location as we explained before. The other reason is the host immune system pressure to which the TSs are exposed [75] since they are targets of both humoral and cell-mediated immune responses [112].

The best-characterized function of this family is the trans-sialidase catalytic activity, which was first described in 1980 [118]. Posterior studies demonstrated that *T. cruzi* is unable to synthesize their own sialic acids and uses the TSs to incorporate sialic acids from host-cell sialoglycoconjugates to acceptor molecules of their membranes as mucins [119–124]. This sialylation confers a negatively charged coat that protects the trypomastigotes from being killed by human anti-α galactosyl antibodies [125]. A neuraminidase was described in TSs, although it is only active when suitable Gal acceptors are present. It was suggested that this neuraminidase activity just represents around 5% of the total activity of the TS enzyme [126].

Other studies suggest that the TS activity has a key role during the *T. cruzi* infection for parasite survival and the establishment of an effective infection [127]. TSs can interact with different cells from the mammalian hosts, as thymocytes, CD4[+] and CD8[+] T cells, B cells, cardiac fibroblasts, endothelial cells, platelets, neurons, and Schwann cells [117]. However, the critical residues that are necessary for the catalytic activity were identified just in a few genes and other roles related to host-ligand interactions and immune regulation were proposed [110,116]. Therefore, renaming this protein family would be advisable since not all their members have TS activity.

Genes encoding TS or TS-like genes were first classified into four groups according to their sequence similarity and functional properties [112,128,129] (Figure 4). TSs of the group I have trans-sialidase and/or neuraminidase catalytic activities [130] and were described in *T. rangeli* too [131]. Interestingly, *T. rangeli* lacks the trans-sialidase activity, retaining only the sialidase [132]. Some of the group I members in *T. cruzi* were the SAPA (shed acute-phase antigen), TS-epi, and TCNA (*T. cruzi* neuraminidase) proteins [116,129], which have active trans-sialidase and neuraminidase activities and are expressed in trypomastigotes [133] (except TS-epi, which is expressed and active in epimastigotes). Both SAPA and TCNA have an N-terminal catalytic region and a C-terminal extension with a tandem repeat of 12 amino acids (SAPA repeats), which consensus sequence is DSSAH [S/G]TPSTP [A/V], and a GPI anchor [134]. Conversely, TS-epi lacks the SAPA repeats and the GPI anchor [135].

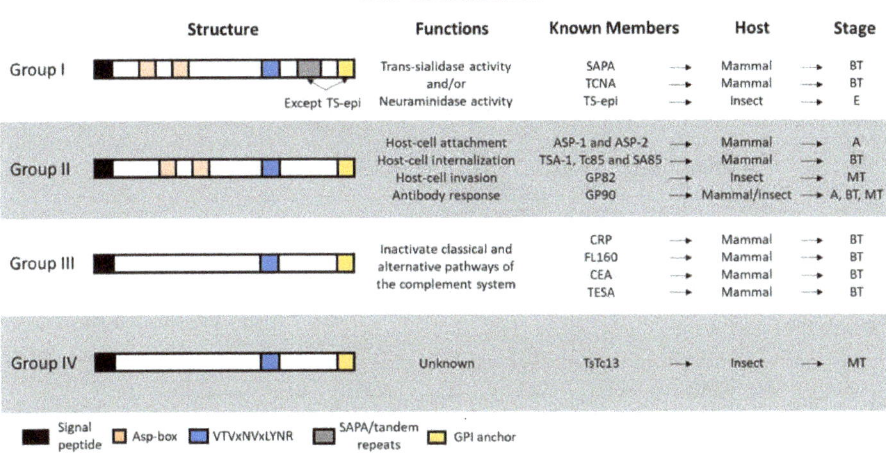

Figure 4. First classification of TS members. Four groups were described according to their sequence similarity and functional properties. The structure and functions of each group are displayed as well as the known members with their host and parasite-stage in which they are expressed. BT: bloodstream trypomastigotes; A: amastigotes; MT: metacyclic trypomastigotes; E: epimastigotes.

TSs of group II are expressed in trypomastigotes and intracellular amastigotes and were also described in *T. rangeli* [136]. This group comprises members of the so-called GP85 glycoproteins (ASP-1, ASP-2, TSA-1, Tc85, SA85, GP82, and GP90 among others) [137] which are related to host-cell attachment [138–142], strong antibody responses in mice and humans [143–145] and *T. cruzi* internalization and invasion [141,146–148]. They shared with the TSs of group I, apart from the common TS motif VTVxNVxLYNR, the motifs known Asp-box (SxDxGxTW) and the C-terminal GPI anchor.

Group III is formed by TSs found in bloodstream trypomastigotes as CRP, FL160, CEA, and Trypomastigote Excreted-Secreted Antigens (TESA), which can inactivate both the classical and the alternative pathways of the complement system protecting the parasite from lysis [149–151]. In addition, the TS group IV have TSs with the characteristic motif VTVxNVxLYNR, but with unknown functions.

However, a study of 2011 with the CL Brener strain established a different classification in eight groups by a sequence cluster analysis [110]. The sequence structure of this classification is displayed in Figure 5. The TSs of each group are defined by specific motifs and show specific activities, being the groups II and V those with more members (around 70% of the TSs in the study). Nevertheless, in databases many members of the TS family are annotated in the *T. cruzi* genomes only as trans-sialidase, without the group they belong to, making more difficult to work with this type of complex sequences.

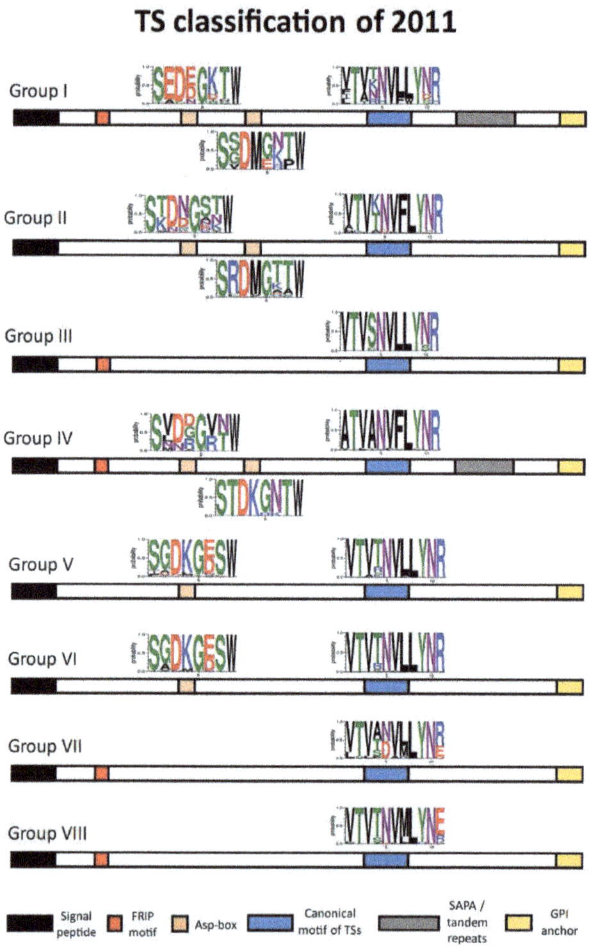

Figure 5. Classification of 2011 of TS members according to a sequence cluster analysis. Each group is defined by specific motifs. Logos of each Asp-box and canonical TS motifs are displayed. Adapted from Freitas, L. M. et al., 2011 [110].

Interestingly, phylogenetic analysis with several species of the *T. cruzi* clade and *T. brucei* showed that the variability of the TS-like sequences seems to be consistent with the aggrupation into eight groups. The detection of each TS group in each specie is displayed in Table 2. Group I TSs were found in all the species, in which two clades were established: the *T. brucei* clade with TSs of *T. brucei*, *T. congolense*, and *T. vivax* among others, and the *T. cruzi* clade with TSs of *T. cruzi*, *T. c. marinkellei* and *T. conorhini* among others [107]. It is important to remember that the TSs of group I are active catalytically, therefore it might be possible that other species just require this type of enzymatic function for their viability, while *T. cruzi* needs more TSs with other functions as interaction with host-cell ligands or immune evasion. Furthermore, other studies revealed that sialidases/sialidase-like proteins similar to all *T. cruzi* TS groups exist in *T. rangeli*, although this parasite exhibits fewer members of the trans-sialidase/sialidase family than *T. cruzi* [152,153]. TSs of group II that belong to different DTU strains were also analyzed in a new phylogenetic tree [113]. The results clustered together strains of the same DTU suggesting that TS group II genes might be used as markers for *T. cruzi* genotyping.

Table 2. Presence of at least one member of each TS group in different *Trypanosoma* species.

		TS Groups of *T. cruzi*							
		Group I	Group II	Group III	Group IV	Group V	Group VI	Group VII	Group VIII
	T. c. marinkellei	✓	✓	✓	✓	✓	✓	✓	✓
	T. rangeli	✓	✓	✓	✓	✓	✓	✓	✓
	T. conorhini	✓	✓			✓			
Trypanosoma	*T. dionisii*	✓	✓						
species with	*T. evansi*	✓							
TS sequence	*T. congolense*	✓							
similarity	*T. vivax*	✓							
	T. grayi	✓							
	T. carassii	✓							
	T. brucei	✓							

In this classification of eight groups, SAPA, TCNA, and TS-epi that are active TSs belonging to the previously defined group I, clustered together in the new group I, in which not all the members displayed the catalytic sites. ASP-2, Tc85, SA85, GP82, and GP90, which belonged to the previously defined group II and are related to host-cell attachment and invasion, were also classified in the new group II. In addition, finally, FL160 and other TSs involved in the complement system inhibition of the previous group III, clustered in the new group III too.

Regarding the common motifs in these TS groups, all of them have the canonical TS motif (VTVxNVxLYNR), although some variations exist (Figure 5). In fact, there is a small motif (FLY) inside the sequence that can act as a virulence factor [154], and it is only present in group II above all and group IV suggesting a host-cell attachment role in these groups. The Asp-box, previously described in viral and bacterial sialidases as SxDxGxTW [155], appears in some TSs of groups I, II, IV, V, and VI with some variations from the consensus sequence. Most of these TSs have one or two Asp-box, but a few displayed three. The function of this motif in *T. cruzi* remains unknown although it was hypothesized that TSs with these Asp-box could be more capable of binding carbohydrate molecules. The FRIP motif (with the pattern xRxP), which is located upstream the Asp-boxes and involved in binding the carboxylate group of sialic acid [156], was found in groups I, III, IV, VII, and VIII. This implies that although some TSs of these groups are enzymatically inactive, they still preserve carbohydrate-binding properties that could be important for the interaction with the host-cell [157,158]. Finally, tandem repeats, as SAPA repeats, were only found in groups I and IV. Interestingly, this classification forms three different patterns of motif occurrence. Groups I and IV have the most complex structure with all the previously described motifs, despite a few variations in the tandem repeats and the VTVxNVxLYNR motif. Groups II, V, and VI have only the Asp-box and the VTVxNVxLYNR motifs, and groups III, VII, and VIII contain only the FRIP and the VTVxNVxLYNR motifs.

A recent study evaluated the presence of each group in different *T. cruzi* strain genomes [47]. Considering these results and the new genomes of Dm28c and TCC (Figure 6), the TS group V is the most expanded, with the only exception of Sylvio X10/1 strain. TS group V was associated with antigenic variation allowing the adaptation of the parasites to the host environment [110]. TS group II is the second most abundant cluster, which contains trans-sialidases with host-interaction functions, and TS group I is the most expanded just in Sylvio X10/1 strain. Interestingly, TS group I that have the enzymatically active trans-sialidases, is much less abundant as predicted among strains. TS group III contains trans-sialidases that inhibit the complement pathways, and the different percentage of these trans-sialidases between strains could explain their different sensibility to the complement lysis. Finally, TS groups IV and VII are the less expanded, being absent in some strains as Sylvio X10/1, B7, or Dm28c. Therefore, this fact, in addition to the distinct distribution of Sylvio X10/1, could be caused by the quality of the assembled genomes and/or the annotation of them, being impossible to discard the presence of those TS groups or a few differences in the percentage of each TS group among strains.

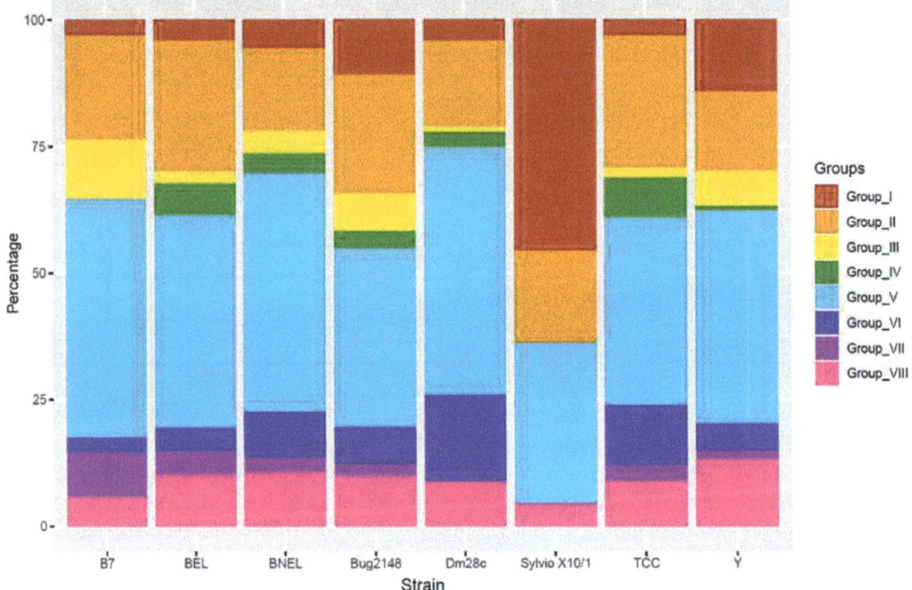

Figure 6. TS group distribution in genomes of different strains of *T. cruzi* and B7 strain of *T. cruzi marinkellei*. The percentage of each TS group is displayed. BNEL: CL Brener Non-Esmeraldo-like; BEL: CL Brener Esmeraldo-like.

6.2. Mucins

This is the most expressed family in the *T. cruzi* membrane and the fourth largest gene family, although 25% of them are non-functional pseudogenes [47,159]. Mucins that bear a dense array of oligosaccharides *O*-linked to serine and/or threonine residues, have two main functions: to protect the parasite from the defensive mechanisms of the host and to ensure the attachment and invasion of specific host cells [160]. These proteins are the principal acceptors of sialic acid in the parasite membrane [161] and they were classified in two subfamilies (TcMUC and TcSMUG) according to structural and biological criteria [162,163]. TcMUC proteins are only expressed in the mammalian stages of the parasite and TcSMUG in the insect-dwelling forms [160,164,165]. TcMUC proteins displayed more diversity than TcSMUG proteins and this is associated with their chromosome localization near to the telomeric regions and the immune system pressure that they suffer in the mammalian hosts [47,165].

TcSMUG (*T. cruzi* small mucin-like genes) subfamily is composed of two groups of genes, named L (large) and S (small), with differences in the genomic structure [166]. Considering the coding region, sequences of TcSMUG S and TcSMUG L display > 80% identity. TcSMUG S genes were identified as the backbone for the GP35/50 mucins that are expressed in the insect-dwelling stages [167]. GP35/50 mucins in metacyclic trypomastigotes bind to target cells to induce a bidirectional Ca^{2+} response which can contribute to the cell invasion [146]. However, the role of GP35/50 mucins in the epimastigotes is associated with protection against proteases of the insect intestinal tract [168]. Interestingly, TcSMUG S members, unlike TcSMUG L ones, are acceptors of the sialic acid residues that the TSs transferred to the parasite membrane. Otherwise, TcSMUG L products might be involved in the attachment to the luminal midgut surface of the vector and are exclusive of the epimastigote form [165,169]. Finally, some researchers saw that the expression of TcSMUG genes is post-transcriptionally regulated by AU-rich motifs of the 3′ UTR that recruit proteins to modulate the stability and translation efficiency of the mRNAs [166,170].

TcMUCs are subclassified in TcMUC I, II, and III genes. Interestingly, mucins from bloodstream trypomastigotes are called tGPI mucins and they suffer sialylation of their *O*-linked oligosaccharides by the TSs. These tGPI mucins are highly heterogeneous due to the simultaneous expression of several TcMUC I and II genes that display differences in their length, sequence, and structure of the attached oligosaccharides [162]. TcMUC II genes are quantitatively predominant over TcMUC I genes for the tGPI mucins [164].

TcMUC I members are more abundant in the amastigotes, whereas TcMUC II members are predominant in membrane lipid rafts of the bloodstream trypomastigotes [115]. TcMUC proteins contain a signal peptide, GPI anchor, and a principal central region. This central region has binding sites for N-acetylglucosamine residues and is rich in threonines. These residues are targets for the *O*-glycosylation and subsequent binding of sialic acid, which may explain why the mucins of mammalian host stages (amastigotes and bloodstream trypomastigotes) show higher glycosylation than those expressed by the epimastigotes [160]. A proportion of the TcMUC II genes are linked in the polycistronic transcription to TS genes [163,171].

In this central region TcMUC I genes have a short hypervariable (HV) section and many tandem repeats of the canonical Thr_8-Lys-Pro_2 sequence, although some degenerations in this sequence were described. Otherwise, TcMUC II genes have a central region with a long HV section and a few tandem repeats that are still rich in Thr and Pro. Some studies suggest that the TcMUC II genes have evolved from TcMUC I genes or vice versa. The common ancestor could be either a TcMUC I gene, which suffered a progressive expansion and diversification of its HV section or a TcMUC II gene which experienced an amplification of their original tandem repeats [163].

There is another type of mucin-like protein named TSSA (Trypomastigote Small Surface Antigen) that belongs to the TcMUC family and is present in the bloodstream trypomastigote membranes. TSSA is encoded by a single-copy gene and seems to have a role in the invasion of the host-cell as an adhesion molecule [172]. Also, it was one of the first immunological markers to allow discrimination between lineages. TSSA forms the called TcMUC group III and, unlike TcMUC I and II genes, it apparently does not display a Thr rich region [173]. Sequence analysis showed a high content of Ser and Thr residues and several signals for *O*-glycosylation [174]. However, another study described TSSA as a hypoglycosylated molecule [175], therefore further research is needed to elucidate its glycan composition and structure.

6.3. MASPs

MASPs (Mucin-Associated Surface Proteins) have a structural similarity to TcMUC II proteins and their expression seems to be up-regulated in mammal-dwelling stages [71]. They are the second largest gene family in the *T. cruzi* genome and received that name from their cluster position among large TS and mucin gene groups. MASPs are characterized by highly conserved N- and C- terminal domains, a GPI anchor, and a variable and repetitive central region [176].

According to some studies the MASP family constitutes about 6% of the parasite haploid genome and comprises between 500 and more than 1000 members varying among strains [47,55]. As in the TS family, the hybrid strain Bug2148 displays around the double of genes that Dm28c, Sylvio X10/1, or Y strains. The high variability of this family is not only due to the telomeric location of some of their members, and some researchers suggested that other mechanisms may exist. The high conservation of some motifs of the UTR sequences of these genes could contribute as sites for homologous recombination. It was suggested that one of the main mechanisms could be the retrotransposition by mobile elements of the TcTREZO type, specific of this gene family with its insertion sites at the conserved 5′ and 3′ ends [177].

MASPs have sites for both *N*- and *O*-glycosylation which undergo extensive post-translational modifications and were detected in trypomastigotes, amastigotes, and epimastigotes [178]. However, they seem to be overexpressed in the infectious stages (metacyclic and bloodstream trypomastigotes) and a critical role in the invasion process favoring the endocytosis was suggested. Other researchers have speculated that changes in the repertoire of MASP antigenic peptides could contribute to the evasion of the host immune system during the acute phase of Chagas disease [179]. Otherwise, antibodies against a specific MASP member can produce a decrease in the parasite internalization. MASP overexpression in the amastigote membrane before the binary fission suggests that some of these proteins play a major biological role in the survival and multiplication of the intracellular amastigotes [180].

7. Conclusions

Genomic studies are essential for the understanding of the *T. cruzi* pathogenicity and biology. The new sequencing technologies have contributed to improve the quality of several genomes of different strains and to elucidate the broad genetic diversity and complexity of this parasite. In this regard, the combination of long- and short-read sequencing methods may overcome the problems in the genome assembly and annotation due to the high intrinsic genome complexity of *T. cruzi*. Therefore, we need to wonder if the CL Brener genome is the best as reference in the databases, considering the new genomes of other strains that were obtained with these sequencing methods that have improved both assembly and annotation processes.

The study of the different repetitive sequences, recombination processes, and gene expansion events of the *T. cruzi* genome shows that genome plasticity plays a key role as a survival strategy during the life cycle of the parasite. Therefore, further research is needed to understand these relevant processes of the parasite biology.

Regarding the principal multi-gene families of *T. cruzi*, this parasite presents a wide variety of surface proteins with important roles in its life cycle. Due to the fact of genome plasticity, these multi-gene families have suffered an expansion and constant evolution that have increased the *T. cruzi* ability of adaptation, survival, and infection of both insect and mammalian hosts.

Altogether, the recent advances in trans-sialidases, mucins, and other multi-gene families can positively increase the current knowledge of host-parasite interactions and will allow the design of effective drugs against the Chagas disease. However, due to the genome complexity of the parasite more studies to unravel the specific structure and functions of those proteins will be needed. In this regard, high-throughput technologies will be useful to establish the development and evolution of multi-gene families of *T. cruzi*.

Author Contributions: M.F. conceived the idea; N.G. and M.F. acquired funding; A.H.-C. wrote the manuscript; M.F., N.G. and F.C.-H. reviewed and edited the manuscript. All authors have read and agreed to the published version of the manuscript.

Funding: This research was funded by: "Ministerio de Economía y competitividad" and "Fondo Europeo de Desarrollo Regional" (SAF2016-75988-R (MINECO/FEDER) to M.F.); "Ministerio de Ciencia, Innovación y Universidades-Agencia Estatal de Investigación" and "Fondo Europeo de Desarrollo Regional" (PGC2018-096132-B-I00 (MICINN/FEDER) to N.G.); "Red de Investigación de Centros de Enfermedades Tropicales" (RICET RD16/0027/0006 to M.F.); and Comunidad de Madrid (S2017/BMD-3671 to M.F.).

Acknowledgments: CBMSO institutional grants from "Fundación Ramón Areces" and "Banco de Santander".

Conflicts of Interest: The authors declare no conflict of interest.

References

1. Maslov, D.A.; Opperdoes, F.R.; Kostygov, A.Y.; Hashimi, H.; Lukeš, J.; Yurchenko, V. Recent advances in trypanosomatid research: Genome organization, expression, metabolism, taxonomy and evolution. *Parasitology* **2019**, *146*, 1–27. [CrossRef]

2. Lukeš, J.; Butenko, A.; Hashimi, H.; Maslov, D.A.; Votýpka, J.; Yurchenko, V. Trypanosomatids Are Much More than Just Trypanosomes: Clues from the Expanded Family Tree. *Trends Parasitol.* **2018**, *34*, 466–480. [CrossRef]

3. Rassi, A.; Rassi, A.; Marcondes de Rezende, J. American trypanosomiasis (Chagas disease). *Infect. Dis. Clin. N. Am.* **2012**, *26*, 275–291. [CrossRef] [PubMed]

4. Andrade, L.O.; Machado, C.R.S.; Chiari, E.; Pena, S.D.J.; Macedo, A.M. *Trypanosoma cruzi*: Role of host genetic background in the differential tissue distribution of parasite clonal populations. *Exp. Parasitol.* **2002**, *100*, 269–275. [CrossRef]

5. Manoel-Caetano, F.S.; Silva, A.E. Implications of genetic variability of *Trypanosoma cruzi* for the pathogenesis of Chagas disease. *Cad. Saúde Pública* **2007**, *23*, 2263–2274. [CrossRef] [PubMed]

6. Rodriguez, H.O.; Guerrero, N.A.; Fortes, A.; Santi-Rocca, J.; Gironès, N.; Fresno, M. *Trypanosoma cruzi* strains cause different myocarditis patterns in infected mice. *Acta Trop.* **2014**, *139*, 57–66. [CrossRef] [PubMed]

7. Santi-Rocca, J.; Fernandez-Cortes, F.; Chillón-Marinas, C.; González-Rubio, M.-L.; Martin, D.; Gironès, N.; Fresno, M. A multi-parametric analysis of *Trypanosoma cruzi* infection: Common pathophysiologic patterns beyond extreme heterogeneity of host responses. *Sci. Rep.* **2017**, *7*, 8893. [CrossRef]

8. Clayton, J. Chagas disease 101. *Nature* **2010**, *465*, S4–S5. [CrossRef]

9. De Souza, W.; de Carvalho, T.M.U.; Barrias, E.S. Review on *Trypanosoma cruzi*: Host Cell Interaction. *Int. J. Cell. Biol.* **2010**, *2010*. [CrossRef]

10. Rodrigues, J.C.F.; Godinho, J.L.P.; de Souza, W. Biology of human pathogenic trypanosomatids: Epidemiology, lifecycle and ultrastructure. *Subcell. Biochem.* **2014**, *74*, 1–42. [CrossRef]

11. Echeverria, L.E.; Morillo, C.A. American Trypanosomiasis (Chagas Disease). *Infect. Dis. Clin. N. Am.* **2019**, *33*, 119–134. [CrossRef] [PubMed]

12. Lukes, J.; Guilbride, D.L.; Votýpka, J.; Zíková, A.; Benne, R.; Englund, P.T. Kinetoplast DNA network: Evolution of an improbable structure. *Eukaryot. Cell* **2002**, *1*, 495–502. [CrossRef] [PubMed]

13. Thomas, S.; Martinez, L.L.I.T.; Westenberger, S.J.; Sturm, N.R. A population study of the minicircles in *Trypanosoma cruzi*: Predicting guide RNAs in the absence of empirical RNA editing. *BMC Genom.* **2007**, *8*, 133. [CrossRef]

14. Gerasimov, E.S.; Zamyatnina, K.A.; Matveeva, N.S.; Rudenskaya, Y.A.; Kraeva, N.; Kolesnikov, A.A.; Yurchenko, V. Common Structural Patterns in the Maxicircle Divergent Region of Trypanosomatidae. *Pathogens* **2020**, *9*, 100. [CrossRef]

15. Aphasizhev, R.; Aphasizheva, I. Mitochondrial RNA editing in trypanosomes: Small RNAs in control. *Biochimie* **2014**, *100*, 125–131. [CrossRef] [PubMed]

16. Westenberger, S.J.; Cerqueira, G.C.; El-Sayed, N.M.; Zingales, B.; Campbell, D.A.; Sturm, N.R. *Trypanosoma cruzi* mitochondrial maxicircles display species- and strain-specific variation and a conserved element in the non-coding region. *BMC Genom.* **2006**, *7*, 60. [CrossRef]

17. Messenger, L.A.; Llewellyn, M.S.; Bhattacharyya, T.; Franzén, O.; Lewis, M.D.; Ramírez, J.D.; Carrasco, H.J.; Andersson, B.; Miles, M.A. Multiple mitochondrial introgression events and heteroplasmy in *Trypanosoma cruzi* revealed by maxicircle MLST and next generation sequencing. *PLoS Negl. Trop. Dis.* **2012**, *6*, e1584. [CrossRef]

18. Gibson, W.; Bingle, L.; Blendeman, W.; Brown, J.; Wood, J.; Stevens, J. Structure and sequence variation of the trypanosome spliced leader transcript. *Mol. Biochem. Parasitol.* **2000**, *107*, 269–277. [CrossRef]

19. Akopyants, N.S.; Kimblin, N.; Secundino, N.; Patrick, R.; Peters, N.; Lawyer, P.; Dobson, D.E.; Beverley, S.M.; Sacks, D.L. Demonstration of genetic exchange during cyclical development of *Leishmania* in the sand fly vector. *Science* **2009**, *324*, 265–268. [CrossRef] [PubMed]

20. Berry, A.S.F.; Salazar-Sánchez, R.; Castillo-Neyra, R.; Borrini-Mayorí, K.; Chipana-Ramos, C.; Vargas-Maquera, M.; Ancca-Juarez, J.; Náquira-Velarde, C.; Levy, M.Z.; Brisson, D.; et al. Sexual reproduction in a natural *Trypanosoma cruzi* population. *PLoS Negl. Trop. Dis.* **2019**, *13*, e0007392. [CrossRef]

21. Schwabl, P.; Imamura, H.; Van den Broeck, F.; Costales, J.A.; Maiguashca-Sánchez, J.; Miles, M.A.; Andersson, B.; Grijalva, M.J.; Llewellyn, M.S. Meiotic sex in Chagas disease parasite *Trypanosoma cruzi*. *Nat. Commun.* **2019**, *10*, 3972. [CrossRef] [PubMed]

22. Tibayrenc, M.; Ayala, F.J. Reproductive clonality of pathogens: A perspective on pathogenic viruses, bacteria, fungi, and parasitic protozoa. *Proc. Natl. Acad. Sci. USA* **2012**, *109*, E3305–E3313. [CrossRef] [PubMed]

23. Tibayrenc, M.; Ayala, F.J. The population genetics of *Trypanosoma cruzi* revisited in the light of the predominant clonal evolution model. *Acta Trop.* **2015**, *151*, 156–165. [CrossRef] [PubMed]

24. Ramírez, J.D.; Llewellyn, M.S. Reproductive clonality in protozoan pathogens–truth or artefact? *Mol. Ecol.* **2014**, *23*, 4195–4202. [CrossRef]

25. Souza, R.T.; Lima, F.M.; Barros, R.M.; Cortez, D.R.; Santos, M.F.; Cordero, E.M.; Ruiz, J.C.; Goldenberg, S.; Teixeira, M.M.G.; da Silveira, J.F. Genome Size, Karyotype Polymorphism and Chromosomal Evolution in *Trypanosoma cruzi*. *PLoS ONE* **2011**, *6*, e23042. [CrossRef]

26. Reis-Cunha, J.L.; Rodrigues-Luiz, G.F.; Valdivia, H.O.; Baptista, R.P.; Mendes, T.A.O.; de Morais, G.L.; Guedes, R.; Macedo, A.M.; Bern, C.; Gilman, R.H.; et al. Chromosomal copy number variation reveals differential levels of genomic plasticity in distinct *Trypanosoma cruzi* strains. *BMC Genom.* **2015**, *16*. [CrossRef]

27. Lima, F.M.; Souza, R.T.; Santori, F.R.; Santos, M.F.; Cortez, D.R.; Barros, R.M.; Cano, M.I.; Valadares, H.M.S.; Macedo, A.M.; Mortara, R.A.; et al. Interclonal Variations in the Molecular Karyotype of *Trypanosoma cruzi*: Chromosome Rearrangements in a Single Cell-Derived Clone of the G Strain. *PLoS ONE* **2013**, *8*, e63738. [CrossRef]

28. Henriksson, J.; Dujardin, J.C.; Barnabé, C.; Brisse, S.; Timperman, G.; Venegas, J.; Pettersson, U.; Tibayrenc, M.; Solari, A. Chromosomal size variation in *Trypanosoma cruzi* is mainly progressive and is evolutionarily informative. *Parasitology* **2002**, *124*, 277–286. [CrossRef]

29. Lewis, M.D.; Llewellyn, M.S.; Gaunt, M.W.; Yeo, M.; Carrasco, H.J.; Miles, M.A. Flow cytometric analysis and microsatellite genotyping reveal extensive DNA content variation in *Trypanosoma cruzi* populations and expose contrasts between natural and experimental hybrids. *Int. J. Parasitol.* **2009**, *39*, 1305–1317. [CrossRef]

30. Zingales, B.; Andrade, S.G.; Briones, M.R.S.; Campbell, D.A.; Chiari, E.; Fernandes, O.; Guhl, F.; Lages-Silva, E.; Macedo, A.M.; Machado, C.R.; et al. A new consensus for *Trypanosoma cruzi* intraspecific nomenclature: Second revision meeting recommends TcI to TcVI. *Mem. Inst. Oswaldo Cruz* **2009**, *104*, 1051–1054. [CrossRef]

31. Brenière, S.F.; Waleckx, E.; Barnabé, C. Over Six Thousand *Trypanosoma cruzi* Strains Classified into Discrete Typing Units (DTUs): Attempt at an Inventory. *PLoS Negl. Trop. Dis.* **2016**, *10*, e0004792. [CrossRef] [PubMed]

32. Anonymous. Recommendations from a satellite meeting. *Mem. Inst. Oswaldo Cruz* **1999**, *94*, 429–432. [CrossRef] [PubMed]

33. Tibayrenc, M. Genetic epidemiology of parasitic protozoa and other infectious agents: The need for an integrated approach. *Int. J. Parasitol.* **1998**, *28*, 85–104. [CrossRef]

34. Brisse, S.; Barnabé, C.; Tibayrenc, M. Identification of six *Trypanosoma cruzi* phylogenetic lineages by random amplified polymorphic DNA and multilocus enzyme electrophoresis. *Int. J. Parasitol.* **2000**, *30*, 35–44. [CrossRef]

35. Brisse, S.; Verhoef, J.; Tibayrenc, M. Characterisation of large and small subunit rRNA and mini-exon genes further supports the distinction of six *Trypanosoma cruzi* lineages. *Int. J. Parasitol.* **2001**, *31*, 1218–1226.

36. Tomazi, L.; Kawashita, S.Y.; Pereira, P.M.; Zingales, B.; Briones, M.R.S. Haplotype distribution of five nuclear genes based on network genealogies and Bayesian inference indicates that *Trypanosoma cruzi* hybrid strains are polyphyletic. *Genet. Mol. Res.* **2009**, *8*, 458–476. [CrossRef]

37. De Freitas, J.M.; Augusto-Pinto, L.; Pimenta, J.R.; Bastos-Rodrigues, L.; Gonçalves, V.F.; Teixeira, S.M.R.; Chiari, E.; Junqueira, Â.C.V.; Fernandes, O.; Macedo, A.M.; et al. Ancestral Genomes, Sex, and the Population Structure of *Trypanosoma cruzi*. *PLoS Pathog.* **2006**, *2*, e24. [CrossRef]

38. Westenberger, S.J.; Barnabé, C.; Campbell, D.A.; Sturm, N.R. Two Hybridization Events Define the Population Structure of *Trypanosoma cruzi*. *Genetics* **2005**, *171*, 527–543. [CrossRef]

39. Marcili, A.; Lima, L.; Cavazzana, M.; Junqueira, A.C.V.; Veludo, H.H.; Maia Da Silva, F.; Campaner, M.; Paiva, F.; Nunes, V.L.B.; Teixeira, M.M.G. A new genotype of *Trypanosoma cruzi* associated with bats evidenced by phylogenetic analyses using SSU rDNA, cytochrome b and Histone H2B genes and genotyping based on ITS1 rDNA. *Parasitology* **2009**, *136*, 641–655. [CrossRef]

40. Lima, L.; Espinosa-Álvarez, O.; Ortiz, P.A.; Trejo-Varón, J.A.; Carranza, J.C.; Pinto, C.M.; Serrano, M.G.; Buck, G.A.; Camargo, E.P.; Teixeira, M.M.G. Genetic diversity of *Trypanosoma cruzi* in bats, and multilocus phylogenetic and phylogeographical analyses supporting Tcbat as an independent DTU (discrete typing unit). *Acta Trop.* **2015**, *151*, 166–177. [CrossRef]

41. Barnabé, C.; Mobarec, H.I.; Jurado, M.R.; Cortez, J.A.; Brenière, S.F. Reconsideration of the seven discrete typing units within the species *Trypanosoma cruzi*, a new proposal of three reliable mitochondrial clades. *Infect. Genet. Evol.* **2016**, *39*, 176–186. [CrossRef] [PubMed]

42. El-Sayed, N.M.; Myler, P.J.; Bartholomeu, D.C.; Nilsson, D.; Aggarwal, G.; Tran, A.-N.; Ghedin, E.; Worthey, E.A.; Delcher, A.L.; Blandin, G.; et al. The Genome Sequence of *Trypanosoma cruzi*, Etiologic Agent of Chagas Disease. *Science* **2005**, *309*, 409–415. [CrossRef] [PubMed]

43. Ivens, A.C.; Peacock, C.S.; Worthey, E.A.; Murphy, L.; Aggarwal, G.; Berriman, M.; Sisk, E.; Rajandream, M.-A.; Adlem, E.; Aert, R.; et al. The genome of the kinetoplastid parasite, *Leishmania major*. *Science* **2005**, *309*, 436–442. [CrossRef] [PubMed]

44. Berriman, M.; Ghedin, E.; Hertz-Fowler, C.; Blandin, G.; Renauld, H.; Bartholomeu, D.C.; Lennard, N.J.; Caler, E.; Hamlin, N.E.; Haas, B.; et al. The genome of the African trypanosome *Trypanosoma brucei*. *Science* **2005**, *309*, 416–422. [CrossRef] [PubMed]

45. Reis-Cunha, J.L.; Bartholomeu, D.C. *Trypanosoma cruzi* Genome Assemblies: Challenges and Milestones of Assembling a Highly Repetitive and Complex Genome. *Methods Mol. Biol.* **2019**, *1955*, 1–22. [CrossRef] [PubMed]

46. El-Sayed, N.M.; Myler, P.J.; Blandin, G.; Berriman, M.; Crabtree, J.; Aggarwal, G.; Caler, E.; Renauld, H.; Worthey, E.A.; Hertz-Fowler, C.; et al. Comparative genomics of trypanosomatid parasitic protozoa. *Science* **2005**, *309*, 404–409. [CrossRef]

47. Callejas-Hernández, F.; Rastrojo, A.; Poveda, C.; Gironès, N.; Fresno, M. Genomic assemblies of newly sequenced *Trypanosoma cruzi* strains reveal new genomic expansion and greater complexity. *Sci. Rep.* **2018**, *8*, 14631. [CrossRef]

48. Baptista, R.P.; Reis-Cunha, J.L.; DeBarry, J.D.; Chiari, E.; Kissinger, J.C.; Bartholomeu, D.C.; Macedo, A.M. Assembly of highly repetitive genomes using short reads: The genome of discrete typing unit III *Trypanosoma cruzi* strain 231. *Microb. Genom.* **2018**, *4*. [CrossRef]

49. Franzén, O.; Ochaya, S.; Sherwood, E.; Lewis, M.D.; Llewellyn, M.S.; Miles, M.A.; Andersson, B. Shotgun Sequencing Analysis of *Trypanosoma cruzi* I Sylvio X10/1 and Comparison with *T. cruzi* VI CL Brener. *PLoS Negl. Trop. Dis.* **2011**, *5*, e984. [CrossRef]

50. Bradwell, K.R.; Koparde, V.N.; Matveyev, A.V.; Serrano, M.G.; Alves, J.M.P.; Parikh, H.; Huang, B.; Lee, V.; Espinosa-Alvarez, O.; Ortiz, P.A.; et al. Genomic comparison of *Trypanosoma conorhini* and *Trypanosoma rangeli* to *Trypanosoma cruzi* strains of high and low virulence. *BMC Genom.* **2018**, *19*. [CrossRef]

51. Franzén, O.; Talavera-López, C.; Ochaya, S.; Butler, C.E.; Messenger, L.A.; Lewis, M.D.; Llewellyn, M.S.; Marinkelle, C.J.; Tyler, K.M.; Miles, M.A.; et al. Comparative genomic analysis of human infective *Trypanosoma cruzi* lineages with the bat-restricted subspecies *T. cruzi marinkellei*. *BMC Genom.* **2012**, *13*, 531. [CrossRef] [PubMed]

52. Camacho, E.; González-de la Fuente, S.; Rastrojo, A.; Peiró-Pastor, R.; Solana, J.C.; Tabera, L.; Gamarro, F.; Carrasco-Ramiro, F.; Requena, J.M.; Aguado, B. Complete assembly of the *Leishmania* donovani (HU3 strain) genome and transcriptome annotation. *Sci. Rep.* **2019**, *9*, 6127. [CrossRef] [PubMed]

53. Callejas-Hernández, F.; Gironès, N.; Fresno, M. Genome Sequence of *Trypanosoma cruzi* Strain Bug2148. *Genome Announc.* **2018**, *6*. [CrossRef] [PubMed]

54. Díaz-Viraqué, F.; Pita, S.; Greif, G.; de Souza, R.C.M.; Iraola, G.; Robello, C. Nanopore Sequencing Significantly Improves Genome Assembly of the Protozoan Parasite *Trypanosoma cruzi*. *Genome Biol. Evol.* **2019**, *11*, 1952–1957. [CrossRef] [PubMed]

55. Berná, L.; Rodriguez, M.; Chiribao, M.L.; Parodi-Talice, A.; Pita, S.; Rijo, G.; Alvarez-Valin, F.; Robello, C. Expanding an expanded genome: Long-read sequencing of *Trypanosoma cruzi*. *Microb. Genom.* **2018**, *4*. [CrossRef]

56. Arner, E.; Kindlund, E.; Nilsson, D.; Farzana, F.; Ferella, M.; Tammi, M.T.; Andersson, B. Database of *Trypanosoma cruzi* repeated genes: 20,000 additional gene variants. *BMC Genom.* **2007**, *8*, 391. [CrossRef]

57. Callejas-Hernández, F.; Gutierrez-Nogues, Á.; Rastrojo, A.; Gironès, N.; Fresno, M. Analysis of mRNA processing at whole transcriptome level, transcriptomic profile and genome sequence refinement of *Trypanosoma cruzi*. *Sci. Rep.* **2019**, *9*. [CrossRef]
58. Weatherly, D.B.; Boehlke, C.; Tarleton, R.L. Chromosome level assembly of the hybrid *Trypanosoma cruzi* genome. *BMC Genom.* **2009**, *10*, 255. [CrossRef]
59. Berná, L.; Pita, S.; Chiribao, M.L.; Parodi-Talice, A.; Alvarez-Valin, F.; Robello, C. Biology of the *Trypanosoma cruzi* Genome. In *Biology of Trypanosoma cruzi*; IntechOpen: London, UK, 2019. [CrossRef]
60. Downing, T.; Imamura, H.; Decuypere, S.; Clark, T.G.; Coombs, G.H.; Cotton, J.A.; Hilley, J.D.; de Doncker, S.; Maes, I.; Mottram, J.C.; et al. Whole genome sequencing of multiple *Leishmania donovani* clinical isolates provides insights into population structure and mechanisms of drug resistance. *Genome Res.* **2011**, *21*, 2143–2156. [CrossRef]
61. Mannaert, A.; Downing, T.; Imamura, H.; Dujardin, J.-C. Adaptive mechanisms in pathogens: Universal aneuploidy in *Leishmania*. *Trends Parasitol.* **2012**, *28*, 370–376. [CrossRef]
62. Dujardin, J.-C.; Mannaert, A.; Durrant, C.; Cotton, J.A. Mosaic aneuploidy in *Leishmania*: The perspective of whole genome sequencing. *Trends Parasitol.* **2014**, *30*, 554–555. [CrossRef]
63. Almeida, L.V.; Coqueiro-Dos-Santos, A.; Rodriguez-Luiz, G.F.; McCulloch, R.; Bartholomeu, D.C.; Reis-Cunha, J.L. Chromosomal copy number variation analysis by next generation sequencing confirms ploidy stability in *Trypanosoma brucei* subspecies. *Microb. Genom.* **2018**, *4*. [CrossRef] [PubMed]
64. Reis-Cunha, J.L.; Baptista, R.P.; Rodrigues-Luiz, G.F.; Coqueiro-Dos-Santos, A.; Valdivia, H.O.; de Almeida, L.V.; Cardoso, M.S.; D'Ávila, D.A.; Dias, F.H.C.; Fujiwara, R.T.; et al. Whole genome sequencing of *Trypanosoma cruzi* field isolates reveals extensive genomic variability and complex aneuploidy patterns within TcII DTU. *BMC Genom.* **2018**, *19*, 816. [CrossRef] [PubMed]
65. Reis-Cunha, J.L.; Valdivia, H.O.; Bartholomeu, D.C. Gene and Chromosomal Copy Number Variations as an Adaptive Mechanism Towards a Parasitic Lifestyle in Trypanosomatids. *Curr. Genom.* **2018**, *19*, 87–97. [CrossRef] [PubMed]
66. Araújo, P.R.; Teixeira, S.M. Regulatory elements involved in the post-transcriptional control of stage-specific gene expression in *Trypanosoma cruzi*: A review. *Memórias Do Inst. Oswaldo Cruz* **2011**, *106*, 257–266. [CrossRef] [PubMed]
67. Pastro, L.; Smircich, P.; Di Paolo, A.; Becco, L.; Duhagon, M.A.; Sotelo-Silveira, J.; Garat, B. Nuclear Compartmentalization Contributes to Stage-Specific Gene Expression Control in *Trypanosoma cruzi*. *Front. Cell Dev. Biol.* **2017**, *5*. [CrossRef]
68. Vargas, N.; Pedroso, A.; Zingales, B. Chromosomal polymorphism, gene synteny and genome size in *T. cruzi* I and *T. cruzi* II groups. *Mol. Biochem. Parasitol.* **2004**, *138*, 131–141. [CrossRef]
69. Herreros-Cabello, A.; Callejas-Hernández, F.; Fresno, M.; Gironès, N. Comparative proteomic analysis of trypomastigotes from *Trypanosoma cruzi* strains with different pathogenicity. *Infect. Genet. Evol.* **2019**, *76*, 104041. [CrossRef]
70. Avila, C.C.; Mule, S.N.; Rosa-Fernandes, L.; Viner, R.; Barisón, M.J.; Costa-Martins, A.G.; de Oliveira, G.S.; Teixeira, M.M.G.; Marinho, C.R.F.; Silber, A.M.; et al. Proteome-Wide Analysis of *Trypanosoma cruzi* Exponential and Stationary Growth Phases Reveals a Subcellular Compartment-Specific Regulation. *Genes* **2018**, *9*, 413. [CrossRef]
71. Atwood, J.A.; Weatherly, D.B.; Minning, T.A.; Bundy, B.; Cavola, C.; Opperdoes, F.R.; Orlando, R.; Tarleton, R.L. The *Trypanosoma cruzi* Proteome. *Science* **2005**, *309*, 473–476. [CrossRef]
72. Godoy, L.M.F.; de Marchini, F.K.; Pavoni, D.P.; Rampazzo, R.C.P.; Probst, C.M.; Goldenberg, S.; Krieger, M.A. Quantitative proteomics of *Trypanosoma cruzi* during metacyclogenesis. *Proteomics* **2012**, *12*, 2694–2703. [CrossRef] [PubMed]
73. Ramirez, J.L. An Evolutionary View of *Trypanosoma cruzi* Telomeres. *Front. Cell. Infect. Microbiol.* **2020**, *9*. [CrossRef] [PubMed]
74. Moraes Barros, R.R.; Marini, M.M.; Antônio, C.R.; Cortez, D.R.; Miyake, A.M.; Lima, F.M.; Ruiz, J.C.; Bartholomeu, D.C.; Chiurillo, M.A.; Ramirez, J.L.; et al. Anatomy and evolution of telomeric and subtelomeric regions in the human protozoan parasite *Trypanosoma cruzi*. *BMC Genom.* **2012**, *13*, 229. [CrossRef]

75. Kim, D.; Chiurillo, M.A.; El-Sayed, N.; Jones, K.; Santos, M.R.M.; Porcile, P.E.; Andersson, B.; Myler, P.; da Silveira, J.F.; Ramírez, J.L. Telomere and subtelomere of *Trypanosoma cruzi* chromosomes are enriched in (pseudo)genes of retrotransposon hot spot and trans-sialidase-like gene families: The origins of *T. cruzi* telomeres. *Gene* **2005**, *346*, 153–161. [CrossRef] [PubMed]

76. Lira, C.B.B.; Giardini, M.A.; Neto, J.L.S.; Conte, F.F.; Cano, M.I.N. Telomere biology of trypanosomatids: Beginning to answer some questions. *Trends Parasitol.* **2007**, *23*, 357–362. [CrossRef] [PubMed]

77. Freitas-Junior, L.H.; Bottius, E.; Pirrit, L.A.; Deitsch, K.W.; Scheidig, C.; Guinet, F.; Nehrbass, U.; Wellems, T.E.; Scherf, A. Frequent ectopic recombination of virulence factor genes in telomeric chromosome clusters of P. falciparum. *Nature* **2000**, *407*, 1018–1022. [CrossRef] [PubMed]

78. Kudla, G.; Helwak, A.; Lipinski, L. Gene Conversion and GC-Content Evolution in Mammalian Hsp70. *Mol. Biol. Evol.* **2004**, *21*, 1438–1444. [CrossRef]

79. Galtier, N. Gene conversion drives GC content evolution in mammalian histones. *Trends Genet.* **2003**, *19*, 65–68. [CrossRef]

80. Chiurillo, M.A.; Cano, I.; Da Silveira, J.F.; Ramirez, J.L. Organization of telomeric and sub-telomeric regions of chromosomes from the protozoan parasite *Trypanosoma cruzi*. *Mol. Biochem. Parasitol.* **1999**, *100*, 173–183. [CrossRef]

81. Kudla, G.; Lipinski, L.; Caffin, F.; Helwak, A.; Zylicz, M. High Guanine and Cytosine Content Increases mRNA Levels in Mammalian Cells. *PLoS Biol.* **2006**, *4*, e180. [CrossRef]

82. Tiengwe, C.; Marcello, L.; Farr, H.; Dickens, N.; Kelly, S.; Swiderski, M.; Vaughan, D.; Gull, K.; Barry, J.D.; Bell, S.D.; et al. Genome-wide Analysis Reveals Extensive Functional Interaction between DNA Replication Initiation and Transcription in the Genome of *Trypanosoma brucei*. *Cell Rep.* **2012**, *2*, 185–197. [CrossRef] [PubMed]

83. Lombraña, R.; Álvarez, A.; Fernández-Justel, J.M.; Almeida, R.; Poza-Carrión, C.; Gomes, F.; Calzada, A.; Requena, J.M.; Gómez, M. Transcriptionally Driven DNA Replication Program of the Human Parasite *Leishmania major*. *Cell Rep.* **2016**, *16*, 1774–1786. [CrossRef] [PubMed]

84. Marques, C.A.; Dickens, N.J.; Paape, D.; Campbell, S.J.; McCulloch, R. Genome-wide mapping reveals single-origin chromosome replication in *Leishmania*, a eukaryotic microbe. *Genome Biol.* **2015**, *16*. [CrossRef]

85. Da Silva, M.S.; Pavani, R.S.; Damasceno, J.D.; Marques, C.A.; McCulloch, R.; Tosi, L.R.O.; Elias, M.C. Nuclear DNA Replication in Trypanosomatids: There Are No Easy Methods for Solving Difficult Problems. *Trends Parasitol.* **2017**, *33*, 858–874. [CrossRef]

86. De Araujo, C.B.; da Cunha, J.P.C.; Inada, D.T.; Damasceno, J.; Lima, A.R.J.; Hiraiwa, P.; Marques, C.; Gonçalves, E.; Nishiyama-Junior, M.Y.; McCulloch, R.; et al. Replication origin location might contribute to genetic variability in *Trypanosoma cruzi*. *BMC Genom.* **2020**, *21*, 414. [CrossRef]

87. De Araujo, C.B.; Calderano, S.G.; Elias, M.C. The Dynamics of Replication in *Trypanosoma cruzi* Parasites by Single-Molecule Analysis. *J. Eukaryot. Microbiol.* **2019**, *66*, 514–518. [CrossRef]

88. De Gaudenzi, J.G.; Noé, G.; Campo, V.A.; Frasch, A.C.; Cassola, A. Gene expression regulation in trypanosomatids. *Essays Biochem.* **2011**, *51*, 31–46. [CrossRef]

89. Martínez-Calvillo, S.; Nguyen, D.; Stuart, K.; Myler, P.J. Transcription Initiation and Termination on *Leishmania major* Chromosome 3. *Eukaryot. Cell.* **2004**, *3*, 506–517. [CrossRef]

90. Smircich, P.; El-Sayed, N.M.; Garat, B. Intrinsic DNA curvature in trypanosomes. *BMC Res. Notes* **2017**, *10*, 585. [CrossRef]

91. Gilinger, G.; Bellofatto, V. Trypanosome spliced leader RNA genes contain the first identified RNA polymerase II gene promoter in these organisms. *Nucleic Acids Res.* **2001**, *29*, 1556–1564. [CrossRef]

92. Siegel, T.N.; Hekstra, D.R.; Kemp, L.E.; Figueiredo, L.M.; Lowell, J.E.; Fenyo, D.; Wang, X.; Dewell, S.; Cross, G.A.M. Four histone variants mark the boundaries of polycistronic transcription units in *Trypanosoma brucei*. *Genes Dev.* **2009**, *23*, 1063–1076. [CrossRef] [PubMed]

93. Respuela, P.; Ferella, M.; Rada-Iglesias, A.; Aslund, L. Histone acetylation and methylation at sites initiating divergent polycistronic transcription in *Trypanosoma cruzi*. *J. Biol. Chem.* **2008**, *283*, 15884–15892. [CrossRef] [PubMed]

94. Padilla-Mejía, N.E.; Florencio-Martínez, L.E.; Figueroa-Angulo, E.E.; Manning-Cela, R.G.; Hernández-Rivas, R.; Myler, P.J.; Martínez-Calvillo, S. Gene organization and sequence analyses of transfer RNA genes in Trypanosomatid parasites. *BMC Genom.* **2009**, *10*, 232. [CrossRef] [PubMed]

95. Günzl, A. The Pre-mRNA Splicing Machinery of Trypanosomes: Complex or Simplified? *Eukaryot Cell* **2010**, *9*, 1159–1170. [CrossRef] [PubMed]

96. Palenchar, J.B.; Bellofatto, V. Gene transcription in trypanosomes. *Mol. Biochem. Parasitol.* **2006**, *146*, 135–141. [CrossRef]

97. Rastrojo, A.; Carrasco-Ramiro, F.; Martín, D.; Crespillo, A.; Reguera, R.M.; Aguado, B.; Requena, J.M. The transcriptome of *Leishmania major* in the axenic promastigote stage: Transcript annotation and relative expression levels by RNA-seq. *BMC Genom.* **2013**, *14*, 223. [CrossRef]

98. Kolev, N.G.; Franklin, J.B.; Carmi, S.; Shi, H.; Michaeli, S.; Tschudi, C. The Transcriptome of the Human Pathogen *Trypanosoma brucei* at Single-Nucleotide Resolution. *PLoS Pathog.* **2010**, *6*, e1001090. [CrossRef]

99. Thomas, S.; Green, A.; Sturm, N.R.; Campbell, D.A.; Myler, P.J. Histone acetylations mark origins of polycistronic transcription in *Leishmania major*. *BMC Genom.* **2009**, *10*, 152. [CrossRef]

100. Minning, T.A.; Weatherly, D.B.; Atwood, J.; Orlando, R.; Tarleton, R.L. The steady-state transcriptome of the four major life-cycle stages of *Trypanosoma cruzi*. *BMC Genom.* **2009**, *10*, 370. [CrossRef]

101. Minning, T.A.; Weatherly, D.B.; Flibotte, S.; Tarleton, R.L. Widespread, focal copy number variations (CNV) and whole chromosome aneuploidies in *Trypanosoma cruzi* strains revealed by array comparative genomic hybridization. *BMC Genom.* **2011**, *12*, 139. [CrossRef]

102. Bartholomeu, D.C.; de Paiva, R.M.C.; Mendes, T.A.O.; DaRocha, W.D.; Teixeira, S.M.R. Unveiling the Intracellular Survival Gene Kit of Trypanosomatid Parasites. *PLoS Pathog.* **2014**, *10*, e1004399. [CrossRef] [PubMed]

103. Lorch, Y.; Maier-Davis, B.; Kornberg, R.D. Role of DNA sequence in chromatin remodeling and the formation of nucleosome-free regions. *Genes Dev.* **2014**, *28*, 2492–2497. [CrossRef] [PubMed]

104. Patino, L.H.; Ramírez, J.D. RNA-seq in kinetoplastids: A powerful tool for the understanding of the biology and host-pathogen interactions. *Infect. Genet. Evol.* **2017**, *49*, 273–282. [CrossRef] [PubMed]

105. Clayton, C. The Regulation of Trypanosome Gene Expression by RNA-Binding Proteins. *PLoS Pathog.* **2013**, *9*, e1003680. [CrossRef] [PubMed]

106. Kelly, S.; Kramer, S.; Schwede, A.; Maini, P.K.; Gull, K.; Carrington, M. Genome organization is a major component of gene expression control in response to stress and during the cell division cycle in trypanosomes. *Open Biol.* **2012**, *2*. [CrossRef]

107. Pech-Canul, Á.D.L.C.; Monteón, V.; Solís-Oviedo, R.-L. A Brief View of the Surface Membrane Proteins from *Trypanosoma cruzi*. *J. Parasitol. Res.* **2017**, *2017*. [CrossRef]

108. Poveda, C.; Herreros-Cabello, A.; Callejas-Hernández, F.; Osuna-Pérez, J.; Maza, M.C.; Chillón-Marinas, C.; Calderón, J.; Stamatakis, K.; Fresno, M.; Gironès, N. Interaction of Signaling Lymphocytic Activation Molecule Family 1 (SLAMF1) receptor with *Trypanosoma cruzi* is strain-dependent and affects NADPH oxidase expression and activity. *PLoS Negl. Trop. Dis.* **2020**, *14*, e0008608. [CrossRef]

109. Medina-Acosta, E.; Franco, A.M.R.; Jansen, A.M.; Sampol, M.; Nevés, N.; Pontes-De-Carvalho, L.; Grimaldi, G.; Nussenzweig, V. Trans-sialidase and Sialidase Activities Discriminate between Morphologically Indistinguishable Trypanosomatids. *Eur. J. Biochem.* **1994**, *225*, 333–339. [CrossRef]

110. Freitas, L.M.; dos Santos, S.L.; Rodrigues-Luiz, G.F.; Mendes, T.A.O.; Rodrigues, T.S.; Gazzinelli, R.T.; Teixeira, S.M.R.; Fujiwara, R.T.; Bartholomeu, D.C. Genomic Analyses, Gene Expression and Antigenic Profile of the Trans-Sialidase Superfamily of *Trypanosoma cruzi* Reveal an Undetected Level of Complexity. *PLoS ONE* **2011**, *6*, e25914. [CrossRef]

111. De Pablos, L.M.; Osuna, A. Multigene Families in *Trypanosoma cruzi* and Their Role in Infectivity. *Infect. Immun.* **2012**, *80*, 2258–2264. [CrossRef]

112. Frasch, A.C.C. Functional Diversity in the Trans-sialidase and Mucin Families in *Trypanosoma cruzi*. *Parasitol. Today* **2000**, *16*, 282–286. [CrossRef]

113. Chiurillo, M.A.; Cortez, D.R.; Lima, F.M.; Cortez, C.; Ramírez, J.L.; Martins, A.G.; Serrano, M.G.; Teixeira, M.M.G.; da Silveira, J.F. The diversity and expansion of the trans-sialidase gene family is a common feature in *Trypanosoma cruzi* clade members. *Infect. Genet. Evol.* **2016**, *37*, 266–274. [CrossRef] [PubMed]

114. Frevert, U.; Schenkman, S.; Nussenzweig, V. Stage-specific expression and intracellular shedding of the cell surface trans-sialidase of *Trypanosoma cruzi*. *Infect. Immun.* **1992**, *60*, 2349–2360. [CrossRef] [PubMed]

115. Lantos, A.B.; Carlevaro, G.; Araoz, B.; Ruiz Diaz, P.; Camara, M.d.L.M.; Buscaglia, C.A.; Bossi, M.; Yu, H.; Chen, X.; Bertozzi, C.R.; et al. Sialic Acid Glycobiology Unveils *Trypanosoma cruzi* Trypomastigote Membrane Physiology. *PLoS Pathog.* **2016**, *12*, e1005559. [CrossRef] [PubMed]

116. Schenkman, S.; Eichinger, D.; Pereira, M.E.A.; Nussenzweig, V. Structural and Functional Properties of Trypanosoma Trans-Sialidase. *Annu. Rev. Microbiol.* **1994**, *48*, 499–523. [CrossRef]

117. Oliveira, I.A.; Freire-de-Lima, L.; Penha, L.L.; Dias, W.B.; Todeschini, A.R. *Trypanosoma cruzi* Trans-Sialidase: Structural Features and Biological Implications. In *Proteins and Proteomics of Leishmania and Trypanosoma*; Santos, A.L.S., Branquinha, M.H., d'Avila-Levy, C.M., Kneipp, L.F., Sodré, C.L., Eds.; Subcellular Biochemistry; Springer: Dordrecht, The Netherlands, 2014; pp. 181–201. ISBN 978-94-007-7305-9.

118. Pereira, M.E.; Loures, M.A.; Villalta, F.; Andrade, A.F. Lectin receptors as markers for *Trypanosoma cruzi*. Developmental stages and a study of the interaction of wheat germ agglutinin with sialic acid residues on epimastigote cells. *J. Exp. Med.* **1980**, *152*, 1375–1392. [CrossRef]

119. Parodi, A.J.; Pollevick, G.D.; Mautner, M.; Buschiazzo, A.; Sanchez, D.O.; Frasch, A.C. Identification of the gene(s) coding for the trans-sialidase of *Trypanosoma cruzi*. *EMBO J.* **1992**, *11*, 1705–1710. [CrossRef]

120. Previato, J.; Andrade, A.F.B.; Pessolani, M.C.V.; Mendonça-Previato, L. Incorporation of sialic acid into *Trypanosoma cruzi* macromolecules. A proposal for a new metabolic route. *Mol. Biochem. Parasitol.* **1985**, *16*, 85–96. [CrossRef]

121. Schauer, R.; Reuter, G.; Mühlpfordt, H.; Andrade, A.F.; Pereira, M.E. The occurrence of N-acetyl- and N-glycoloylneuraminic acid in *Trypanosoma cruzi*. *Hoppe-Seyler's Z Physiol. Chem.* **1983**, *364*, 1053–1057. [CrossRef]

122. Schenkman, S.; Mortara, R.A. HeLa cells extend and internalize pseudopodia during active invasion by *Trypanosoma cruzi* trypomastigotes. *J. Cell. Sci.* **1992**, *101*, 895–905.

123. Scudder, P.; Doom, J.P.; Chuenkova, M.; Manger, I.D.; Pereira, M.E. Enzymatic characterization of β-D-galactoside α 2,3-trans-sialidase from *Trypanosoma cruzi*. *J. Biol. Chem.* **1993**, *268*, 9886–9891. [PubMed]

124. Zingales, B.; Carniol, C.; de Lederkremer, R.M.; Colli, W. Direct sialic acid transfer from a protein donor to glycolipids of trypomastigote forms of *Trypanosoma cruzi*. *Mol. Biochem. Parasitol.* **1987**, *26*, 135–144. [CrossRef]

125. Pereira-Chioccola, V.L.; Acosta-Serrano, A.; Correia de Almeida, I.; Ferguson, M.A.; Souto-Padron, T.; Rodrigues, M.M.; Travassos, L.R.; Schenkman, S. Mucin-like molecules form a negatively charged coat that protects *Trypanosoma cruzi* trypomastigotes from killing by human anti-α-galactosyl antibodies. *J. Cell. Sci.* **2000**, *113*, 1299–1307. [PubMed]

126. Pereira, M.E. A developmentally regulated neuraminidase activity in *Trypanosoma cruzi*. *Science* **1983**, *219*, 1444–1446. [CrossRef]

127. Nardy, A.F.F.R.; Freire-de-Lima, C.G.; Pérez, A.R.; Morrot, A. Role of *Trypanosoma cruzi* Trans-sialidase on the Escape from Host Immune Surveillance. *Front. Microbiol.* **2016**, *7*, 348. [CrossRef]

128. Colli, W. Trans-sialidase: A unique enzyme activity discovered in the protozoan *Trypanosoma cruzi*. *FASEB J.* **1993**, *7*, 1257–1264. [CrossRef] [PubMed]

129. Cross, G.A.; Takle, G.B. The surface trans-sialidase family of *Trypanosoma cruzi*. *Annu. Rev. Microbiol.* **1993**, *47*, 385–411. [CrossRef] [PubMed]

130. Schenkman, S.; Jiang, M.S.; Hart, G.W.; Nussenzweig, V. A novel cell surface trans-sialidase of *Trypanosoma cruzi* generates a stage-specific epitope required for invasion of mammalian cells. *Cell* **1991**, *65*, 1117–1125. [CrossRef]

131. Buschiazzo, A.; Campetella, O.; Frasch, A.C.C. *Trypanosoma rangeli* sialidase: Cloning, expression and similarity to *T. cruzi* trans-sialidase. *Glycobiology* **1997**, *7*, 1167–1173. [CrossRef]

132. Amaya, M.F.; Buschiazzo, A.; Nguyen, T.; Alzari, P.M. The high resolution structures of free and inhibitor-bound *Trypanosoma rangeli* sialidase and its comparison with *T. cruzi* trans-sialidase. *J. Mol. Biol.* **2003**, *325*, 773–784. [CrossRef]

133. Schenkman, S.; Pontes de Carvalho, L.; Nussenzweig, V. *Trypanosoma cruzi* trans-sialidase and neuraminidase activities can be mediated by the same enzymes. *J. Exp. Med.* **1992**, *175*, 567–575. [CrossRef] [PubMed]

134. Pollevick, G.D.; Affranchino, J.L.; Frasch, A.C.; Sánchez, D.O. The complete sequence of a shed acute-phase antigen of *Trypanosoma cruzi*. *Mol. Biochem. Parasitol.* **1991**, *47*, 247–250. [CrossRef]

135. Briones, M.R.; Egima, C.M.; Schenkman, S. *Trypanosoma cruzi* trans-sialidase gene lacking C-terminal repeats and expressed in epimastigote forms. *Mol. Biochem. Parasitol.* **1995**, *70*, 9–17. [CrossRef]

136. Añez-Rojas, N.; Peralta, A.; Crisante, G.; Rojas, A.; Añez, N.; Ramírez, J.L.; Chiurillo, M.A. *Trypanosoma rangeli* expresses a gene of the group II trans-sialidase superfamily. *Mol. Biochem. Parasitol.* **2005**, *142*, 133–136. [CrossRef]

137. Mattos, E.C.; Tonelli, R.R.; Colli, W.; Alves, M.J.M. The Gp85 surface glycoproteins from *Trypanosoma cruzi*. *Subcell. Biochem.* **2014**, *74*, 151–180. [CrossRef]

138. Colli, W.; Alves, M.J. Relevant glycoconjugates on the surface of *Trypanosoma cruzi*. *Mem. Inst. Oswaldo Cruz* **1999**, *94*, 37–49. [CrossRef]

139. Claser, C.; Espíndola, N.M.; Sasso, G.; Vaz, A.J.; Boscardin, S.B.; Rodrigues, M.M. Immunologically relevant strain polymorphism in the Amastigote Surface Protein 2 of *Trypanosoma cruzi*. *Microbes Infect.* **2007**, *9*, 1011–1019. [CrossRef]

140. Giordano, R.; Fouts, D.L.; Tewari, D.; Colli, W.; Manning, J.E.; Alves, M.J. Cloning of a surface membrane glycoprotein specific for the infective form of *Trypanosoma cruzi* having adhesive properties to laminin. *J. Biol. Chem.* **1999**, *274*, 3461–3468. [CrossRef]

141. Magdesian, M.H.; Giordano, R.; Ulrich, H.; Juliano, M.A.; Juliano, L.; Schumacher, R.I.; Colli, W.; Alves, M.J. Infection by *Trypanosoma cruzi*. Identification of a parasite ligand and its host cell receptor. *J. Biol. Chem.* **2001**, *276*, 19382–19389. [CrossRef]

142. Tonelli, R.R.; Giordano, R.J.; Barbu, E.M.; Torrecilhas, A.C.; Kobayashi, G.S.; Langley, R.R.; Arap, W.; Pasqualini, R.; Colli, W.; Alves, M.J.M. Role of the gp85/trans-sialidases in *Trypanosoma cruzi* tissue tropism: Preferential binding of a conserved peptide motif to the vasculature in vivo. *PLoS Negl. Trop. Dis.* **2010**, *4*, e864. [CrossRef]

143. Santos, M.A.; Garg, N.; Tarleton, R.L. The identification and molecular characterization of *Trypanosoma cruzi* amastigote surface protein-1, a member of the trans-sialidase gene super-family. *Mol. Biochem. Parasitol.* **1997**, *86*, 1–11. [PubMed]

144. Wizel, B.; Nunes, M.; Tarleton, R.L. Identification of *Trypanosoma cruzi* trans-sialidase family members as targets of protective CD8+ TC1 responses. *J. Immunol.* **1997**, *159*, 6120–6130. [PubMed]

145. Wizel, B.; Palmieri, M.; Mendoza, C.; Arana, B.; Sidney, J.; Sette, A.; Tarleton, R. Human infection with *Trypanosoma cruzi* induces parasite antigen-specific cytotoxic T lymphocyte responses. *J. Clin. Investig.* **1998**, *102*, 1062–1071. [CrossRef] [PubMed]

146. Yoshida, N. Molecular basis of mammalian cell invasion by *Trypanosoma cruzi*. *Acad. Bras. Cienc.* **2006**, *78*, 87–111. [CrossRef] [PubMed]

147. Favoreto, S.; Dorta, M.L.; Yoshida, N. *Trypanosoma cruzi* 175-kDa protein tyrosine phosphorylation is associated with host cell invasion. *Exp. Parasitol.* **1998**, *89*, 188–194. [CrossRef]

148. Nogueira, N. Host and parasite factors affecting the invasion of mononuclear phagocytes by *Trypanosoma cruzi*. *Ciba Found. Symp.* **1983**, *99*, 52–73. [CrossRef]

149. Matsumoto, T.K.; Cotrim, P.C.; da Silveira, J.F.; Stolf, A.M.S.; Umezawa, E.S. *Trypanosoma cruzi*: Isolation of an immunodominant peptide of TESA (Trypomastigote Excreted-Secreted Antigens) by gene cloning. *Diagn. Microbiol. Infect. Dis.* **2002**, *42*, 187–192. [CrossRef]

150. Beucher, M.; Norris, K.A. Sequence diversity of the *Trypanosoma cruzi* complement regulatory protein family. *Infect. Immun.* **2008**, *76*, 750–758. [CrossRef]

151. Kipnis, T.L.; David, J.R.; Alper, C.A.; Sher, A.; da Silva, W.D. Enzymatic treatment transforms trypomastigotes of *Trypanosoma cruzi* into activators of alternative complement pathway and potentiates their uptake by macrophages. *Proc. Natl. Acad. Sci. USA* **1981**, *78*, 602–605. [CrossRef]

152. Grisard, E.C.; Stoco, P.H.; Wagner, G.; Sincero, T.C.M.; Rotava, G.; Rodrigues, J.B.; Snoeijer, C.Q.; Koerich, L.B.; Sperandio, M.M.; Bayer-Santos, E.; et al. Transcriptomic analyses of the avirulent protozoan parasite *Trypanosoma rangeli*. *Mol. Biochem. Parasitol.* **2010**, *174*, 18–25. [CrossRef]

153. Stoco, P.H.; Wagner, G.; Talavera-Lopez, C.; Gerber, A.; Zaha, A.; Thompson, C.E.; Bartholomeu, D.C.; Lückemeyer, D.D.; Bahia, D.; Loreto, E.; et al. Genome of the avirulent human-infective trypanosome—*Trypanosoma rangeli*. *PLoS Negl. Trop. Dis.* **2014**, *8*, e3176. [CrossRef] [PubMed]

154. Tonelli, R.R.; Torrecilhas, A.C.; Jacysyn, J.F.; Juliano, M.A.; Colli, W.; Alves, M.J.M. In vivo infection by *Trypanosoma cruzi*: The conserved FLY domain of the gp85/trans-sialidase family potentiates host infection. *Parasitology* **2011**, *138*, 481–492. [CrossRef] [PubMed]

155. Roggentin, P.; Rothe, B.; Kaper, J.B.; Galen, J.; Lawrisuk, L.; Vimr, E.R.; Schauer, R. Conserved sequences in bacterial and viral sialidases. *Glycoconj. J.* **1989**, *6*, 349–353. [CrossRef] [PubMed]

156. Gaskell, A.; Crennell, S.; Taylor, G. The three domains of a bacterial sialidase: A β-propeller, an immunoglobulin module and a galactose-binding jelly-roll. *Structure* **1995**, *3*, 1197–1205. [CrossRef]

157. Cremona, M.L.; Campetella, O.; Sánchez, D.O.; Frasch, A.C. Enzymically inactive members of the trans-sialidase family from *Trypanosoma cruzi* display β-galactose binding activity. *Glycobiology* **1999**, *9*, 581–587. [CrossRef] [PubMed]

158. Todeschini, A.R.; Dias, W.B.; Girard, M.F.; Wieruszeski, J.-M.; Mendonça-Previato, L.; Previato, J.O. Enzymatically inactive trans-sialidase from *Trypanosoma cruzi* binds sialyl and β-galactopyranosyl residues in a sequential ordered mechanism. *J. Biol. Chem.* **2004**, *279*, 5323–5328. [CrossRef] [PubMed]

159. Acosta-Serrano, A.; Almeida, I.C.; Freitas-Junior, L.H.; Yoshida, N.; Schenkman, S. The mucin-like glycoprotein super-family of *Trypanosoma cruzi*: Structure and biological roles. *Mol. Biochem. Parasitol.* **2001**, *114*, 143–150. [CrossRef]

160. Buscaglia, C.A.; Campo, V.A.; Frasch, A.C.C.; Di Noia, J.M. *Trypanosoma cruzi* surface mucins: Host-dependent coat diversity. *Nat. Rev. Microbiol.* **2006**, *4*, 229–236. [CrossRef]

161. Schenkman, S.; Ferguson, M.A.; Heise, N.; de Almeida, M.L.; Mortara, R.A.; Yoshida, N. Mucin-like glycoproteins linked to the membrane by glycosylphosphatidylinositol anchor are the major acceptors of sialic acid in a reaction catalyzed by trans-sialidase in metacyclic forms of *Trypanosoma cruzi*. *Mol. Biochem. Parasitol.* **1993**, *59*, 293–303. [CrossRef]

162. Mucci, J.; Lantos, A.B.; Buscaglia, C.A.; Leguizamón, M.S.; Campetella, O. The *Trypanosoma cruzi* Surface, a Nanoscale Patchwork Quilt. *Trends Parasitol.* **2017**, *33*, 102–112. [CrossRef]

163. Campo, V.; Di Noia, J.M.; Buscaglia, C.A.; Agüero, F.; Sánchez, D.O.; Frasch, A.C.C. Differential accumulation of mutations localized in particular domains of the mucin genes expressed in the vertebrate host stage of *Trypanosoma cruzi*. *Mol. Biochem. Parasitol.* **2004**, *133*, 81–91. [CrossRef] [PubMed]

164. Buscaglia, C.A.; Campo, V.A.; Di Noia, J.M.; Torrecilhas, A.C.T.; De Marchi, C.R.; Ferguson, M.A.J.; Frasch, A.C.C.; Almeida, I.C. The surface coat of the mammal-dwelling infective trypomastigote stage of *Trypanosoma cruzi* is formed by highly diverse immunogenic mucins. *J. Biol. Chem.* **2004**, *279*, 15860–15869. [CrossRef] [PubMed]

165. Urban, I.; Boiani Santurio, L.; Chidichimo, A.; Yu, H.; Chen, X.; Mucci, J.; Agüero, F.; Buscaglia, C.A. Molecular diversity of the *Trypanosoma cruzi* TcSMUG family of mucin genes and proteins. *Biochem. J.* **2011**, *438*, 303–313. [CrossRef] [PubMed]

166. Di Noia, J.M.; D'Orso, I.; Sánchez, D.O.; Frasch, A.C. AU-rich elements in the 3′-untranslated region of a new mucin-type gene family of *Trypanosoma cruzi* confers mRNA instability and modulates translation efficiency. *J. Biol. Chem.* **2000**, *275*, 10218–10227. [CrossRef]

167. Cámara, M.d.l.M.; Balouz, V.; Cameán, C.C.; Cori, C.R.; Kashiwagi, G.A.; Gil, S.A.; Macchiaverna, N.P.; Cardinal, M.V.; Guaimas, F.; Lobo, M.M.; et al. *Trypanosoma cruzi* surface mucins are involved in the attachment to the Triatoma infestans rectal ampoule. *PLoS Negl. Trop. Dis.* **2019**, *13*, e0007418. [CrossRef]

168. Mortara, R.A.; da Silva, S.; Araguth, M.F.; Blanco, S.A.; Yoshida, N. Polymorphism of the 35- and 50-kilodalton surface glycoconjugates of *Trypanosoma cruzi* metacyclic trypomastigotes. *Infect. Immun.* **1992**, *60*, 4673–4678. [CrossRef]

169. Nogueira, N.F.S.; Gonzalez, M.S.; Gomes, J.E.; de Souza, W.; Garcia, E.S.; Azambuja, P.; Nohara, L.L.; Almeida, I.C.; Zingales, B.; Colli, W. *Trypanosoma cruzi*: Involvement of glycoinositolphospholipids in the attachment to the luminal midgut surface of *Rhodnius prolixus*. *Exp. Parasitol.* **2007**, *116*, 120–128. [CrossRef]

170. D'Orso, I.; Frasch, A.C. TcUBP-1, a developmentally regulated U-rich RNA-binding protein involved in selective mRNA destabilization in trypanosomes. *J. Biol. Chem.* **2001**, *276*, 34801–34809. [CrossRef]

171. Freitas-Junior, L.H.; Briones, M.R.; Schenkman, S. Two distinct groups of mucin-like genes are differentially expressed in the developmental stages of *Trypanosoma cruzi*. *Mol. Biochem. Parasitol.* **1998**, *93*, 101–114. [CrossRef]

172. Cánepa, G.E.; Degese, M.S.; Budu, A.; Garcia, C.R.S.; Buscaglia, C.A. Involvement of TSSA (trypomastigote small surface antigen) in *Trypanosoma cruzi* invasion of mammalian cells. *Biochem. J.* **2012**, *444*, 211–218. [CrossRef]

173. Di Noia, J.M.; D'Orso, I.; Aslund, L.; Sánchez, D.O.; Frasch, A.C. The *Trypanosoma cruzi* mucin family is transcribed from hundreds of genes having hypervariable regions. *J. Biol. Chem.* **1998**, *273*, 10843–10850. [CrossRef] [PubMed]

174. Di Noia, J.M.; Buscaglia, C.A.; De Marchi, C.R.; Almeida, I.C.; Frasch, A.C.C. A *Trypanosoma cruzi* small surface molecule provides the first immunological evidence that Chagas' disease is due to a single parasite lineage. *J. Exp. Med.* **2002**, *195*, 401–413. [CrossRef] [PubMed]

175. Cámara, M.d.l.M.; Cánepa, G.E.; Lantos, A.B.; Balouz, V.; Yu, H.; Chen, X.; Campetella, O.; Mucci, J.; Buscaglia, C.A. The Trypomastigote Small Surface Antigen (TSSA) regulates *Trypanosoma cruzi* infectivity and differentiation. *PLoS Negl. Trop. Dis.* **2017**, *11*, e0005856. [CrossRef] [PubMed]

176. Bartholomeu, D.C.; Cerqueira, G.C.; Leão, A.C.A.; daRocha, W.D.; Pais, F.S.; Macedo, C.; Djikeng, A.; Teixeira, S.M.R.; El-Sayed, N.M. Genomic organization and expression profile of the mucin-associated surface protein (masp) family of the human pathogen *Trypanosoma cruzi*. *Nucleic Acids Res.* **2009**, *37*, 3407–3417. [CrossRef]

177. Souza, R.T.; Santos, M.R.M.; Lima, F.M.; El-Sayed, N.M.; Myler, P.J.; Ruiz, J.C.; da Silveira, J.F. New *Trypanosoma cruzi* Repeated Element That Shows Site Specificity for Insertion. *Eukaryot. Cell* **2007**, *6*, 1228–1238. [CrossRef]

178. Atwood, J.A.; Minning, T.; Ludolf, F.; Nuccio, A.; Weatherly, D.B.; Alvarez-Manilla, G.; Tarleton, R.; Orlando, R. Glycoproteomics of *Trypanosoma cruzi* trypomastigotes using subcellular fractionation, lectin affinity, and stable isotope labeling. *J. Proteome Res.* **2006**, *5*, 3376–3384. [CrossRef]

179. dos Santos, S.L.; Freitas, L.M.; Lobo, F.P.; Rodrigues-Luiz, G.F.; Mendes, T.A.d.O.; Oliveira, A.C.S.; Andrade, L.O.; Chiari, É.; Gazzinelli, R.T.; Teixeira, S.M.R.; et al. The MASP Family of *Trypanosoma cruzi*: Changes in Gene Expression and Antigenic Profile during the Acute Phase of Experimental Infection. *PLoS Negl. Trop. Dis.* **2012**, *6*, e1779. [CrossRef]

180. De Pablos, L.M.; González, G.G.; Solano Parada, J.; Seco Hidalgo, V.; Díaz Lozano, I.M.; Gómez Samblás, M.M.; Cruz Bustos, T.; Osuna, A. Differential Expression and Characterization of a Member of the Mucin-Associated Surface Protein Family Secreted by *Trypanosoma cruzi*. *Infect. Immun.* **2011**, *79*, 3993–4001. [CrossRef]

Publisher's Note: MDPI stays neutral with regard to jurisdictional claims in published maps and institutional affiliations.

 *genes*

Article

Genomic Organization and Generation of Genetic Variability in the RHS (Retrotransposon Hot Spot) Protein Multigene Family in *Trypanosoma cruzi*

Werica P. Bernardo [1], Renata T. Souza [1], André G. Costa-Martins [2] , Eden R. Ferreira [1], Renato A. Mortara [1], Marta M. G. Teixeira [2], José Luis Ramirez [3,*] and José F. Da Silveira [1,*]

[1] Department of Microbiology, Immunology and Parasitology, Escola Paulista de Medicina, Universidade Federal de São Paulo, São Paulo 04023-062, SP, Brazil; wericabernardo@gmail.com (W.P.B.); renataepm@hotmail.com (R.T.S.); edendearaujo@gmail.com (E.R.F.); ramortara@unifesp.br (R.A.M.)

[2] Department of Parasitology, Instituto de Ciências Biomédicas, Universidade de São Paulo, São Paulo 05508-000, SP, Brazil; andreguilherme@usp.br (A.G.C.-M.); mmgteix@icb.usp.br (M.M.G.T.)

[3] Fundación Instituto de Estudios Avanzados (IDEA), Universidad Central de Venezuela, Caracas 1080, Venezuela

* Correspondence: ramsjoseluis@gmail.com (J.L.R); jose.franco@unifesp.br (J.F.D.S.)

Received: 7 August 2020; Accepted: 14 September 2020; Published: 17 September 2020

Abstract: Retrotransposon Hot Spot (RHS) is the most abundant gene family in *Trypanosoma cruzi*, with unknown function in this parasite. The aim of this work was to shed light on the organization and expression of RHS in *T. cruzi*. The diversity of the RHS protein family in *T. cruzi* was demonstrated by phylogenetic and recombination analyses. Transcribed sequences carrying the RHS domain were classified into ten distinct groups of monophyletic origin. We identified numerous recombination events among the RHS and traced the origins of the donors and target sequences. The transcribed RHS genes have a mosaic structure that may contain fragments of different RHS inserted in the target sequence. About 30% of RHS sequences are located in the subtelomere, a region very susceptible to recombination. The evolution of the RHS family has been marked by many events, including gene duplication by unequal mitotic crossing-over, homologous, as well as ectopic recombination, and gene conversion. The expression of RHS was analyzed by immunofluorescence and immunoblotting using anti-RHS antibodies. RHS proteins are evenly distributed in the nuclear region of *T. cruzi* replicative forms (amastigote and epimastigote), suggesting that they could be involved in the control of the chromatin structure and gene expression, as has been proposed for *T. brucei*.

Keywords: *Trypanosoma cruzi*; Retrotransposon Hot Spot (RHS) multigene family; chromosome distribution; recombination; gene mosaic structure; evolution; nuclear protein

---

1. Introduction

The flagellate protozoan *Trypanosoma cruzi* is the etiologic agent of Chagas disease or American trypanosomiasis, which affects 6–7 million people mainly in Latin America, with an increasing number of cases in non-endemic countries such as Canada, the United States of America, and some European countries [1]. When compared with other members of the genus Trypanosoma, the *T. cruzi* genome was expanded, being 2.3-fold larger than that of *T. brucei* and *T. rangeli*. Repetitive DNA sequences comprise about 52% of the *T. cruzi* genome [2–4]. The dramatic expansion and diversification of repetitive sequences, particularly of multigene family encoding proteins, such as surface proteins (TS (Trans-Sialidase), MASP (Mucin-Associated Surface Protein), mucins, gp63, Retrotransposon Hot Spot (RHS), and DGF-1 (Dispersed Gene Family-1)) may have contributed to the speciation of the *T. cruzi* taxon [2,5]. RHS proteins are coded by a multigene family found in the genus Trypanosoma.

RHS refers to a hot spot for retrotransposon insertion within the RHS gene. When retrotransposons are inserted in this site, they generate RHS pseudogenes carrying one or more retroelements flanked by two separate halves of RHS [6]. Multiple RHS genes have been annotated in the genomes of mammalian trypanosomes (African trypanosomes—*T. brucei*, *T. congolense*, and *T. vivax*; American trypanosomes—*T. cruzi*, *T. cruzi marinkellei*, and *T. rangeli*; and cosmopolitan trypanosomes—*T. theileri*, *T. evansi*, *T. conorhini*) and *T. grayi* isolated from reptiles.

RHS proteins were first identified in *T. brucei* and were classified into six subfamilies (RHS1 to RHS6) based on the C-terminal region sequence [6]. The RHS proteins of *T. brucei* share a highly conserved amino-terminal (N-terminal) region, while the carboxy-terminal (C-terminal) portion is highly variable [6]. The N-terminal region has an ATP/GTP binding motif encoded by five codons located upstream of the hot spot insertion site for the retrotransposons Ingi (an autonomous long interspersed element—LINE) and RIME (a non-autonomous short interspersed element—SINE). The pseudogene may be the result of homologous recombination between two RHS variants by crossing-over involving the 5' region upstream of the retroelement insertion site. Retrotransposon insertion generates nonsense mutations or frameshifts within the coding sequence, resulting in truncated RHS proteins [6].

The role of the RHS family has been investigated in *T. brucei*, and it has been suggested that RHSs are involved in the control of the expansion of the retroelements in this organism [6,7]. Earlier studies in *T. brucei* showed an increase in the level of RHS transcripts after the ablation of argonaute protein, suggesting that the RHS family may be under the control of siRNA (small interfering RNA) [8]. High throughput analysis of small non-coding RNAs showed that a large number of pseudogene-derived siRNAs originated from pseudogene–pseudogene pairs, in which the major components were RHS pseudogenes [9], and it has been hypothesized that RHS pseudogenes in *T. brucei* are a source of antisense siRNAs, which regulate the expression of the RHS family. More recent studies proposed that the RHS family could be involved in the chromatin modeling [10], transcription elongation, and mRNA export in *T. brucei* [11].

Beyond an initial genomic analysis showing multiple RHS (gene) pseudogenes, little is known about the organization, structure, and expression of these genes and their products in *T. cruzi*. In the current study, we aimed to investigate the structure, evolution, and expression of the RHS multigene family in *T. cruzi*. We also provide insights into the strategies used by *T. cruzi* for preserving complete and functional RHS genes.

2. Material and Methods

2.1. Parasites

Trypanosome isolates used in this study were the *T. cruzi* clone CL Brener (CLB) (TRYCC426, [12], and G strain [13]), *T. cruzi marinkellei* (TCC344), *T. rangeli* SC58 [14] and *T. brucei rhodesiense* YTAT 1.1. The epimastigotes of *T. cruzi*, *T. cruzi marinkellei*, and *T. rangeli* were grown in axenic cultures at 28 °C in liver-infusion tryptose (LIT) medium [15] supplemented with 10–20% heat-inactivated fetal calf serum. Procyclic forms of *T. brucei rhodesiense* YTAT 1.1 were cultured in a semi-defined medium (SDM-79) supplemented with 10% heat-inactivated fetal bovine serum at 27 °C. *T. cruzi* extracellular amastigotes were obtained by culture tissue trypomastigote differentiation in a LIT medium, as previously described [16].

2.2. Identification of RHS Sequences in T. cruzi and T. cruzi marinkellei Genome Databases

The search for homologous RHS genes in the TriTrypDB and GenBank databases was performed using the algorithms BLASTp, tBLASTn, BLASTx, and the presence of RHS domain architecture was confirmed using rpsBLAST [17]. RHS transcripts of CLB were used as queries to identify homologous sequences in other *Trypanosoma* species using the tBLASTn (e-value of 1×10^{-3}) search program. The retrieved sequences were evaluated for the presence of RHS domains with the rpsBLAST algorithm (e-value of 1×10^{-5}) against the database of conserved domains [18]. An extra round of tBLASTn was performed using found RHS sequences as a query to improve genome survey sensibility. Figure S1 shows the flowchart of this analysis. Sequence alignments were carried out with RHS of clone CLB excluding truncated sequences. The nucleotide and amino acid sequences were aligned using the MUSCLE program [19] and the poorly conserved regions were removed using the Gblocks program [20].

2.3. Classification and Phylogenetic Analyses of RHS

For these analyses, we selected RHS transcripts of the *T. cruzi* clone CLB [21]. Transcribed genes were analyzed for the presence of RHS domains with the rpsBLAST algorithm using 1×10^{-5} e-value against the NCBI Conserved Domain Database (CDD) [18] (Figure S1). Sequences that showed false-positive RHS domains and pseudogenes were excluded. In the phylogenetic analysis, the global multiple alignment was carried out with the MUSCLE algorithm [19]. Phylogenetic trees were generated using the "Maximum likelihood method" using the RaxML v 8.2.9 program [22], with an automatic search for substitution models (PROTGAMMAAUTO) selected by the Akaike information criterion (AIC) (auto-prot = AIC) information criterion, with 1000 bootstrap replicas. The phylogenetic tree was visualized with the program FigTree V 1.4.2 [23].

2.4. Detection of Potential Recombination Events in RHS Sequences

The RHS sequences selected for the phylogenetic study were also used to identify recombination events in the clone CLB using the RDP4 program (Recombination Detection Program) [24], which allows the identification and statistical analysis of recombination events from a set of aligned sequences. It uses non-parametric recombination detection methods (algorithms RDP, GENECONV, MaxChi, Chimera, Bootscan, 3Seq, and SiSscan) to identify breakpoints in the genomic sequences where recombination begins and ends, in addition to the donor parental sequences of the recombinant fragment. For recombination events, sequences detected by at least 6 of the 7 algorithms in the RDP4 package were considered recombinant.

2.5. Expression and Purification of Recombinant RHS

An 877-bp fragment encoding a 292-aa region of the carboxy-terminal domain of the RHS (TcCLB.511055.20) was amplified by PCR from CLB genomic DNA, cloned into pGEM-T, and sequenced to confirm gene identity. Then, it was subcloned into pGEX-1λT to produce the RHS-GST fusion protein as described by Martins et al., 2015 [24]. *E. coli* BL21 bacteria were transformed with the RHS-GST construct, grown in LB medium, and protein expression was induced with 1 mM isopropyl β-D-1-thiogalactopyranoside (IPTG). The RHS recombinant protein was extracted from the insoluble fraction of bacterial lysates with Laemmli's sample buffer and separated on 10% SDS-PAGE. The band W to the recombinant protein was excised from the gel and extracted by dialysis against ammonium bicarbonate and distilled water [24]. The purity of recombinant RHS was checked by SDS-PAGE stained with colloidal Coomassie Blue and immunoblotting (Figure S2). Purified protein was quantified with Coomassie Plus (Pierce, Thermo Fisher Scientific, Waltham, MA, USA) in 96-well plates at 620 nm.

2.6. Antibody Production, Western Blot, and Immunofluorescence Analyses

About two mg of the purified RHS recombinant protein were sent to Rheabiotech Research and Development Laboratory, SP, Brazil, for the production of polyclonal anti-RHS antibodies in mice. The specificity and reactivity of the anti-RHS antibodies were determined by ELISA and Western blot assays using the recombinant protein RHS.

Epimastigotes (10^8 cells) of *T. cruzi* (clone CLB, strain G), *T. cruzi marinkellei*, and *T. rangeli*, and procyclic forms (10^7 cells) of *T. brucei* were washed in PBS and lysed with $4 \times$ Laemmli's sample buffer, and the extracts were subjected to SDS-PAGE (10% for separation gel and 3% for packaging gel) at 120 V for 45 min. Proteins were transferred to Hybond ECL membranes (Amersham, GE Healthcare Life Sciences). For the Western blot reaction, the membrane was blocked in $1\times$ PBS solution containing

7.5% skimmed milk powder (PBS/milk solution) for 1 h at room temperature. The membrane was then incubated with PBS/milk solution anti-RHS1 (dilution 1:500) for 1 h, at room temperature. Subsequently, the membrane was washed three times (3 × 5 min) in PBS containing 0.05% Tween 20 (PBS/Tween solution). Secondary antibodies (Sigma Aldrich) were incubated for 1 h at room temperature at a dilution of 1:10,000. Bound antibody signals were amplified with ECL (Enhanced Chemiluminescence) substrate (GE Healthcare) and luminescent bands visualized in an Alliance 2.7 photo documenter (UVItec).

For indirect immunofluorescence assay, *T. cruzi* epimastigotes (10^7 cells) were harvested from the culture medium, washed with PBS, and fixed with 2% paraformaldehyde in PBS for 15 min at room temperature. Then, the parasites were washed with PBS and incubated with anti-RHS antibodies (1:1000 dilution) in the presence of 0.1% saponin and 1% PBS/BSA for 1 h at room temperature. The parasites were washed once more with PBS and incubated for 1 h with an Alexa Flour 568 anti-mouse IgG antibody raised in goat diluted 1:100 in 1% PBS/BSA and 1 mM DAPI (4′,6 -diamino-2-phenylindole, Molecular Probes). Subsequently, epimastigotes were washed with PBS and the slides were mounted using Glycerol-PPD (p-Phenylenediamine). Images were acquired with a TCS SP5 II TandemScanner confocal microscope (Leica Microsystems, Wetzlar, Germany) using a 63 × NA 1.40 PlanApo oil immersion objective and processed with Imaris software 7.0 (Bitplane).

3. Results

3.1. Mapping of RHS Sequences on the Chromosomes of Clone CLB of T. cruzi

Natural populations of *T. cruzi* reproduce predominantly by binary fission, therefore they exhibit a clonal population structure [25–28]. However, the occurrence of hybridization has been demonstrated in vitro [29] and also in natural populations of *T. cruzi* [28,30–36]. Based on several genetic markers, *T. cruzi* isolates were classified into six discrete typing units (DTU) named lineages TcI to TcVI [37–39]. The isolates from lineages V and VI have a hybrid evolutionary origin from at least two hybridization events between lineages TcII and TcIII [28,33,34,39].

The clone CL Brener (CLB) is a hybrid strain grouped in lineage TcVI, and sequence analysis of its genome revealed the presence of two haplotypes [2], one of which has contigs similar to the Esmeraldo strain of lineage TcII. The sequence divergence between the two haplotypes is 5.4% [2]. The genomic sequences generated in the Genome Project of *T. cruzi* clone CLB have been organized in 41 pairs of homologous chromosomes (TcChr), with the smallest having 77,958 bp (TcChr1) and the largest 2,371,736 bp (TcChr41) [2,40,41]. Due to the hybrid nature of CLB, each pair of homologous chromosomes consists of one homolog, which is an Esmeraldo-like-haplotype (S), and another homolog, which is a non-Esmeraldo-like haplotype (P), totaling 82 in silico chromosomes (TcChr) [2,40]. A search for RHS sequences in the CLB genome deposited in the TriTrypDB database resulted in 525 RHS sequences (111 genes, 384 pseudogenes, 30 truncated sequences), which are distributed in the haplotypes as follows: 48 complete genes, 177 pseudogenes, and 8 truncated sequences in the Esmeraldo haplotype (S), and 63 complete genes, 207 pseudogenes and 22 truncated sequences in the non-Esmeraldo haplotype (P) (Table S1). Besides these sequences, we found 42 complete RHS genes, 175 pseudogenes, and 11 truncated sequences among the unallocated contigs, totaling 753 RHS sequences in the CLB genome. RHS gene sizes range from 351 to 3014 bp. The estimated RHS content of the CLB genome was 3,271,841 bp, comprising about 5.4% of the *T. cruzi* genome sequence.

The distribution of RHS sequences along the CLB chromosomes is shown in Figure S3. Among 82 chromosomes, three chromosomes, TcChr1-S, TcChr4-S, and TcChr34-S, did not show RHS sequences. Larger chromosomes, such as TcChr40 and TcChr41, have predominantly RHS pseudogenes (Table S1), suggesting that RHS and other repetitive sequences could be involved in the expansion of the chromosome size. It is important to highlight that the total number of RHS sequences present in the genome of the CLB may be even greater than that obtained in this analysis. When non-transcribed sequences were included in our analysis, the total number of RHS sequences was larger than one thousand, showing the presence of fragments dispersed in the genome, which are reminiscent of RHS genes. These results reflect the complexity of the *T. cruzi* genome and RHS family [2,6,42]. The haploid genome of *T. cruzi* is about 2- and 5-fold larger than that of *T. brucei* and *Leishmania* spp., respectively. In addition, multigenic families (trans-sialidases, mucins, DGF-1, MASP, RHS, and GP63 proteases) underwent a very pronounced expansion process in *T. cruzi* [2,3,6,42–44].

The frequency of RHS sequences in each chromosome of CLB was plotted as a heatmap in Figure 1, and the proportion of total RHS length in each chromosome is shown in Figure S4. RHS sequences comprise 0.34% to 6.14% of the entire length of each CLB chromosome. Overall, the frequency of RHS was similar in most pairs of homologous chromosomes. However, in some homologous pairs, this proportion was quite different, e.g., between the haplotypes S and P of the chromosome TcChr20 or TcChr21.

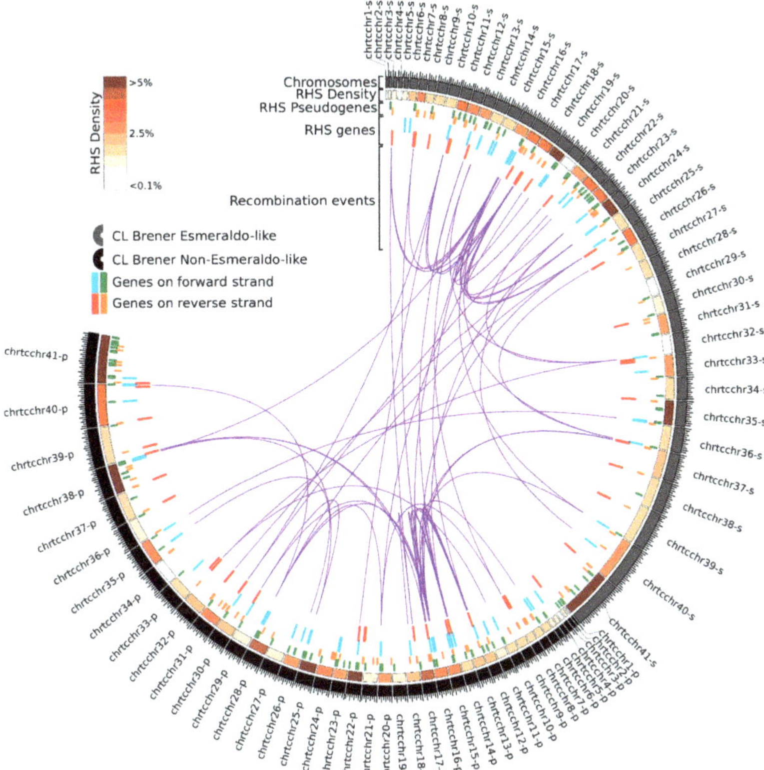

Figure 1. Circos diagram depicting the genomic organization and recombination events of the RHS family in the whole genome of *T. cruzi* clone CLB. Inner track 1 represents the recombination between RHS genes. The recombinant sequences are linked to putative major and minor parental, using purple and green lines, respectively. Track 2 shows the genomic organization of RHS genes in chromosomes. Genes on forward and reverse strands are colored in blue and red, respectively. Track 3 shows the genomic organization of RHS pseudogenes in chromosomes. Pseudogenes on forward and reverse strands are colored in green and orange, respectively. Track 4 depicts a heat map of RHS genes' and pseudogenes' density for each chromosome. Values were obtained by summing the length (bp) of RHS genes and pseudogenes and were divided by the chromosome size. Outer track 5 shows the representation of *T. cruzi* CLB chromosomes for Esmeraldo (haplotype S) and non-Esmeraldo (haplotype P) allelic loci.

3.2. Phylogeny and Classification of the RHS Multigene Family of Clone CLB

In the phylogenetic analysis, the transcribed RHS genes were examined for the presence of RHS domains by rpsBLAST using an e-value of 1×10^{-5} against the database of conserved domains [18]. Aiming to reveal the real extension of recombination events within RHS genes, in this analysis, we excluded non-LTR retrotransposons or other protein families with which RHS are commonly associated. The presence of conserved RHS domains (pfam07999, PTZ00209, and TIGRO1631) was also confirmed in other databases (CDD, Pfam, SMART, KOG, COG, PRK, and TIGR). The analysis of 139 RHS amino acid sequences was carried out using the maximum likelihood method in the RxML v 8.2.9 program by replacement models (PROTGAMMAAUTO). One thousand bootstrap replicas were processed to confirm the degree of reliability of the groups, assuming bootstrap values >75. Seventy-four RHS sequences can be categorized into groups 1 to 10 with values above the cutoff (indicated in colors), while three groups comprising 65 sequences with bootstrap values below the cutoff (indicated in black) were designated as unclassified groups. The number of sequences per group ranged from two RHS sequences in group 10 (light blue) to 15 sequences in group 3 (red) (Figure 2 and Table 1). Phylogenetic analysis showed that each RHS group consists of a monophylogenetic group. The results were also shown in the format rooted in the midpoint (Figure S5), where all the sequences with their respective TriTrypDB access numbers can be appreciated [41].

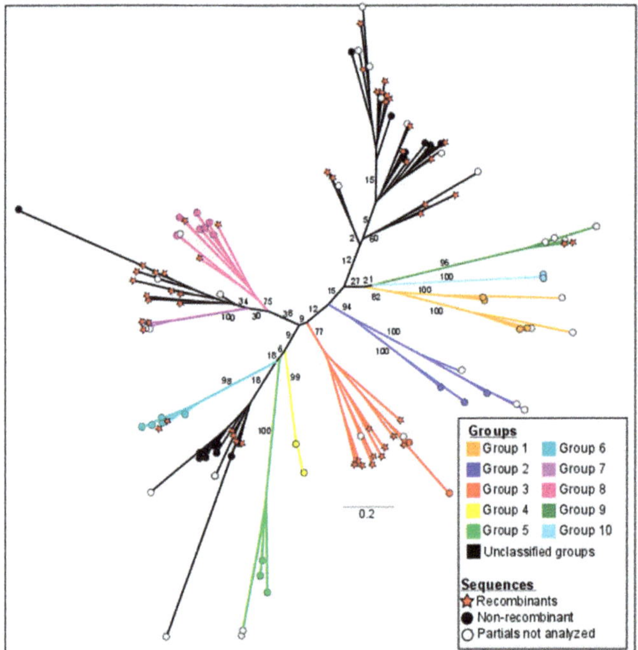

Figure 2. Phylogeny and classification of transcribed RHS sequences. Phylogenetic analysis was carried out using the RaxML v 8.2.9 program with an automatic search for substitution models (PROTGAMMAAUTO) selected using the Akaike information criterion (AIC) (auto-prot = AIC), with 1000 bootstrap replicates. Groups 1–10 comprise RHS sequences, with supported values separated by colors, and RHS sequences with bootstrap values below the cutoff (unclassified groups) are indicated in black.

Table 1. Distribution of the members of RHS groups across the chromosomes of clone CLB.

Group	Gene ID TriTrypDB [1]	CDS (bp) [2]	Peptide (aa) [3]	Direction of Transcription [4]	Subtelomeric Region [5]	Chromosome [6]
1	TcCLB.511845.10	270	90	Sense	-	TcChr20-P (580,762–581,031)
	TcCLB.509717.176	402	134	Sense	-	TcChr4-P (157,230–157,631)
	TcCLB.509295.90	771	256	Sense	Tel 6	TcChr28-P (746,714–747,484)
	TcCLB.510479.11	1701	567	Sense	-	TcChr38-P (1,335,682–1,337,382)
	TcCLB.506961.10	1929	642	Anti-Sense	-	TcChr18-S (118–2046)
	TcCLB.506001.90	2763	920	Sense	-	TcChr4-P (166,550–169,312)
	TcCLB.507167.70	2772	923	Sense	Tel 6	TcChr28-P (837,994–840,765)
	TcCLB.508479.500	2892	963	Anti-Sense	-	TcChr40-P (1,914,173–1,917,064)
2	TcCLB.509875.11	819	273	Sense	Tel 13	TcChr26-P (793,295–794,113)
	TcCLB.509873.10	831	276	Sense	Tel 13	TcChr26-P (794,215–795,045)
	TcCLB.508285.10	1767	588	Sense	Tel 3	TcChr19-S (653,962–655,728)
	TcCLB.506421.10	1038	345	Anti-Sense	Tel 49	TcChr31-P (53,479–54,51)
	TcCLB.509915.60	1767	588	Anti-Sense	Tel 49	TcChr31-P (64,469–66,235)
	TcCLB.506443.150	2400	799	Sense	Tel 24	TcChr11-P (510,464–512,863)
	TcCLB.507555.80	2757	918	Anti-Sense	Tel 35	TcChr35-S (510,464–512,863)
3	TcCLB.459199.10	2820	939	Anti-Sense	Tel 28	TcChr15-P (5578–8397)
	TcCLB.506047.20	1815	604	Sense	Tel 9	TcChr35-S (1,183,688–1,185,502)
	TcCLB.506017.51	1122	374	Sense	-	TcChr29-P (869,711–870,832)
	TcCLB.507167.20	2835	944	Sense	Tel 6	TcChr28-P (849,015–851,849)
	TcCLB.507611.10	2841	946	Anti-Sense	Tel 17	TcChr37-S (1391–4231)
	TcCLB.506393189	2274	758	Sense	-	TcChr14-P (596,251–598,524)
	TcCLB.506323.30	2790	929	Anti-Sense	Tel 4	TcChr22-P (62,292–65,081)
	TcCLB.509429.4	2613	871	Sense	-	TcChr6-P (364,778–367,390)
	TcCLB.511773.110	2472	995	Anti-Sense	-	TcChr17-P (301–2772)
	TcCLB.508037.10	1146	381	Anti-Sense	Tel 48	TcChr27-S (1297–2442)
	TcCLB.511929.30	2781	926	Sense	-	TcChr25-P (736,933–739,713)
	TcCLB.504109.200	3294	1097	Anti-Sense	-	TcChr39-P (599–3892)
	TcCLB.508473.10	4512	1503	Sense	Tel 30	TcChr39-S (1,847,980–1,852,491)
	TcCLB.507625.10	4149	1382	Sense	Tel 45	TcChr40-S (1,133,828–1,137,976)
	TcCLB.39997.10	1053	350	Anti-Sense	-	TcChr37-P (33,320–34,372)
4	TcCLB.504343.30	1779	592	Anti-Sense	-	TcChr7-S (60,071–61,849)
	TcCLB.507.907.30	1779	592	Anti-Sense	-	TcChr7-S (73,533–75,311)
	TcCLB.507.907.60	1779	592	Anti-Sense	-	TcChr7-S (62,859–64,637)
	TcCLB.505207.30	1626	541	Anti-Sense	-	TcChr41-P (8244–9869)
5	TcCLB.511019.80 *	1500	499	Sense	-	TcChr35-P (101,616–103,187)
	TcCLB.503881.30	1509	502	Sense	-	TcChr33-S (730,729–732,237)
	TcCLB.508119.140	1503	500	Anti-Sense	-	TcChr33-P (724,554–726,056)
	TcCLB.511907.330	1503	500	Sense	-	TcChr26-P (250,686–252,188)
	TcCLB.506529.680	444	148	Sense	-	TcChr6-S (201,683–202,126)
	TcCLB.510889.352	510	170	Sense	-	TcChr6-P (201,577–202,086)
6	TcCLB.509085.120	1896	631	Anti-Sense	-	TcChr15-P (164,566–166,461)
	TcCLB.509437.110	1896	631	Sense	-	TcChr15-P (256,827–258,722)
	TcCLB.509349.20	1893	630	Anti-Sense	Tel 2	TcChr11-S (115,973–117,865)
	TcCLB.508479.80	1947	648	Sense	-	TcChr40-P (1,993,454–1,995,400)
	TcCLB.509163.110	1962	653	Sense	-	TcChr35-P (1,138,639–1,140,600)
	TcCLB.511871.130	1896	631	Sense	-	TcChr15-S (101,636–103,531)
	TcCLB.511861.90	1896	631	Sense	-	TcChr15-P (118,605–120,500)
	TcCLB.511863.4	1572	524	Sense	-	TcChr15-P (101,616–103,187)
7	TcCLB.506809.5	354	117	Sense	-	TcChr16-P (453,466–453,819)
	TcCLB.509575.10	2763	920	Sense	-	TcChr16-P (389,424–392,186)
	TcCLB.424771.10	873	290	Sense	-	TcChr16-P (477,440–478,312)
	TcCLB.507843.10	1779	592	Sense	-	TcChr16-S (390,540–392,318)
	TcCLB.509827.4	962	320	Sense	-	TcChr16-S (389,477–390,438)
	TcCLB.507841.14	2562	854	Sense	-	TcChr16-S (452,820–455,381)
8	TcCLB.511019.13	1548	516	Sense	-	TcChr35-P (446,350–447,897)
	TcCLB.509219.20	3633	1210	Sense	-	TcChr20-P (567,813–571,445)
	TcCLB.506271.30	324	108	Sense	-	TcChr20-P (586,959–587,282)
	TcCLB.510643.190	2496	831	Sense	-	TcChr16-P (642,806–645,301)
	TcCLB.505997.60	2316	771	Anti-Sense	Tel 1	TcChr9-P (12,280–14,595)
	TcCLB.506595.149	2465	821	Anti-Sense	-	TcChr33-P (101–2565)
	TcCLB.511371.10	1785	594	Sense	-	TcChr5-S (200,095–201,879)
	TcCLB.511415.11	1095	365	Anti-Sense	-	TcChr9-S (30,121–31,215)
	TcCLB.508559.90	1821	606	Sense	Tel 21	TcChr25-S (700,188–702,008)
	TcCLB.511585.320	1932	643	Anti-Sense	Tel 14	TcChr33-S (31,331–33,262)
	TcCLB.507015.10 *	2988	995	Anti-Sense	Tel 10	TcChr13-P (1626–4613)
	TcCLB.509917.19	1815	605	Anti-Sense	Tel 49	TcChr31-P (54,619–56,433)

Table 1. *Cont.*

Group	Gene ID TriTrypDB [1]	CDS (bp) [2]	Peptide (aa) [3]	Direction of Transcription [4]	Subtelomeric Region [5]	Chromosome [6]
9	TcCLB.503401.11	243	81	Sense	-	TcChr22-S (214,572–214,814)
	TcCLB.506629.240	327	109	Anti-Sense	-	TcChr39-P (389,442–389,768)
	TcCLB.509829.9	909	303	Anti-Sense	-	TcChr39-S 392,244–393)
	TcCLB.509329.9	752	250	Sense	-	TcChr22-P (339,264–340,015)
	TcCLB.509463.41 *	1209	403	Anti-Sense	-	TcChr22-P (391,811–393,019)
	TcCLB.509843.10	1503	500	Sense	-	TcChr22-S (214,918–216,420)
10	TcCLB.506139.200	1674	557	Sense	-	TcChr18-P (357,746–359,419)
	TcCLB.510845.10	1824	608	Anti-Sense	-	TcChr19-S (28,739–30,562)

[1] TriTrypDB [41]. [2] CDS (coding DNA sequence), size in bp. [3] The translated peptide, size in amino acid (aa). [4] The direction of transcription. [5] RHS is located in the subtelomeric regions of the chromosomes of clone CLB [45]. [6] Genomic coordinates at the in silico chromosome of clone CLB (TcChr) [40]. * The other allele at the same locus is a pseudogene.

The bulk of detailed information of the RHS groups of the CLB genome, such as chromosome mapping, genomic location including the subtelomeric region, the sizes of the coding sequence, and the predicted translated protein, is shown in Table 1. Most of RHS transcribed genes (70%) encode proteins of approximately 60 to 180 kDa, and the remainder encode peptides of 38 to 10 kDa. The RHS sequences selected for phylogenetic analysis were those assigned to CLB chromosomes (TcChr). Out of 74 RHS sequences, 58 genes have only one copy located in haplotype S or P, resulting in a hemizygous condition. Twenty-two of the hemizygotes are located in the subtelomere, a polymorphic region susceptible to homologous recombination, including ectopic recombination [5,45,46].

Our results showed that RHS hemizygotes can also be found in the interstitial chromosome regions in which the synteny is interrupted by a set of RHS sequences [47,48]. It has been proposed that the *T. cruzi* genome is organized in two compartments: a core compartment comprising conserved and hypothetical conserved genes, and a non-syntenic region (disruptive compartment) enriched by repetitive sequences such as members of multigene families TS, MASP, and mucins [3]. Other multigene families (GP63, DGF-1, and RHS) are dispersed throughout both compartments [3].

The subtelomeres of *T. cruzi* could be included in the disruptive compartment since they are enriched by genes encoding surface proteins (TS, MASP and DGF-1), retrotransposon hot spot genes (RHS), retrotransposon elements, satellite DNA, RNA-helicase and N-acetyltransferase genes [45,48–51]. Twenty-five chromosomal ends of CLB chromosomes (TcChr) are composed mostly of RHS genes and pseudogenes [45]. The disruptive compartment including the subtelomeric regions could act as sites for homologous recombination [2,3,5,26,28–30,32–35].

The members of the RHS groups are organized in multiple clusters at various genomic locations on different chromosomes, including the core and disruptive compartments and subtelomeres. (Table 1 and Figure 1). The distance between two contiguous RHS genes ranged from 2 to 50,000 bp and the identity from 55 to 98%, suggesting the occurrence of gene duplication by homologous mitotic recombination, as has been described in fungi [52,53]. Some rearrangements could be explained by unequal crossing-over between homologous chromatids (interhomolog crossover) leading to the loss of the tandem counterparts in one of the haplotypes. For example, the RHS genes of groups 1 and 7 located on chromosomes TcChr4-P and TcChr7-S, respectively, were mapped in only one haplotype, indicating the loss of these genes in the corresponding haplotype (Figure 3A,B).

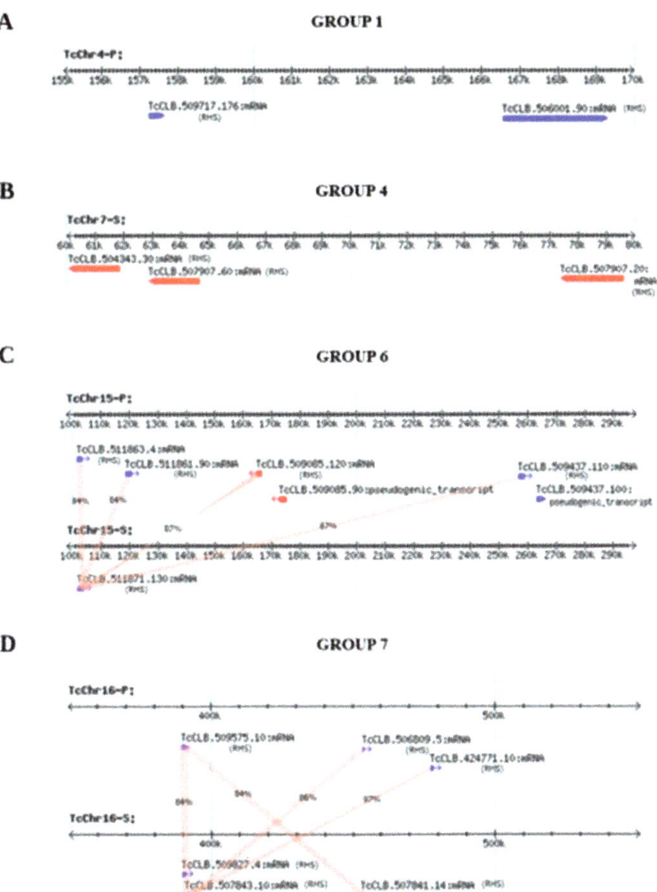

Figure 3. Gene duplication events in the RHS sequences of clone CLB. The figure shows the physical map of the chromosome regions involved in the recombination event. For clarity, only RHS sequences are shown. The direction of transcription is indicated by blue (sense) and red (anti-sense) arrows. (**A**,**B**) Groups 1 and 4: duplication of RHS genes by unequal crossing-over with loss of tandem counterparts in one of the haplotypes (TcCh4-S and TcChr7-P). (**C**) Group 6: duplication of the RHS genes by unequal crossing-over with the conservation of one of the RHS counterparts in the TcChr15-S haplotype. (**D**) Group 7: duplication followed by genetic conversion between paralogous genes located in the TcChr16-P and TcChr16-S haplotypes (interlocus nonallelic gene conversion). The identity between homologous RHS proteins of the P and S haplotypes is indicated in the figure. The identity between paralogous RHS proteins ranged from 93 to 100%. The physical maps showing the position of RHS sequences were downloaded from the public genome database TriTrypDB [41].

The RHS genes of group 6 were mapped to the chromosomes TcChr15-P and TcChr15-S, and only the first gene (TcCLB.511871.130) of the cluster was present on the TcChr15-S haplotype, the remainder was lost by unequal crossing-over-recombination between homologous chromatids (Figure 3C). The homologous RHS genes of the TcChr15-P encode proteins with >93% identity with each other, and they share 84% identity with the paralogous RHS (TcCLB.511871.130) of the TcChr15-S haplotype. These results showed that duplications gave rise to RHS sequences in tandem that maintained the structure of the functional gene.

The RHS genes of group 7 located on the chromosomes TcChr16-P and ThChr16-S share 84–97% identity (Figure 3D), and this arrangement could be explained by genetic duplication followed by genetic conversion between non-alleles (interlocus nonallelic gene conversion), e.g., between the RHS genes TcCLB.507843.10 (TcChr16-S) and TcCLB.506809.5

3.3. Generation of Genetic Variability by Recombination between T. cruzi RHS Sequences

In the phylogenetic analysis, we found sixty-five RHS sequences distributed in branches with low bootstrap values, which were included in the unclassified groups. Due to the high number of unclassified sequences, we investigated whether recombination events had also occurred in these sequences. We used the Circos plot to map the recombination events between RHS with a single link connecting each pair of paralogs (Figure 1). We identified 53 recombination events in 139 RHS sequences that were confirmed by at least six of the seven algorithms of the RDP4 package (Figure 4). We found that about 60% of the recombination events occurred in the unclassified sequences. Thirty-two unclassified RHS sequences were involved in the recombination events. The size of the fragment inserted into the target sequence by recombination is quite variable, and it may represent approximately 4% of the entire RHS gene. The recombination between the RHS genes results in mosaic structures that can contain up to three fragments of different RHSs inserted in the target sequence.

The recombination events occurred in different regions of RHS including the coding regions of the amino- and carboxy-terminal portions, as well as in the central region of the protein. Most recombination events were detected in the RHS sequences of group 3 that served as donors into unclassified sequences and eventually into sequences from other RHS groups. The recombination events occurred in specific regions, e.g., the amino-terminal coding region of RHS genes. As an example, the insertion of the same RHS sequence TcCLB.507841.14 of group 7 into the amino-terminal coding region of unclassified RHS sequences is shown (Figure 4, see recombination events 46 to 53).

3.4. Expression and Subcellular Localization of RHS in T. cruzi

The expression of RHS in *T. cruzi* and other trypanosomes was analyzed by Western blot using anti-RHS antibodies raised against a recombinant protein carrying a 292-amino acid region from the carboxy-terminal domain of RHS (TcCLB.511055.20) of CLB. This region is conserved among RHS of some *T. cruzi* strains (Dm28c, Sylvio X10/1, Y, Bug2148, Tulahuen, TCC) and *T. cruzi marinkellei*. The location of RHS (TcCLB.511055.20) in the nucleus has been experimentally demonstrated in the nuclear subproteome of clone CLB [54].

The anti-RHS polyclonal antibodies identified different protein profiles among *T. cruzi* strains and trypanosome species. They reacted strongly with two bands of 118 kDa and 112 kDa in the *T. cruzi* clone CLB and G strain, and weakly with two additional bands of 65 kDa and 29 kDa in CLB. A single band of 65 kDa was detected in *T. cruzi marinkellei* and *T. rangeli*, and a band of 82 kDa in *T. brucei* (Figure 5A). The sizes of RHS proteins identified by Western blot are consistent with those predicted RHS ORFs in the *T. cruzi* strains and *T. cruzi marinkellei*. These results suggest that the RHS genes encoding the 118 kDa and 112 kDa proteins are expressed in the CLB and G strain, whereas the lower molecular weight (65 kDa and 29 kDa) RHS proteins are expressed only in lower amounts in CLB. *T. cruzi marinkellei* and *T. rangeli* showed a similar expression profile consisting of a single 65 kDa band. The presence of an 82 kDa RHS in *T. brucei* is in agreement with the RHS protein profile (85 to 110 kDa) described in this trypanosome [6].

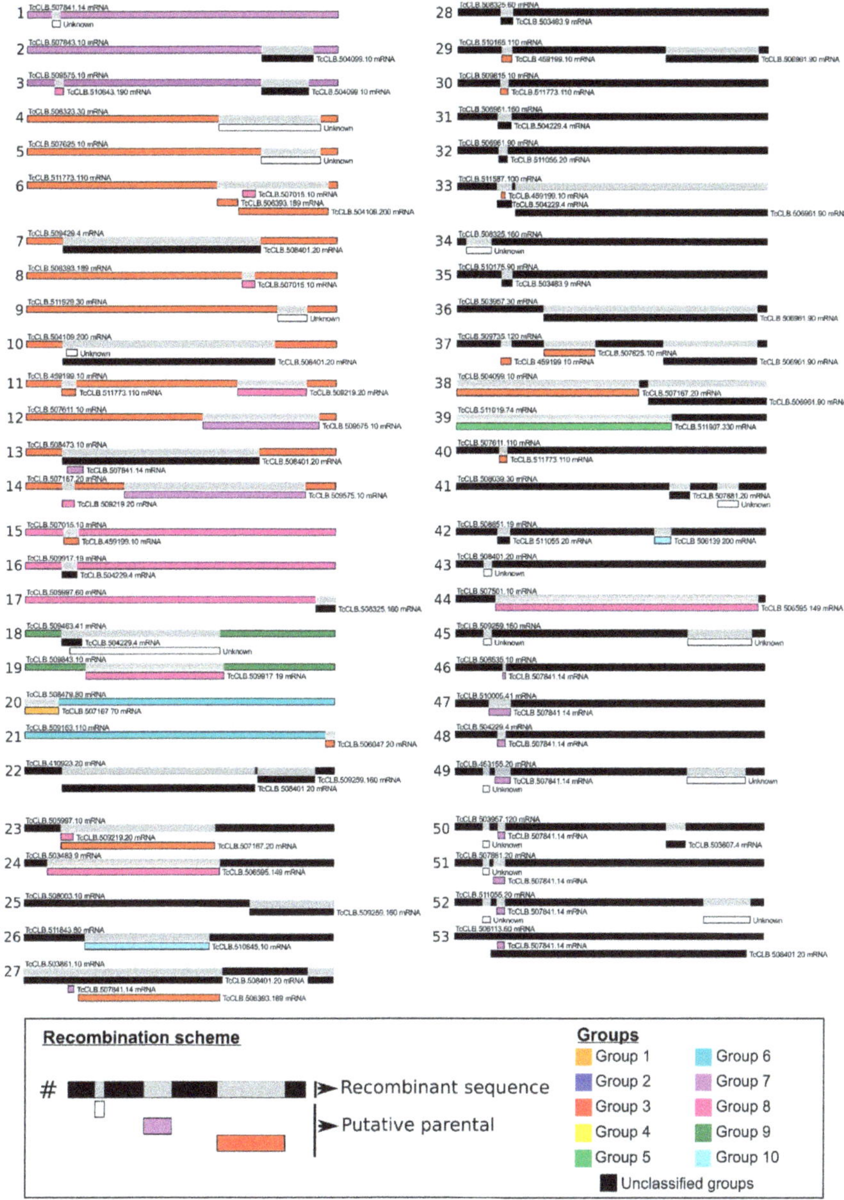

Figure 4. Detection of potential recombination events in *T. cruzi* RHS sequences. Recombination analysis was performed using the RDP4 program composed of non-parametric recombination detection methods by the algorithms: RDP, GENECONV, MaxChi, Chimera, Bootscan, SiSscan, and 3Seq. RHS sequences of groups 1–10 (parental sequences) are highlighted in different colors and unclassified groups (recombinant sequences) are presented in black. All RHS sequences are also indicated by their access number in the TriTrypDB [41].

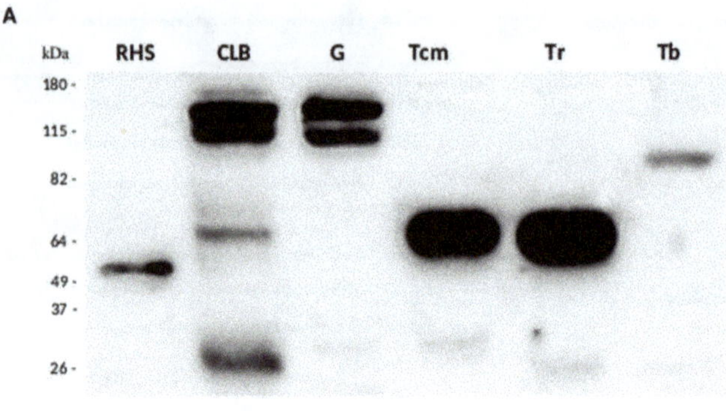

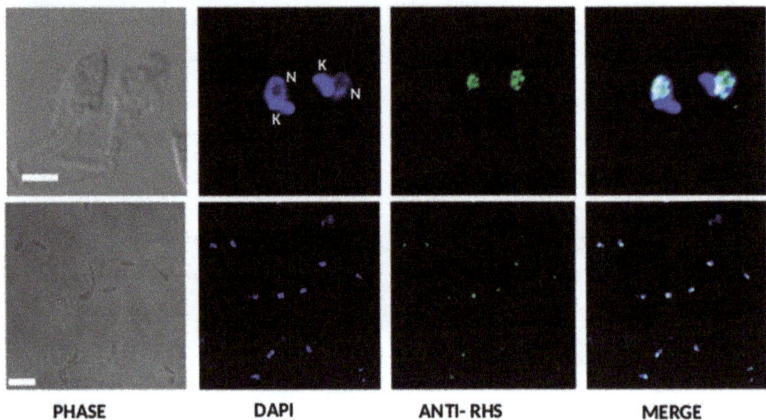

Figure 5. Analysis of the expression of RHS by Western blot in *T. cruzi* and other trypanosomes and cellular localization in *T. cruzi* by indirect immunofluorescence. (**A**) Protein extracts of epimastigotes of *T. cruzi* (CLB and strain G), *T. cruzi marinkellei* (Tcm), and *T. rangeli* (Tr) and procyclic forms of *T. brucei* (Tb) were separated by SDS-PAGE, transferred to nitrocellulose membranes, and incubated with anti-RHS polyclonal antibodies (diluted 1:500). The RHS recombinant protein was included as a positive control. The molecular masses of the reference proteins are indicated on the left in kDa. (**B**) Confocal microscopy images from indirect immunofluorescence reaction with anti-RHS antibodies (diluted 1:1000) in permeabilized epimastigotes of clone CLB. The labeling of the nucleus and kinetoplast DNA (DAPI) and RHS proteins is shown in blue and green, respectively. At the top, the reaction with two epimastigotes is shown at 3 μm scale. In the lower panel, the image shows epimastigotes (scale bar 10 μm). N, nucleus; K, kinetoplast.

Permeabilized parasites were analyzed by indirect immunofluorescence, using anti-RHS antibodies (Figure 5B). Nuclear and kinetoplast DNA was labeled with DAPI, and the RHS proteins were detected with fluorescent anti-RHS antibodies (shown in blue and green, respectively). The fluorescence distribution in the permeabilized parasites is concentrated at the nuclear region, confirmed by its colocalization with DAPI (Figure 5B merge). RHS distribution was concentrated in spots within the nucleus. Anti-RHS also reacted within the nucleus of intracellular amastigote (Figure 6), but no reaction was found in trypomastigotes. Taken together, these results suggest that RHS proteins of clone CLB have a predominantly nuclear location.

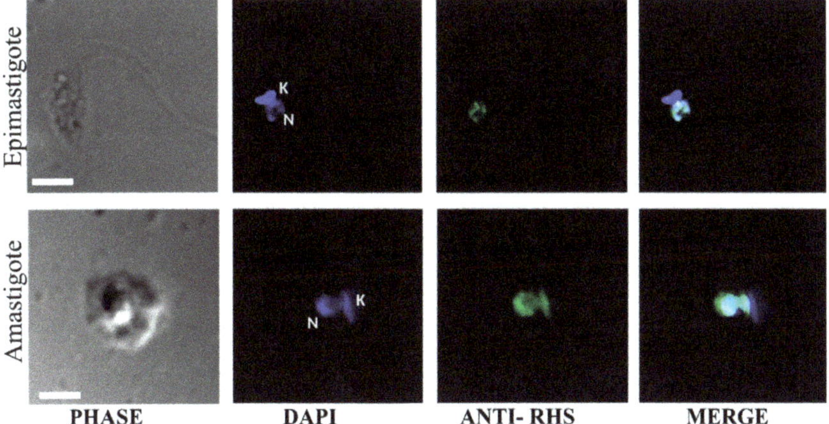

Figure 6. Cellular localization of RHS in the amastigote of T. cruzi. Confocal microscopy images from indirect immunofluorescence reaction with anti-RHS antibodies (diluted 1:1000) in permeabilized epimastigotes and amastigotes of clone CLB. The labeling of the nucleus and kinetoplast DNA (DAPI) and RHS proteins is shown in blue and green, respectively. Scale bar 3 µm. N, nucleus; K, kinetoplast.

4. Discussion

4.1. Genomic Organization and Generation of Genetic Variability in the RHS Multigene Family in T. cruzi

RHS is a genus-specific multigene family identified in the genome of all trypanosomes sequenced so far. RHS genes have a retrotransposon insertion site in their 5′ coding region, which is predicted to disrupt more than 50% of the members of this family. Therefore, our phylogenetic analysis was restricted to transcribed RHS sequences with an uninterrupted ORF encoding the RHS domain. RHS proteins of clone CLB were categorized into 10 groups with significant bootstrap (Figure 2), suggesting that each RHS subfamily is a monophyletic group, as previously reported in *T. brucei* [6]. Regarding the unclassified RHS sequences, they were separated from the rest of the groups, suggesting some structural differentiation among these sequences, and they evolved together with other RHS groups. Our search showed that *T. cruzi* RHS paralogous genes shared 75–100% identity at the amino acid level, whereas they shared 30–47% identity with orthologous genes from other trypanosome species, such as *T. rangeli*, *T. grayi*, *T. evansi*, *T. vivax*, *T. brucei*, *T. theileri* and *T. conorhini*. From these results, we may infer that RHS genes evolved from a common ancestor and started diverging by speciation.

Once we defined the RHS sequence groups of *T. cruzi* CLB, the next question was whether recombination events occurred among the members of the various RHS groups including the unclassified ones. The comparison of transcribed RHS sequences showed the occurrence of one to three recombinational events resulting in a mosaic structure, which contains up to three fragments derived from different RHSs. The RHS sequences of unclassified groups comprised ~47% of total transcribed RHS, being involved in ~60% of the recombinational events in which they were used as a template to generate new RHS sequences. Our results suggest that the RHS family has been subjected to rapid gene turnover, resulting in different paralogous groups that are conserved for functional reasons. We believe that the unclassified RHSs may act as sequence reservoirs that can recombine with functional paralogs to generate diversity, and at the same time preserve intact copies in the RHS gene family. The lack of ancestral sequences could be explained by a continuous process of gene turnover mediated by gene conversion (allelic or ectopic) and unequal crossing-over.

The complexity of the RHS family may also be related to the large number of pseudogenes that comprise more than 50% of the family [2,6,7,42]. In *T. cruzi* and *T. brucei*, the repertoire of pseudogenes is of great importance in the generation of variants of multigenic families involved in parasitic virulence [6,55–59]. Taken together, these results suggest that trypanosomes developed alternative mechanisms for achieving genetic diversity in the multigene families, one of which uses incomplete genes (pseudogenes) in the generation of functional genes, while others promote recombination between functional genes. These mechanisms acting together may lead to the generation of multiple RHS sequences, resulting in the diversity within this family but preserving intact RHS copies in the genome.

Sequence diversity in the RHS multigene family of *T. cruzi* may be generated by unequal crossing-over (sister chromatid exchange and interhomolog crossover), segmental gene conversion, and interlocus nonallelic gene conversion. Tandem duplication generated by unequal crossing-over over between non-sister homologous chromatids (interhomolog crossover) may occur with the loss of tandem allelic counterparts in one of the haplotypes, leading to a condition called hemizygosity. Out of 139 transcribed RHS genes of CLB, 58 genes (~42%) have only one allele with no counterpart in the other haplotype (S or P), resulting in a hemizygous condition. We identified 22 RHS hemizygotes mapped in the subtelomere, which is a polymorphic region that is susceptible to homologous and ectopic recombination [5,45,46,49,51]. Callejas et al., 2006 [60] identified a large hemizygous subtelomere region in the chromosome I of *T. brucei*. This region accounted for three-quarters of the length of chromosome I and resulted in the amplification and divergence of gene families such as VSG (Variant Surface Glycoprotein) [60].

There is some evidence in the genome of *T. cruzi* that segmental gene conversion is involved in the generation of sequence diversity for multigene families organized in tandem array repeats [61–64]. In addition to segmental genetic conversion, we also found evidence of interlocus nonallelic gene conversion (IGC) among gene duplicates between loci. Gene conversion has been proposed as an active force in the evolution of trypanosomes [65]. Araujo et al., 2020 [66] showed that DNA replication origins in *T. cruzi* are preferentially located at the subtelomeric region, which is a site of conflict between transcription and replication that may lead to DNA double-strand breaks and generation of diversity. Wier et al., 2016 [67] suggested that gene conversion is the mechanism used by *T. brucei gambiensis* to avoid the Meselson effect of accumulation of mutations on the chromosomes for lack of sexual recombination in this species. The proposed mechanism is based on the repair of a defective gene copy on a chromosome by copying and pasting the functional gene from the homologous chromosome.

4.2. The Role of RHS Proteins in T. cruzi

We found that RHS proteins are located in the nucleus of epimastigotes and amastigotes of *T. cruzi*. This is in agreement with previous work [54] that identified the presence of 74 RHS proteins with apparent molecular masses of 12 to 111 kDa in the nuclear proteome of *T. cruzi* epimastigotes [54]. These data were corroborated by Western blot analysis, in which we identified RHS proteins from 29 to 118 kDa in CLB. Despite the large number of RHSs expressed in *T. cruzi*, the profile of proteins recognized by anti-RHS antibodies is relatively simple, composed of 2–3 strongly reactive proteins. A similar profile was described in *T. brucei*, and it may be due to the absence of cross-reactivity between RHSs of different families [6].

Proteomic studies revealed that RHS proteins are expressed in epimastigotes of *T. cruzi* [68,69]. More recently, approximately 39 RHS isoforms expressed in *T. cruzi* trypomastigotes have been identified [70]. However, the diversity of RHS proteins detected by immunoblotting was more restricted, since only eight RHS isoforms were observed in this study [71]. The absence of reactivity of anti-RHS antibodies generated against the carboxy-terminal domain of RHS (TcCLB.511055.20) of CLB with *T. cruzi* trypomastigotes suggests that RHS proteins carrying the epitopes used in the mice immunization were not expressed in this developmental form. RHS proteins seem to be constitutively expressed in *T. brucei*, but they are more abundant in the procyclic forms of this parasite [6]. More recently, it has been reported that several RHSs are stage-specific regulated [10].

Since RHS is a target for the insertion of retrotransposons, the participation of RHS in controlling the expansion of these mobile elements has been proposed. Other functions for RHS have been related to *T. brucei*. TbRRM, a modulator of the chromatin structure in *T. brucei*, interacts with RHS transcripts, proteins and histones, suggesting that the RHS family could be involved in chromatin modeling [10]. Recently, it has been reported that several RHS proteins (RHS2, RHS4, and RHS6) may act as factors involved in the transcription elongation and mRNA export in *T. brucei* [11].

Little is known about the role of RHS in the *T. cruzi* life cycle. *T. cruzi* RHS proteins have been identified in the secretome of epimastigotes, trypomastigotes, and amastigotes, indicating that they are exported to the extracellular medium [71–74]. Bautista-Lopez et al., 2017 [71] showed that RHS proteins were present in the extracellular vesicles (EVs) released by *T. cruzi* trypomastigotes and amastigotes in infected Vero cells. The secreted RHS proteins reacted with sera from chronic chagasic patients ranging from asymptomatic to advanced cardiomyopathy. EVs are important modulators of the mammalian host—*T. cruzi* relationships, such as heart parasitism, susceptibility to infection of mammalian cells, and inflammatory response [72,75]. The immunoreactivity of RHSs from EVs suggests that they could participate, possibly as adjuvants, in the interaction of *T. cruzi* with the mammalian host. In this context, it is noteworthy that RHS is more abundant in the *T. cruzi* strains infective for humans (Bug2148, Y, and Sylvio X10) than in B7, which is not infective in humans [44].

In conclusion, our data suggest that unequal mitotic crossing-over and gene conversion play a significant role in shaping the patterns of homology between the RHS paralogous repeats that accelerate the generation of diversity within this multigene family. Recombination among transcribed RHS genes leads to the generation of multiple chimeric functional RHS genes. Finally, we showed the nuclear location of RHS in the replicative forms of *T. cruzi*. Although evidence for the functions of RHS in *T. cruzi* has been elusive, we suggest that these proteins could play a role in modulating the chromatin structure at the transcriptional and posttranscriptional levels, as has been suggested in *T. brucei* [10,11].

Supplementary Materials: The following are available online at http://www.mdpi.com/2073-4425/11/9/1085/s1, Figure S1: Flowchart of RHS sequences' identification, and quality validation; Figure S2: Integrity and purity of RHS recombinant protein; Figure S3: Distribution of RHS sequences across the chromosomes of clone CLB of *T. cruzi*; Figure S4: Proportion of total RHS length in each chromosome of clone CLB; Figure S5: Phylogeny and classification of transcribed RHS sequences of clone CLB; Table S1. Mapping of RHS sequences on the chromosomes of clone CLB of *Trypanosoma cruzi*.

Author Contributions: J.F.D.S., R.T.S., A.G.C.-M., and M.M.G.T. conceived the study; W.P.B., R.T.S., E.R.F., and A.G.C.-M. designed and performed the experiments; W.P.B., R.T.S., E.R.F., A.G.C.-M., J.L.R., and J.F.D.S. contributed to the analysis and interpretation of the data; J.F.D.S. and J.L.R. wrote and edited the manuscript, M.M.G.T. and R.A.M. revised it critically; R.A.M. and J.F.D.S. were responsible for the project management and funding acquisition. All authors have read and agreed to the published version of the manuscript.

Funding: The authors wish to thank the following funding sources: Fundação de Amparo à Pesquisa do Estado de São Paulo (FAPESP), Brasil, post-doctoral fellowship to E.R.F. (2016/16918-5), research thematic grant to R.A.M. and J.F.D.S. (2016/15000-4); Conselho Nacional de Desenvolvimento Científico e Tecnológico (CNPq), Brazil, master fellowship to W.P.B. (134397/2015-0), Young Talent Researcher fellowship/CNPq to R.T.S. (314048/20 13-8) and P.Q. fellowship to J.F.D.S. (306591/2015-4), R.A.M. (302068/2016-3); Coordenação de Aperfeiçoamento de Pessoal de Nível Superior (CAPES), Brazil, post-doctoral fellowship to A.G.C.-M. from CAPES (PROTAX). J.L.R. was supported by IDEA Plan Operativo Anual # 0012, Venezuela.

Acknowledgments: We thank Fernanda Sycko Uliana Marchiano e Rafaela Andrade do Carmo for assistance in the genomic analysis and with the figures.

Conflicts of Interest: The authors declare no conflict of interest.

References

1. WHO. Chagas Disease (American Trypanosomiasis). 2020. Available online: https://www.who.int/health-topics/chagas-disease (accessed on 27 July 2020).

2. El-Sayed, N.M.; Myler, P.J.; Bartholomeu, D.C.; Nilsson, D.; Aggarwal, G.; Tran, A.N.; Ghedin, E.; Worthey, E.A.; Delcher, A.L.; Blandin, G.; et al. The Genome Sequence of *Trypanosoma cruzi*, Etiologic Agent of Chagas Disease. *Science* **2005**, *309*, 409–415. [CrossRef]

3. Berná, L.; Rodriguez, M.; Chiribao, M.L.; Parodi-Talice, A.; Pita, S.; Rijo, G.; Alvarez-Valin, F.; Robello, C. Expanding an expanded genome: Long-read sequencing of *Trypanosoma cruzi*. *Microb. Genomics* **2018**, *4*. [CrossRef]

4. Pita, S.; Díaz-Viraqué, F.; Iraola, G.; Robello, C. The Tritryps Comparative Repeatome: Insights on Repetitive Element Evolution in Trypanosomatid Pathogens. *Genome Biol. Evol.* **2019**, *11*, 546–551. [CrossRef] [PubMed]

5. Chiurillo, M.A.; Barros, R.R.M.; Souza, R.T.; Marini, M.M.; Antonio, C.R.; Cortez, D.R.; Curto, M.; Lorenzi, H.A.; Schijman, A.G.; Ramirez, J.L.; et al. Subtelomeric I-SceI-mediated double-strand breaks are repaired by homologous recombination in *Trypanosoma cruzi*. *Front. Microbiol.* **2016**, *7*, 1–8. [CrossRef] [PubMed]

6. Bringaud, F.; Biteau, N.; Melville, S.E.; Hez, S.; El-Sayed, N.M.; Leech, V.; Berriman, M.; Hall, N.; Donelson, J.E.; Baltz, T. A New, Expressed Multigene Family Containing a Hot Spot for Insertion of Retroelements Is Associated with Polymorphic Subtelomeric Regions of Trypanosoma brucei. *Eukaryot. Cell* **2002**, *1*, 137–151. [CrossRef]

7. Bringaud, F.; Bartholomeu, D.C.; Blandin, G.; Delcher, A.; Baltz, T.; El-Sayed, N.M.A.; Ghedin, E. The *Trypanosoma cruzi* L1Tc and NARTc Non-LTR Retrotransposons Show Relative Site Specificity for Insertion. *Mol. Biol. Evol.* **2006**, *23*, 411–420. [CrossRef] [PubMed]

8. Durand-Dubief, M.; Absalon, S.; Menzer, L.; Ngwabyt, S.; Ersfeld, K.; Bastin, P. The Argonaute protein TbAGO1 contributes to large and mini-chromosome segregation and is required for control of RIME retroposons and RHS pseudogene-associated transcripts. *Mol. Biochem. Parasitol.* **2007**, *156*, 144–153. [CrossRef]

9. Wen, Y.-Z.; Zheng, L.-L.; Liao, J.-Y.; Wang, M.-H.; Wei, Y.; Guo, X.-M.; Qu, L.-H.; Ayala, F.J.; Lun, Z.-R. Pseudogene-derived small interference RNAs regulate gene expression in African Trypanosoma brucei. *Proc. Natl. Acad. Sci. USA* **2011**, *108*, 8345–8350. [CrossRef]

10. Naguleswaran, A.; Gunasekera, K.; Schimanski, B.; Heller, M.; Hemphill, A.; Ochsenreiter, T.; Roditi, I. Trypanosoma brucei RRM1 Is a Nuclear RNA-Binding Protein and Modulator of Chromatin Structure. *MBio* **2015**, *6*, 1–11. [CrossRef]

11. Florini, F.; Naguleswaran, A.; Gharib, W.H.; Bringaud, F.; Roditi, I. Unexpected diversity in eukaryotic transcription revealed by the retrotransposon hotspot family of Trypanosoma brucei. *Nucleic Acids Res.* **2019**, *47*, 1725–1739. [CrossRef]

12. Zingales, B.; Pereira, M.E.S.; Almeida, K.A.; Umezawa, E.S.; Nehme, N.S.; Oliveira, R.P.; Macedo, A.; Souto, R.P. Biological Parameters and Molecular Markers of Clone CL Brener—The Reference Organism of the *Trypanosoma cruzi* Genome Project. *Mem. Inst. Oswaldo Cruz* **1997**, *92*, 811–814. [CrossRef] [PubMed]

13. Yoshida, N. Surface antigens of metacyclic trypomastigotes of *Trypanosoma cruzi*. *Infect. Immun.* **1983**, *40*, 836–839. [CrossRef] [PubMed]

14. Steindel, M.; Pinto, J.C.C.; Toma, H.K.; Mangia, R.H.R.; Ribeiro-Rodrigues, R.; Romanha, A.J. Trypanosoma rangeli (Tejera, 1920) isolated from a sylvatic rodent (*Echimys dasythrix*) in Santa Catarina island, Santa Catarina state: First report of this trypanosome in southern Brazil. *Mem. Inst. Oswaldo Cruz* **1991**, *86*, 73–79. [CrossRef] [PubMed]

15. Camargo, E.P. Growth and differentiation in *Trypanosoma cruzi*. I. Origin of metacyclic trypanosomes in liquid media. *Rev. Inst. Med. Trop. Sao Paulo* **1964**, *6*, 93–100.

16. Vieira da Silva, C.; Luquetti, A.O.; Rassi, A.; Mortara, R.A. Involvement of Ssp-4-related carbohydrate epitopes in mammalian cell invasion by *Trypanosoma cruzi* amastigotes. *Microbes Infect.* **2006**, *8*, 2120–2129. [CrossRef]

17. Altschul, S.F.; Gish, W.; Miller, W.; Myers, E.W.; Lipman, D.J. Basic local alignment search tool. *J. Mol. Biol.* **1990**, *215*, 403–410. [CrossRef]

18. Conserved Domains and Protein Classification. Available online: https://www.ncbi.nlm.nih.gov/cdd (accessed on 10 February 2017).

19. Edgar, R.C. MUSCLE: Multiple sequence alignment with high accuracy and high throughput. *Nucleic Acids Res.* **2004**, *32*, 1792–1797. [CrossRef]

20. Castresana, J. Selection of Conserved Blocks from Multiple Alignments for Their Use in Phylogenetic Analysis. *Mol. Biol. Evol.* **2000**, *17*, 540–552. [CrossRef]

21. TriTrypDB-31_TcruziCLBrener_AnnotatedTranscripts.fast. Available online: https://tritrypdb.org/common/downloads/release-31/TcruziCLBrener/fasta/data/ (accessed on 12 April 2017).

22. Stamatakis, A. RAxML version 8: A tool for phylogenetic analysis and post-analysis of large phylogenies. *Bioinformatics* **2014**, *30*, 1312–1313. [CrossRef]

23. FigTree GitHub Repository. Available online: http://tree.bio.ed.ac.uk/software/figtree/ (accessed on 15 March 2017).

24. Martins, N.O.; de Souza, R.T.; Cordero, E.M.; Maldonado, D.C.; Cortez, C.; Marini, M.M.; Ferreira, E.R.; Bayer-Santos, E.; de Almeida, I.C.; Yoshida, N.; et al. Molecular Characterization of a Novel Family of *Trypanosoma cruzi* Surface Membrane Proteins (TcSMP) Involved in Mammalian Host Cell Invasion. *PLoS Negl. Trop. Dis.* **2015**, *9*, e0004216. [CrossRef]

25. Tibayrenc, M.; Ward, P.; Moya, A.; Ayala, F.J. Natural populations of *Trypanosoma cruzi*, the agent of Chagas disease, have a complex multiclonal structure. *Proc. Natl. Acad. Sci. USA.* **1986**, *83*, 115–119. [CrossRef] [PubMed]

26. Oliveira, R.P.; Broude, N.E.; Macedo, A.M.; Cantor, C.R.; Smith, C.L.; Pena, S.D.J. Probing the genetic population structure of *Trypanosoma cruzi* with polymorphic microsatellites. *Proc. Natl. Acad. Sci. USA* **1998**, *95*, 3776–3780. [CrossRef] [PubMed]

27. Tibayrenc, M.; Ayala, F.J. The clonal theory of parasitic protozoa: 12 years on. *Trends Parasitol.* **2002**, *18*, 405–410. [CrossRef]

28. Pena, S.D.J.; Machado, C.R.; Macedo, A.M. *Trypanosoma cruzi*: Ancestral genomes and population structure. *Mem. Inst. Oswaldo Cruz* **2009**, *104*, 108–114. [CrossRef] [PubMed]

29. Gaunt, M.W.; Yeo, M.; Frame, I.A.; Stothard, J.R.; Carrasco, H.J.; Taylor, M.C.; Mena, S.S.; Veazey, P.; Miles, G.A.J.; Acosta, N.; et al. Mechanism of genetic exchange in American trypanosomes. *Nature* **2003**, *421*, 936–939. [CrossRef]

30. Machado, C.A.; Ayala, F.J. Nucleotide sequences provide evidence of genetic exchange among distantly related lineages of *Trypanosoma cruzi*. *Proc. Natl. Acad. Sci. USA* **2001**, *98*, 7396–7401. [CrossRef]

31. Augusto-Pinto, L.; Teixeira, S.M.R.; Pena, S.D.J.; Machado, C.R. Single-nucleotide polymorphisms of the *Trypanosoma cruzi* MSH2 gene support the existence of three phylogenetic lineages presenting differences in mismatch-repair efficiency. *Genetics* **2003**, *164*, 117–126.

32. Brisse, S.; Henriksson, J.; Barnabé, C.; Douzery, E.J.P.; Berkvens, D.; Serrano, M.; De Carvalho, M.R.C.; Buck, G.A.; Dujardin, J.-C.; Tibayrenc, M. Evidence for genetic exchange and hybridization in *Trypanosoma cruzi* based on nucleotide sequences and molecular karyotype. *Infect. Genet. Evol.* **2003**, *2*, 173–183. [CrossRef]

33. Westenberger, S.J.; Barnabé, C.; Campbell, D.A.; Sturm, N.R. Two Hybridization Events Define the Population Structure of *Trypanosoma cruzi*. *Genetics* **2005**, *171*, 527–543. [CrossRef]

34. de Freitas, J.M.; Augusto-Pinto, L.; Pimenta, J.R.; Bastos-Rodrigues, L.; Gonçalves, V.F.; Teixeira, S.M.R.; Chiari, E.; Junqueira, Â.C.V.; Fernandes, O.; Macedo, A.M.; et al. Ancestral Genomes, Sex, and the Population Structure of *Trypanosoma cruzi*. *PLoS Pathog.* **2006**, *2*, e24. [CrossRef]

35. Berry, A.S.F.; Salazar-Sánchez, R.; Castillo-Neyra, R.; Borrini-Mayorí, K.; Chipana-Ramos, C.; Vargas-Maquera, M.; Ancca-Juarez, J.; Náquira-Velarde, C.; Levy, M.Z.; Brisson, D. Sexual reproduction in a natural *Trypanosoma cruzi* population. *PLoS Negl. Trop. Dis.* **2019**, *13*, e0007392. [CrossRef] [PubMed]

36. Schwabl, P.; Imamura, H.; Van den Broeck, F.; Costales, J.A.; Maiguashca-Sánchez, J.; Miles, M.A.; Andersson, B.; Grijalva, M.J.; Llewellyn, M.S. Meiotic sex in Chagas disease parasite *Trypanosoma cruzi*. *Nat. Commun.* **2019**, *10*, 3972. [CrossRef] [PubMed]

37. Tibayrenc, M. Genetic subdivisions within *Trypanosoma cruzi* (Discrete Typing Units) and their relevance for molecular epidemiology and experimental evolution. *Kinetoplastid Biol. Dis.* **2003**, *2*, 1–6. [CrossRef]

38. Zingales, B.; Andrade, S.; Briones, M.; Campbell, D.; Chiari, E.; Fernandes, O.; Guhl, F.; Lages-Silva, E.; Macedo, A.; Machado, C.; et al. A new consensus for *Trypanosoma cruzi* intraspecific nomenclature: Second revision meeting recommends TcI to TcVI. *Mem. Inst. Oswaldo Cruz* **2009**, *104*, 1051–1054. [CrossRef]

39. Zingales, B.; Miles, M.A.; Campbell, D.A.; Tibayrenc, M.; Macedo, A.M.; Teixeira, M.M.G.; Schijman, A.G.; Llewellyn, M.S.; Lages-Silva, E.; Machado, C.R.; et al. The revised *Trypanosoma cruzi* subspecific nomenclature: Rationale, epidemiological relevance and research applications. *Infect. Genet. Evol.* **2012**, *12*, 240–253. [CrossRef]

40. Weatherly, D.B.; Boehlke, C.; Tarleton, R.L. Chromosome level assembly of the hybrid *Trypanosoma cruzi* genome. *BMC Genom.* **2009**, *10*, 255. [CrossRef]

41. TriTrypDB. The Kinetoplastid Genomics Resource. Available online: https://tritrypdb.org/tritrypdb/ (accessed on 15 February 2017).

42. El-Sayed, N.M.; Myler, P.J.; Blandin, G.; Berriman, M.; Crabtree, J.; Aggarwal, G.; Caler, E.; Renauld, H.; Worthey, E.A.; Hertz-Fowler, C.; et al. Comparative genomics of trypanosomatid parasitic protozoa. *Science* **2005**, *309*, 404–409. [CrossRef]

43. Azuaje, F.J.; Ramirez, J.L.; Da Silveira, J.F. In silico, biologically-inspired modelling of genomic variation generation in surface proteins of *Trypanosoma cruzi*. *Kinetoplastid Biol. Dis.* **2007**, *6*, 6. [CrossRef]

44. Callejas-Hernández, F.; Rastrojo, A.; Poveda, C.; Gironès, N.; Fresno, M. Genomic assemblies of newly sequenced *Trypanosoma cruzi* strains reveal new genomic expansion and greater complexity. *Sci. Rep.* **2018**, *8*, 14631. [CrossRef]

45. Moraes Barros, R.R.; Marini, M.M.; Antônio, C.; Cortez, D.R.; Miyake, A.M.; Lima, F.M.; Ruiz, J.C.; Bartholomeu, D.C.; Chiurillo, M.A.; Ramirez, J.; et al. Anatomy and evolution of telomeric and subtelomeric regions in the human protozoan parasite *Trypanosoma cruzi*. *BMC Genomics* **2012**, *13*, 229. [CrossRef]

46. Chiurillo, M.A.; Regina Antonio, C.; Mendes Marini, M.; de Souza, R.T.; Franco da Silveira, J. Chromosomes Ends and Telomere Biology of Trypanosomatids. In *Frontiers in Parasitology*; Bentham Science United Arab Emirates: Sharjah, UAE, 2017; pp. 104–133.

47. Ghedin, E.; Bringaud, F.; Peterson, J.; Myler, P.; Berriman, M.; Ivens, A.; Andersson, B.; Bontempi, E.; Eisen, J.; Angiuoli, S.; et al. Gene synteny and evolution of genome architecture in trypanosomatids. *Mol. Biochem. Parasitol.* **2004**, *134*, 183–191. [CrossRef] [PubMed]

48. Kim, D.; Chiurillo, M.A.; El-Sayed, N.; Jones, K.; Santos, M.R.M.; Porcile, P.E.; Andersson, B.; Myler, P.; da Silveira, J.F.; Ramírez, J.L. Telomere and subtelomere of *Trypanosoma cruzi* chromosomes are enriched in (pseudo)genes of retrotransposon hot spot and trans-sialidase-like gene families: The origins of *T. cruzi* telomeres. *Gene* **2005**, *346*, 153–161. [CrossRef] [PubMed]

49. Bartholomeu, D.C.; Cerqueira, G.C.; Leão, A.C.A.; DaRocha, W.D.; Pais, F.S.; Macedo, C.; Djikeng, A.; Teixeira, S.M.R.; El-Sayed, N.M. Genomic organization and expression profile of the mucin-associated surface protein (masp) family of the human pathogen *Trypanosoma cruzi*. *Nucleic Acids Res.* **2009**, *37*, 3407–3417. [CrossRef]

50. Martins, C.; Baptista, C.S.; Ienne, S.; Cerqueira, G.C.; Bartholomeu, D.C.; Zingales, B. Genomic organization and transcription analysis of the 195-bp satellite DNA in *Trypanosoma cruzi*. *Mol. Biochem. Parasitol.* **2008**, *160*, 60–64. [CrossRef]

51. Ramirez, J.L. An Evolutionary View of *Trypanosoma cruzi* Telomeres. *Front. Cell. Infect. Microbiol.* **2020**, *9*, 1–7. [CrossRef]

52. Zolan, M.E. Chromosome-length polymorphism in fungi. *Microbiol. Rev.* **1995**, *59*, 686–698. [CrossRef]

53. Symington, L.S.; Rothstein, R.; Lisby, M. Mechanisms and Regulation of Mitotic Recombination in Saccharomyces cerevisiae. *Genetics* **2014**, *198*, 795–835. [CrossRef]

54. dos Santos Júnior, A.D.C.M.; Kalume, D.E.; Camargo, R.; Gómez-Mendoza, D.P.; Correa, J.R.; Charneau, S.; de Sousa, M.V.; de Lima, B.D.; Ricart, C.A.O. Unveiling the *Trypanosoma cruzi* Nuclear Proteome. *PLoS ONE* **2015**, *10*, e0138667. [CrossRef]

55. Thon, G.; Baltz, T.; Eisen, H. Antigenic diversity by the recombination of pseudogenes. *Genes Dev.* **1989**, *3*, 1247–1254. [CrossRef]

56. Barry, J.D.; Ginger, M.L.; Burton, P.; McCulloch, R. Why are parasite contingency genes often associated with telomeres? *Int. J. Parasitol.* **2003**, *33*, 29–45. [CrossRef]

57. Marcello, L.; Barry, J.D. Analysis of the VSG gene silent archive in Trypanosoma brucei reveals that mosaic gene expression is prominent in antigenic variation and is favored by archive substructure. *Genome Res.* **2007**, *17*, 1344–1352. [CrossRef] [PubMed]

58. Hall, J.P.J.; Wang, H.; Barry, J.D. Mosaic VSGs and the Scale of Trypanosoma brucei Antigenic Variation. *PLoS Pathog.* **2013**, *9*, e1003502. [CrossRef] [PubMed]

59. Roth, C.; Bringaud, F.; Layden, R.E.; Baltz, T.; Eisen, H. Active late-appearing variable surface antigen genes in Trypanosoma equiperdum are constructed entirely from pseudogenes. *Proc. Natl. Acad. Sci. USA* **1989**, *86*, 9375–9379. [CrossRef] [PubMed]

60. Callejas, S.; Leech, V.; Reitter, C.; Melville, S. Hemizygous subtelomeres of an African trypanosome chromosome may account for over 75% of chromosome length. *Genome Res.* **2006**, *16*, 1109–1118. [CrossRef] [PubMed]

61. Takle, G.B.; O'Connor, J.; Young, A.J.; Cross, G.A.M. Sequence homology and absence of mRNA defines a possible pseudogene member of the *Trypanosoma cruzi* gp85/sialidase multigene family. *Mol. Biochem. Parasitol.* **1992**, *56*, 117–127. [CrossRef]

62. Taylor, M.C.; Muhia, D.K.; Baker, D.A.; Mondragon, A.; Schaap, P.; Kelly, J.M. *Trypanosoma cruzi* adenylyl cyclase is encoded by a complex multigene family. *Mol. Biochem. Parasitol.* **1999**, *104*, 205–217. [CrossRef]

63. Allen, C.L.; Kelly, J.M. *Trypanosoma cruzi*: Mucin Pseudogenes Organized in a Tandem Array. *Exp. Parasitol.* **2001**, *97*, 173–177. [CrossRef]

64. Cerqueira, G.C.; Bartholomeu, D.C.; DaRocha, W.D.; Hou, L.; Freitas-Silva, D.M.; Machado, C.R.; El-Sayed, N.M.; Teixeira, S.M.R. Sequence diversity and evolution of multigene families in *Trypanosoma cruzi*. *Mol. Biochem. Parasitol.* **2008**, *157*, 65–72. [CrossRef] [PubMed]

65. Jackson, A.P. Tandem gene arrays in Trypanosoma brucei: Comparative phylogenomic analysis of duplicate sequence variation. *BMC Evol. Biol.* **2007**, *7*, 54. [CrossRef]

66. de Araujo, C.B.; da Cunha, J.P.C.; Inada, D.T.; Damasceno, J.; Lima, A.R.J.; Hiraiwa, P.; Marques, C.; Gonçalves, E.; Nishiyama-Junior, M.Y.; McCulloch, R.; et al. Replication origin location might contribute to genetic variability in *Trypanosoma cruzi*. *BMC Genomics* **2020**, *21*, 414. [CrossRef]

67. Weir, W.; Capewell, P.; Foth, B.; Clucas, C.; Pountain, A.; Steketee, P.; Veitch, N.; Koffi, M.; de Meeûs, T.; Kaboré, J.; et al. Population genomics reveals the origin and asexual evolution of human infective trypanosomes. *Elife* **2016**, *5*, 1–14. [CrossRef] [PubMed]

68. Parodi-Talice, A.; Durán, R.; Arrambide, N.; Prieto, V.; Piñeyro, M.D.; Pritsch, O.; Cayota, A.; Cerveñansky, C.; Robello, C. Proteome analysis of the causative agent of Chagas disease: *Trypanosoma cruzi*. *Int. J. Parasitol.* **2004**, *34*, 881–886. [CrossRef] [PubMed]

69. Parodi-Talice, A.; Monteiro-Goes, V.; Arrambide, N.; Avila, A.R.; Duran, R.; Correa, A.; Dallagiovanna, B.; Cayota, A.; Krieger, M.; Goldenberg, S.; et al. Proteomic analysis of metacyclic trypomastigotes undergoing *Trypanosoma cruzi* metacyclogenesis. *J. Mass Spectrom.* **2007**, *42*, 1422–1432. [CrossRef] [PubMed]

70. Brunoro, G.V.F.; Caminha, M.A.; da Silva Ferreira, A.T.; da Veiga Leprevost, F.; Carvalho, P.C.; Perales, J.; Valente, R.H.; Menna-Barreto, R.F.S. Reevaluating the *Trypanosoma cruzi* proteomic map: The shotgun description of bloodstream trypomastigotes. *J. Proteomics* **2015**, *115*, 58–65. [CrossRef]

71. Bautista-López, N.L.; Ndao, M.; Camargo, F.V.; Nara, T.; Annoura, T.; Hardie, D.B.; Borchers, C.H.; Jardim, A. Characterization and Diagnostic Application of *Trypanosoma cruzi* Trypomastigote Excreted-Secreted Antigens Shed in Extracellular Vesicles Released from Infected Mammalian Cells. *J. Clin. Microbiol.* **2017**, *55*, 744–758. [CrossRef]

72. Trocoli Torrecilhas, A.C.; Tonelli, R.R.; Pavanelli, W.R.; da Silva, J.S.; Schumacher, R.I.; de Souza, W.; e Silva, N.C.; de Almeida Abrahamsohn, I.; Colli, W.; Manso Alves, M.J. *Trypanosoma cruzi*: Parasite shed vesicles increase heart parasitism and generate an intense inflammatory response. *Microbes Infect.* **2009**, *11*, 29–39. [CrossRef]

73. Bayer-Santos, E.; Aguilar-Bonavides, C.; Rodrigues, S.P.; Cordero, E.M.; Marques, A.F.; Varela-Ramirez, A.; Choi, H.; Yoshida, N.; Da Silveira, J.F.; Almeida, I.C. Proteomic analysis of *Trypanosoma cruzi* secretome: Characterization of two populations of extracellular vesicles and soluble proteins. *J. Proteome Res.* **2013**, *12*, 883–897. [CrossRef]

74. Brossas, J.Y.; Gulin, J.E.N.; Bisio, M.M.C.; Chapelle, M.; Marinach-Patrice, C.; Bordessoules, M.; Ruiz, G.P.; Vion, J.; Paris, L.; Altcheh, J.; et al. Secretome analysis of *Trypanosoma cruzi* by proteomics studies. *PLoS ONE* **2017**, *12*, e0185504. [CrossRef]

75. Garcia-Silva, M.R.; Cura Das Neves, R.F.; Cabrera-Cabrera, F.; Sanguinetti, J.; Medeiros, L.C.; Robello, C.; Naya, H.; Fernandez-Calero, T.; Souto-Padron, T.; De Souza, W.; et al. Extracellular vesicles shed by *Trypanosoma cruzi* are linked to small RNA pathways, life cycle regulation, and susceptibility to infection of mammalian cells. *Parasitol. Res.* **2014**, *113*, 285–304. [CrossRef]

 *genes*

Article

Identification of Novel Interspersed DNA Repetitive Elements in the *Trypanosoma cruzi* Genome Associated with the 3'UTRs of Surface Multigenic Families

Simone Guedes Calderano [1,2,*,†], Milton Yutaka Nishiyama Junior [2,3,†], Marjorie Marini [4,5], Nathan de Oliveira Nunes [2,3], Marcelo da Silva Reis [2,6], José Salvatore Leister Patané [2,6], José Franco da Silveira [4], Julia Pinheiro Chagas da Cunha [2,6] and Maria Carolina Elias [2,6,*]

1   Laboratório de Parasitologia, Instituto Butantan, São Paulo 05503-900, Brazil
2   Center of Toxins, Immune Response and Cell Signaling (CeTICS), Instituto Butantan, São Paulo 05503-900, Brazil; milton.nishiyama@butantan.gov.br (M.Y.N.J.); nathannunes@usp.br (N.d.O.N.); marcelo.reis@butantan.gov.br (M.d.S.R.); jose.patane@butantan.gov.br (J.S.L.P.); julia.cunha@butantan.gov.br (J.P.C.d.C.)
3   Laboratório de Toxinologia Aplicada, Instituto Butantan, São Paulo 05503-900, Brazil
4   Departamento de Micro, Imuno e Parasitologia, Universidade Federal de São Paulo, São Paulo 04023-062, Brazil; marjorie.marini@gmail.com (M.M.); jose.franco@unifesp.br (J.F.d.S.)
5   Biomedicina, Centro Universitário São Camilo, São Paulo 04263-200, Brazil
6   Laboratório de Ciclo Celular, Instituto Butantan, São Paulo 05503-900, Brazil
*   Correspondence: simone.calderano@butantan.gov.br (S.G.C.); carolina.eliassabbaga@butantan.gov.br (M.C.E.)
†   These authors contributed equally.

Received: 8 September 2020; Accepted: 14 October 2020; Published: 21 October 2020

Abstract: *Trypanosoma cruzi* is the etiological agent of Chagas disease, which affects millions of people in Latin America. No transcriptional control of gene expression has been demonstrated in this organism, and 50% of its genome consists of repetitive elements and members of multigenic families. In this study, we applied a novel bioinformatics approach to predict new repetitive elements in the genome sequence of *T. cruzi*. A new repetitive sequence measuring 241 nt was identified and found to be interspersed along the genome sequence from strains of different DTUs. This new repeat was mostly on intergenic regions, and upstream and downstream regions of the 241 nt repeat were enriched in surface protein genes. RNAseq analysis revealed that the repeat was part of processed mRNAs and was predominantly found in the 3' untranslated regions (UTRs) of genes of multigenic families encoding surface proteins. Moreover, we detected a correlation between the presence of the repeat in the 3'UTR of multigenic family genes and the level of differential expression of these genes when comparing epimastigote and trypomastigote transcriptomes. These data suggest that this sequence plays a role in the posttranscriptional regulation of the expression of multigenic families.

Keywords: *Trypanosoma cruzi*; genome; repeats; 3'UTR; multigenic family

---

1. Introduction

The protozoan *T. cruzi* is the causative agent of Chagas disease and affects approximately 7 million people, mostly in Central and South America, where another 18 million people live at risk of infection [1]. This parasite exhibits a complex life cycle varying between the nonreplicative/infective form, known as the trypomastigote (bloodstream in mammalian host and metacyclic inside the vector), and the replicative forms, known as the amastigote (in the mammalian host) and epimastigote (in the invertebrate vector). Morphological and metabolic changes are observed among these life forms with the presence of distinct proteomic [2,3] and transcriptomic [4] profiles. However, the control of gene expression in *T. cruzi* relies largely on posttranscriptional and translation levels since transcription does not occur from a specific RNA pol II promoter for each gene but, rather, nonrelated genes are transcribed as a unique polycistron and then trans-spliced into individual mature mRNA molecules [5]. Therefore, other levels of gene expression regulation stand out, such as mRNA processing [6], translational repression [7–9], polysome recruitment [10], and codon adaptation [11,12]. In this scenario, noncoding DNA may also be involved as a regulatory element in mRNA expression [10–13].

Among coding and noncoding DNA, the *T. cruzi* genome presents at least 50% repetitive sequences, which include multigenic families, retrotransposons and subtelomeric repeats [14–16]. Of the repetitive DNA elements found within intergenic regions, most have no identified function to date. For example, satellite DNA is a 195 bp repetitive element that can be used as a *T. cruzi* infection marker in molecular diagnostics [17]; however, no function has been attributed to this sequence. Conversely, multigenic families mostly encode surface proteins involved in cell invasion as well as immune system evasion by *T. cruzi* [18]. The expression levels of these genes vary along the *T. cruzi* life cycle according to their function, but little is known about how they are regulated.

The 3'UTR (untranslated region) from mRNA and RNA binding proteins (RNA-BP) has emerged as a key factor in mRNA stability and protein expression level regulation in *T. cruzi*, including some proteins from multigenic families [19–21]. In this study, we developed a new strategy to identify new repetitive elements in the *T. cruzi* genome and found an intergenic repetitive sequence located downstream of many genes of multigenic families, such as mucin-associated proteins (MASPs) and trans-sialidases. Using RNAseq analysis, we confirmed that this sequence is present on the 3'UTRs of these mRNAs and is correlated with gene expression regulation, indicating that this repetitive sequence may have a cis-regulatory function on the expression of multigenic family mRNAs.

2. Materials and Methods

2.1. Filtering Steps for 150 bp Fragments

All of the 150 nucleotide sequences, obtained by a sliding window all over the genome sequence, were filtered through 4 different parameters: (1) Any sequence with at least one undefined nucleotide (N) was excluded; (2) Sequences with less than 10 copies were excluded; (3) Fragments with a significant match against the repetitive elements using RepeatMasker software (version 4.0.7) [22] were excluded (parameters "-species trypanosoma -pa 60 -u -xm -engine ncbi -excln"); and (4) Any repeat from multigenic family genes was excluded. Fragments were submitted to Blast-n alignment against an in house multigenic family database composed of 4999 genes (surface protease (GP63), mucin-associated surface protein (MASP), retrotransposons and trans-sialidase) from CL Brener (-S and -P haplotypes) using the parameters "-e-value 1e-72 -dust no -qcov_hsp_perc 100".

2.2. Search Terms in the TriTryp Database

To establish the number of specific genes on *T. cruzi* strains Dm28c, Y, TCC, CL Brener haplotypes Esmeraldo-like (S), and non-Esmeraldo like (P), the following terms were searched in TriTrypDB [23]: "trans-sialidase", "mucin-associated surface protein", "TcMUC", "mucin like", "surface protease GP63", "hypothetical protein", "90 kDa surface protein", "serine-alanine and proline-rich protein", "dispersed gene family protein 1", elongation factor 1-γ" and "retrotransposon hot spot protein".

The searched terms "TcMUC" and "mucin like" were considered to be one category, "mucin". Additionally, the terms "90 kDa surface protein" and "serine-alanine and proline-rich protein" were categorized under "90 kDa surface protein".

2.3. Statistical Analysis

Student's t-test was used for comparisons between samples using GraphPad Prism software (GraphPad Software Inc., San Diego, CA, USA).

2.4. Genomes Analyzed

Genome sequences from *T. cruzi* strains were downloaded from TritrypDB [23]: Dm28c 2018 (version 2018-05-30 release 46); Y C6 (version 2019-08-26 release 46); TCC (version 2018-05-30 release 46); CL Brener_S (version 2015-12-07 release 46); CL Brener_P (version 2015-12-07 release 46); marinkellei (version 1.0); Brazil A4 (version 2019-08-26 version 46); and Sylvio X10/1 (version 2017-03-18).

2.5. Monte Carlo Test of a 241 nt Repeat

To test whether the frequency of repeats associated with trans-sialidases was higher than expected by chance, we conducted a Monte-Carlo test [24], in which the total number of repeats found in the CL Brener S genome sequence (334) and in Dm28c strain (1117) were randomly re-inserted in the genome sequence. For each replicate, first a "fake" long and single chromosome was generated by concatenating the 41 chromosomes end-to-end from CL Brener_S and all 636 contigs of DM28c. Then, the repeats were randomly re-inserted between the first and last nucleotide of this "fake" chromosome. Finally, we determined how many of these re-inserted repeats had a trans-sialidase in their surroundings (either with the repeat falling into it or at 5′ and/or 3′). The rationale is that if the original number of repeats were to be inserted from scratch along chromosomes/contigs (for instance, by a natural biological process), then the distribution among the pseudoreplicates (334 draws per peudoreplicate for CL Brener S and 1117 for Dm28c) can be considered a reasonable indicator of the variability of the probability of random insertions near trans-sialidases. If the original value of repeats associated with such genes falls within the 95% highest density interval (HDI) of that distribution of the pseudoreplicates, then the hypothesis of random association with trans-sialidases in the original chromosomes/contigs cannot be discarded; on the contrary, if the original value of trans-sialidase associations is outside the 95% HDI (either higher or lower), then the hypothesis of random association with trans-sialidases can be ruled out at an $\alpha = 0.05$ level.

2.6. RNA-seq Analysis

Coverage analyses were performed to investigate the relationships between the identified 241 nt repeat regions in the *T. cruzi* strains as well as their closest upstream and downstream genes and respective range regions. The expression profile for the repeat region, upstream and downstream genes and the region between them were quantified based on the RNA-Seq data from *T. cruzi* strains (CL_Brener Esmeraldo-Like, CL_Brener Non-Esmeraldo-Like, and YC6) from NCBI, for the two different stages of the parasite (epimastigote and trypomastigote) with two biological replicates for *T. cruzi* CL_Brener and three biological replicates for the other ones.

The accession numbers of SRAs from the NCBI of *T. cruzi* strains are as follows: CL_Brener epimastigote (SRX1643253, SRX1643239) and trypomastigote (SRX1643235, SRX1643234); Y epimastigote (SRX574896, SRX574895, SRX574894) and trypomastigote (SRX574893, SRX574892, SRX574891, SRX574890).

To quantify the expression profile of these regions, we aligned the samples to the reference genome using Hisat2 version 2.1.0 [25] with "−k1" parameter, which allows only one alignment per read. Then, the raw counts were quantified based on the alignment of reads to each genome strain with the tool multiBamCov, and the coverage of the region was estimated with coverageBed from bedtools version 2.26.0 [26]. The EdgeR version 3.28.1 [27] was used to estimate the average expression profile between the replicates calculated and the log2 fold change between the stages trypomastigote and epimastigote for every strain and for every region.

2.7. TPM Statistical Test

To test for a possible association of the repeat with the closest gene upstream in transcripts, in opposition to the closest gene downstream, we used previously available RNA-seq data from [databank]. Reads from three different lineages were used (CL Brener-P, CL Brener-S, and Y) for both epi- and trypomastigote forms, with transcripts per million (TPM) values averaged between forms within the same genome for each set of genic regions containing 5′ gene/repeat/3′ gene. We used the smallest difference between the TPM of the repeat to either the upstream or downstream gene's TPM as a sign of the most likely posttranscriptional genic association of the repeat ("−1" if the smallest TPM difference was upstream, "+1" if downstream). The one-tailed sign test was then employed in R [28] to test the hypothesis of a stronger association to the upstream gene. This procedure was carried in two subsets of genes: (1) the set of four gene families to which repeats were found to be more associated with (trans-sialidases, etc.) appearing at either of the neighboring positions and (2) the remaining genic regions ("background regions") in which none of the flanking genes was a member of those four gene families. Cases in which there was a second repeat flanking at the 5′ end of the upstream gene or at the 3′ end of the downstream gene were discarded.

3. Results

3.1. Identification and Distribution of a Novel DNA Repeat on the T. cruzi *Genome*

The *T. cruzi* genome is composed of diverse repetitive elements that vary in size and copy number among different strains. The smallest high copy number element found in the *T. cruzi* genome is satellite DNA, which is 195 bp long with approximately 20,000 copies [29]. Therefore, we established a 150-nucleotide length to screen for new repetitive elements. We decided to start investigating both haplotypes (Esmeraldo like-S and non-Esmeraldo like-P) of the clone CL Brener genome sequence. This strain was used for the *T. cruzi* genome sequence project, and therefore, a considerable amount of information is available allowing future co-relation analysis. Moreover, the CL Brener strain has a hybrid origin containing haplotypes from different DTUs which could increase the robustness of our observations. Therefore, once the parental strains are from DTUs II and III, any DNA element found in both haplotypes would be more likely to be found in the genome sequences of other DTUs. In our approach, we used a sliding 150-nucleotide window along all chromosome sequences in each haplotype, moving it one nucleotide at a time, resulting in millions of 150-nucleotide fragments covering the entire genome of *T. cruzi* CL Brener (Figure 1A). A list with 52 million fragments was obtained and summarized, showing the frequency of 100% identical fragments that appeared during the window screening.

Furthermore, four sequential filtering steps were used to clean these data and isolate potential repetitive sequences. The first two steps excluded any fragment with at least one undefined (N) nucleotide, and then only the ones that appeared at least ten times on the list were selected. Next, the two additional filtering steps excluded the fragments with a significant match against the repetitive elements using RepeatMasker software and excluded any fragment from multigenic family genes. Therefore, the final list of 67 unique 150-nucleotide fragments was obtained (Figure 1A). Once all of the fragments were obtained by a sliding window, where sequential fragments were only one nucleotide apart, the final 67 fragments were aligned to determine whether they were independent repetitive sequences and/or part of a longer sequence. As shown in Figure 1B, all 67 sequences present 100% identity and aligned together, resulting in a consensus sequence composed of 241 nucleotides (Supplementary File S1). Therefore, using the 150-nucleotide sliding window and filtering and alignment steps, we identified a novel repetitive sequence on the *T. cruzi* genome that has not been described to date.

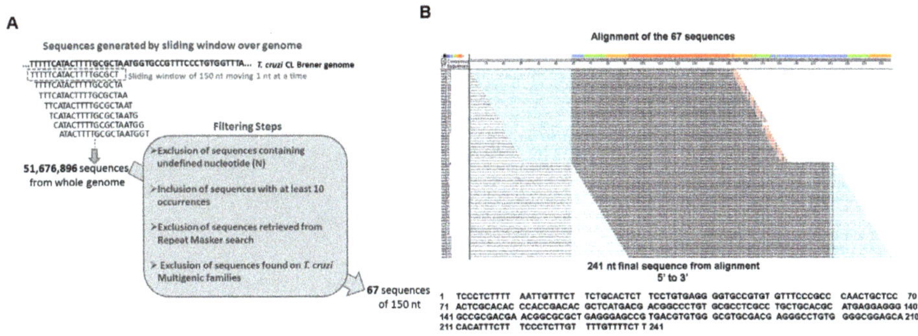

Figure 1. Strategy to identify a novel repetitive element in the *T. cruzi* CL Brener genome. (**A**) Schematic representation of the 150 nucleotide (nt) sliding window used to generate sequences covering all of the CL Brener genome and filtering steps used to exclude known repetitive elements. (**B**) Alignment of the 67 sequences of 150 bp obtained after the filtering steps that resulted in the consensus sequence of 241 nucleotides of the repetitive element.

3.2. The 241 nt Repeat Is Enriched at Intergenic Regions of T. cruzi Genome Sequences

Since the 241 nt repeat was present in both CL Brener haplotypes, we next wanted to determine whether this repeat (i) is present in others *T. cruzi* strains and (ii) is present in other trypanosomatids. Blast-n search revealed that the 241 nt sequence is present in all searched *T. cruzi* strains, including the bat strain *T. cruzi marinkellei*, but it is absent in *Leishmania* and *Trypanosoma brucei* (Supplementary File S2).

To characterize this new repetitive element found in *T. cruzi*, we first checked the genome sequences of *T. cruzi* strains from different DTUs available in TritrypDB that were sequenced using long reads (PacBio technology) that provide more reliable assembly of these genomes (Table 1). Then, we chose one strain of each DTU to be analyzed, named as follows: Dm28c (TcI), Y (TcII), and TCC (TcVI). There are no strains from TcIII and TcIV sequenced by PacBio (Pacific Bioscience of California, Inc. Menlo Park, CA, USA) technology, and the TcV strain genome (Bug 2148) lacks annotation (Table 1). Even though the *T. cruzi marinkellei* genome was not sequenced by PacBio, we decided to perform some analyses on this strain in order to gain evolutionary insights into this repeat.

Table 1. *T. cruzi* genomes available in TritrypDB. With the exception of *T. cruzi marinkellei* (sequenced by Illumina technology-Illumina Inc., San Diego, CA, USA) and the CL Brener strain (reference genome sequenced by whole genome shotgun assembly), the strains were sequenced by PacBio technology. NA: Not Annotated/* chosen strains for formal analysis.

DTU	B7	TcI			TcII	TcV	TcVI		
T. cruzi strain	Marinkellei	Dm28c *	Brazil A4	Sylvio X10/1	Y C6 *	Bug2148	TCC *	CL Brener S *	CL Brener P *
Chr count	0	0	43	47	40	0	0	41	41
Contig count	16783	636	359	0	226	929	1236	0	0
Genome size (Mbp)	38.65	53.27	45.56	41.38	47.22	55.16	87.06	32.53	32.53
Total gene count	10282	19112	18779	20684	17713	NA	29302	10596	11106

To identify the locations of this repetitive sequence on the genome, Blast-n searches on the *T. cruzi* genome sequences were performed. From the retrieved regions, only those with a minimum length of 140 bp and 95% identity to the 241 nt repeat were selected. The great majority of this repetitive element was found distributed on intergenic regions of the analyzed strains (Supplementary File S3): 100% in Dm28c (1117 of 1117), 99.5% in Y (742 of 746), 100% in TCC (1171 of 1171), 96.7% on CL Brener S (322 of 334), and 97.5% on CL Brener P (398 of 408). The repeats found in the genic regions (Supplementary File S4) were in genes of a hypothetical protein (four genes on Y, eight genes on CL Brener_S, and eight genes on CL Brener_P), ATPase (1 on CL Brener_S and 2 on CL Brener_P) and trans-sialidase (two on CL Brener_S). In fact, the identity between these two trans-sialidase sequences and the 241 nucleotide consensus sequence is the cause of the alignment break observed in Figure 1B (nucleotides 30–60 and 180–210), where the 150 nt fragments 100% identical to these trans-sialidases were eliminated after the multigenic family filtering step. Then, we further investigated the repeats located on intergenic regions.

The selected sequences from Blast-n were at least 140 nucleotides long, but approximately 90% of repetitive sequences found on the genome ranged from 231 to 244 nucleotides in DM28c, TCC and CL Brener and 87.5% in the Y strain (Figure 2A). Since the consensus sequence of this new repetitive element is 241 nucleotides long, we call these repeats a 241 nt repeat. These repeats located on intergenic regions were found interspersed throughout the genome sequences rather than organized in tandem in a head-to-tail fashion (Supplementary Files S5–S7), and they were present in most chromosomes from assembled genomes of CL Brener and Y strains (Table 2) (the DM28c and TCC genome sequences are not chromosome assembled). Even though larger chromosomes showed more copies of the repeat, its distribution was not proportional to chromosome size, as seen on Y strain chromosomes 2 and 3 (Table 2), where 3 copies are found on chromosome 2 and 73 copies are found on chromosome 3. Different copy numbers were also observed between CL Brener haplotypes, as seen on chromosome 40 (Table 2), where 34 repeats were found on the S haplotype and 8 were found on the P. In addition, the repeat distribution showed different profiles among the chromosomes, as it was observed at some chromosome edges in some chromosomes and concentrated in the middle in others (Supplementary Files S5–S7). Therefore, there was no preferential location along all chromosomes from the CL Brener and Y strains.

3.3. Regions Upstream and Downstream of the 241 nt Repeats Are Enriched in Surface Protein Genes

The interspersed distribution pattern of the 241 nt repeat and its intergenic location led us to investigate possible signs of correlation of this DNA repetitive element to nearby genes. First, these genes were classified according to their transcription orientation, where genes whose transcription orientation moved in the direction of the repeat were termed "upstream genes" and those whose transcription direction moved away from the repeat were termed "downstream genes" (Figure 2B), regardless of the strain being considered. From this analysis, different patterns could be observed, as shown in Figure 2B. Most 241 nt repeats were located between genes on the same polycistronic transcription unit (PTU) on the sense strand (indicated by ++) and anti-sense strand (indicated by −−), as shown on Table 3. In both cases (++ and −−), there were two genes, one gene upstream and the other downstream, flanking the 241 nt sequence. Fewer 241 nt repeats were located between convergent PTUs (indicated by +− in Table 3), and in this case, both adjacent genes were considered upstream. The 241 nt repeats were also located between divergent PTUs (indicated by −+ in Table 3), when both adjacent genes were denominated downstream. Additionally, some 241 nt copies had only one gene adjacent to it (indicated by +* and *− in Table 3), and these genes were always upstream genes. No 241 nt repeat was found with a single downstream gene close to it in the CL Brener strain (indicated by −* and *+ in Table 3); however, a few copies of this pattern were found on Dm28c, Y and TCC strains (Table 3).

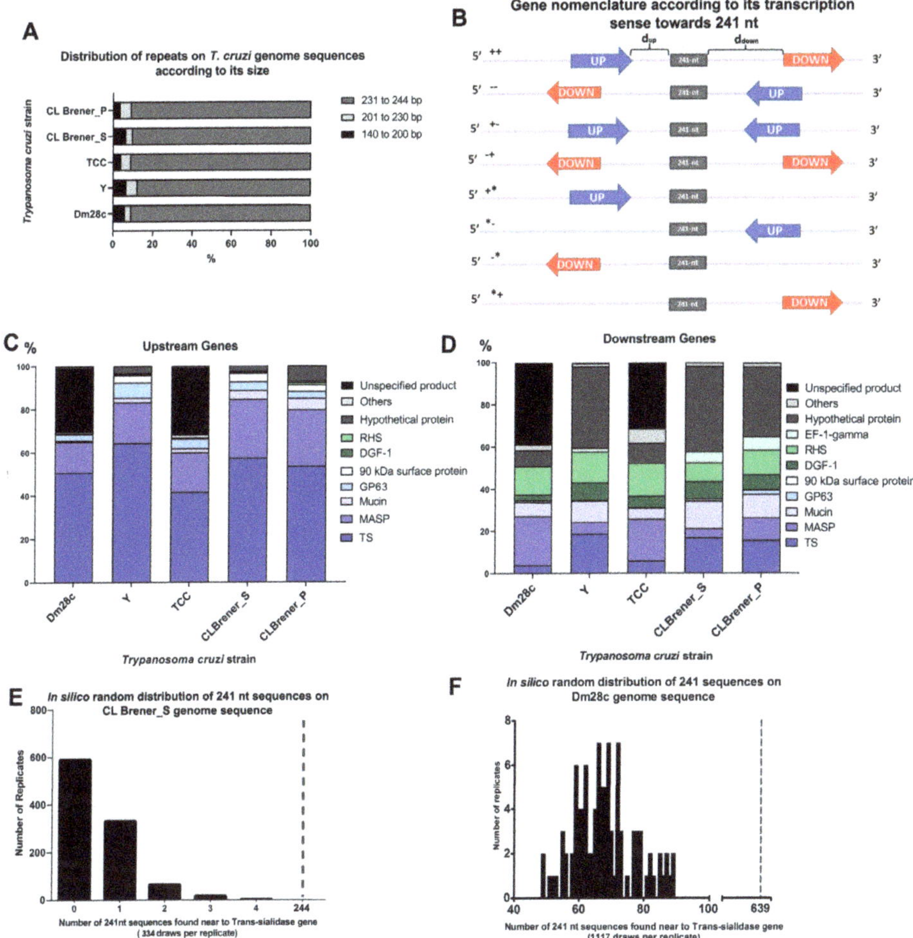

Figure 2. Localization of the 241 nt repetitive element in the *T. cruzi* CL Brener genome. (**A**) Consensus sequence of the 241 nt repeat that was submitted to Blast-n against the CL Brener genome, and the retrieved sequences with at least 95% identity and sizes from 140–244 nucleotides were selected. The graphic shows the frequency distribution of the retrieved sequences by their size in the genome sequence of *T. cruzi* strains Dm28c (TcI), Y (TcII), TCC (TcVI), and CL Brener S and P (TcVI). (**B**) Schematic representation of genes surrounding the 241 nt found inside the intergenic region. Genes were classified as upstream (UP) or downstream (DOWN) of the 241 nt repeat according to their transcription orientation. Distinct patterns of upstream and downstream genes in relation to the repeat are observed and indicated by the symbols ++, −−, +−, −+. In some cases, only one gene is associated with the repeat and is indicated by *+, +*, *−, and −*. Letter "d" indicates the distance between the repeat and upstream (d_{up})/downstream (d_{down}) genes. (**C**) Percentage of genes upstream and (**D**) downstream of the 241 nt repeat on the *T. cruzi* genome sequences of Dm28c, Y, TCC, and CL Brener strains. (**E**) A total of 334 sequences of 241 nt were randomly distributed in the CL Brener_S genome sequence. The graph shows the number of repeats found near a trans-sialidase gene. The dashed line represents the number of repeats identified close to a trans-sialidase gene in the CL Brener genome (S haplotype). (**F**) A total of 1117 sequences of 241 nt were randomly distributed in the Dm28c genome sequence. The graph shows the number of repeats found near a trans-sialidase gene. The dashed line represents the number of repeats identified close to a trans-sialidase gene in the Dm28c genome Abbreviations: TS-trans-sialidase, MASP-mucin-associated surface protein, GP-glycoprotein, DGF-1-dispersed gene family-1, RHS-retrotransposon hot spot and EF-1 γ-elongation factor-1 γ.

Table 2. Number of 241 nt repeats found on each chromosome of *T. cruzi* CL Brener.

| | Y C6 | | CL Brener | | |
| | | | S and P | S | P |
Chr	Chr Size (kb)	N° of Repeats	Chr Size (kb)	N° of Repeats	N° of Repeats
1	2,950,016	146	77,958	0	0
2	1,943,341	3	151.74	2	1
3	1608.8	73	196,644	1	0
4	1,578,048	2	200,401	3	1
5	1,465,819	45	227,319	0	1
6	1,365,397	24	389,024	3	2
7	1238.82	2	391,095	9	3
8	1,238,493	2	393,493	1	2
9	1,233,391	8	509,634	0	1
10	1,196,034	6	518,846	0	0
11	1,179,968	2	526.14	4	1
12	1,154,569	3	533,093	2	4
13	1,073,329	10	558,364	1	0
14	1041.73	6	598,625	5	7
15	973,991	10	612,853	4	17
16	931,817	0	646,207	11	16
17	919,065	3	648,584	8	4
18	889,019	10	655,081	28	23
19	879,731	15	671,453	7	1
20	835,455	23	656,799	15	5
21	802.19	3	704,149	4	9
22	794,882	13	710,778	3	2
23	771,598	18	655,477	5	7
24	748,092	8	779,922	6	10
25	747,041	8	822,374	9	24
26	713.53	17	801,422	9	6
27	704,292	0	850,241	0	0
28	683,656	4	853,233	12	27
29	683,261	6	870,934	12	18
30	618,893	9	863,882	2	2
31	613,739	9	947,473	4	6
32	587,789	6	968,069	0	1
33	572,88	19	1,041,172	3	13
34	565,606	1	1,065,764	4	1
35	563,146	6	1,186,946	1	4
36	542,602	21	1,180,744	2	1
37	354,446	3	1,355,803	3	2
38	332,206	0	1,444,805	24	67
39	241,231	2	1,854,104	3	3
40	239,696	5	2,036,759	34	8
41			2,371,736	90	108
unplaced contig		201			\
Total		752		334	408

Table 3. Number of repeats found in the intergenic region of each *T. cruzi* strain. The + symbol indicates gene transcription from the sense strand, and the − symbol indicates gene transcription from the anti-sense strand. The left symbol represents the gene at the left side of the repeat, and the right symbol represents the gene at the right side of the repeat.

	TcI	TcII		TcVI	
	Dm28c	YC6	TCC	CLBrener_S	CLBrener_P
++	472	349	472	150	190
− −	512	351	516	148	163
+−	94	38	103	22	45
−+	2	1	11	2	3
+*	11	4	24	4	1
*−	18	6	30	8	6
−*	5	1	5	0	0
*+	3	2	10	0	0
Total	1117	752	1171	334	408

The next step was to identify which genes surround the 241 nt repeat. Among upstream genes, the large majority were from multigenic families (trans-sialidase, MASP, mucin and GP63), that is, representing 96.6% of the genes in CL Brener_S, 91.2% in CL Brener_P, 95.6% in the Y strain, and 68.8% and 66.4% in the Dm28c and TCC strains, respectively. The remaining genes were mostly hypothetical proteins (representing 1.1% in Dm28c, 3.3% in Y, 0.8% in TCC, 2.5% and 6.9% in CL Brener S and P haplotypes, respectively) and some other genes (listed in Supplementary File S3) that collectively represent 1% in Dm28c, 0.9% in Y, 1.9% in TCC, and 0.8% and 1.7% in CL Brener S and P haplotypes, respectively (Figure 2C). When analyzing the percentage of these genes on the genome, these genes of multigenic families collectively represent 18.07% of the genes in Dm28c, 32.19% in Y, 16.15% in TCC, and 15.19% and 14.52% in CL Brener S and P, respectively. Additionally, the hypothetical protein genes from the genome correspond to 37.33% of the genes in Dm28c, 58.24% in Y, 37.14% in TCC, and 38.44% and 38.43% in CL Brener S and P, respectively, but only a small percentage was found upstream of the 241 nt repeat. Taken together, these data suggest that the 241 nt repeat is preferentially located near multigenic families along the genome and is not randomly distributed.

Once it was determined that the upstream genes are mostly composed of multigenic family genes and trans-sialidase (TS) genes are the most abundant among them, we tested whether the association between the 241 nt repeat and TS had biological relevance or was just a consequence of the random distribution of the 241 nt repeat in the genome. To address this question, we conducted a Monte Carlo test on genome sequences of CL Brener S (TcVI) and DM28c (TcI) strains that are highly divergent [30]. This test consists of random re-insertions of 241 nt repeats in the genome sequence according to the number of repeats originally identified in genome sequences (334 for CL Brener S and 1117 for Dm28c). In each replicate, random re-insertion was performed, and the number of trans-sialidase genes found flanking this repeat was counted. Monte Carlos analysis of CL Brener S showed that up to four TS genes were located close to the repeat after its random reinsertion into the CL Brener S genome (Figure 2E). As indicated by the dashed line in Figure 2E, the total number of TS genes found close to the 241 nt repeat in the *T. cruzi* CL Brener_S was 244, which is significantly higher than expected for the random distribution of the repeat ($p < 0.01$). For the Sylvio X10/1 strain, the Monte Carlo analysis showed that up to 89 repeats were found close to TS genes after the random re-insertion of the repeat, as shown in Figure 2F. Again, the number of TS genes flanking the 241 nt repeat (dashed line in Figure 2F) in the genome sequence of Sylvio X10/1 was significantly higher ($p < 0.01$) than that expected by chance distribution of the repeat in the genome. Therefore, these findings indicate that the proximity of repeats and the TS genes was not randomly distributed in the genomes analyzed and may have biological function.

We then analyzed the pattern of gene distribution in downstream genes, which was found to differ from that in upstream genes. The same multigenic family enriched upstream of the repeat (trans-sialidase, MASP, mucin and GP63) represented 35.5% of the downstream genes in the CL Brener_S, 39.8% in the CL Brener_P, 34.9% in the Y strain and 34.7% and 31.3% in the Dm28c and TCC strains, respectively. Additionally, two other multigenic family genes were among the downstream genes: DGF-1 (2.7% in Dm28c, 8.2% in Y, 5.4% in TCC, 8% and 7% in CL Brener S and P, respectively) and RHS (13.5% in Dm28c, 14.7% in Y, 15.6% in TCC, 9% and 11.7% in CL Brener S and P, respectively). The remaining genes were mostly genes for hypothetical proteins (7.6% in DM28c, 38.8% in Y, 9.5% in TCC, 40.5% in CL Brener_S and 33.1% on CL Brener_P). The Dm28c and TCC strains also presented "unspecific product" genes that represented 38.8% of the downstream genes in the first strain and 31.5% in the latter (Figure 2D). The higher amount of hypothetical protein genes and the lower amount of multigenic family genes among the downstream genes corroborate the closer relation of the repeat to upstream genes than to downstream genes.

In addition to the strains analyzed above, we also verified the repertoire of genes located upstream and downstream to the 241 nt repeats in *T. cruzi* Brazil A4 and Sylvio X10/1 strains (PacBio sequenced) as well as in the ancestral *T. cruzi marinkellei* strain. The Brazil A4 strain and *T. cruzi marinkellei* presented similar repertoires in their upstream genes (Supplementary Files S8 and S9), where trans-sialidase genes represented the great majority (73.7% in Brazil A4 and 75% in *T. cruzi marinkellei*), followed by MASP genes (13% in Brazil A4 and 6.3% in *T. cruzi marinkellei*). Other multigenic family genes were also observed among the upstream genes (Supplementary Files S8 and S9) and, collectively, the multigenic family genes (TS, MASP, mucin and GP63) represented 93.7% and 85% of the upstream genes from the Brazil A4 strain and *T. cruzi marinkellei*, respectively. When analyzing genes found downstream to the repeat, the repertoires found in Brazil A4 and *T. cruzi marinkellei* were similar to those found in previously analyzed strains but differed in the amount of multigenic family genes (Supplementary Files S8 and S9). In the ancestral *T. cruzi marinkellei*, MASP genes comprised 36.8%, followed by trans-sialidase and hypothetical protein genes (both representing 13.2% of the downstream genes). In the Brazil A4 strain, the most abundant genes among downstream genes were hypothetical protein genes (35.8%) and trans-sialidase (24.1%) (Supplementary Files S8 and S9). Surprisingly, *T. cruzi* Sylvio X10/1 strain analysis revealed different genes flanking the 241 nt repeat (Supplementary Files S8 and S9). Bacterial neuraminidase repeat (BNR)-like domain genes were the most abundant genes among the upstream genes (49.82%) and the second most abundant among the downstream genes (27.86%). The concanavalin A-like lectin/glucanases superfamily represented 20% of the upstream genes and 14.5% of the downstream genes, while leishmanolysin represented 7.64% of the upstream genes and 8.4% of the downstream genes. RHS (5.09 and 3.44%, of the upstream and downstream genes respectively), EF1-γ (4.73% and 2.29% of the upstream and downstream genes respectively), trans-sialidase (1.09% and 0.38% of the upstream and downstream genes respectively) and DGF-1 (4.73% of the upstream genes) also flanked the repeat. Genes identified as "unspecific products" comprised 35% of the downstream genes (Supplementary Files S8 and S9). The fact that the 241 nt repeat is exclusive to *T. cruzi* and that the bat subspecies *T. cruzi marinkellei* presented similar composition illustrates how ancient this repeat found among *T. cruzi* is and reinforces its potential biological role.

3.4. The 241 nt Repeat Is Found Closer to Upstream Genes and May be Part of the 3' UTR of Trans-Sialidase Gene mRNA

The intergenic location of the 241 nt repeat and the different gene profiles of upstream and downstream genes of the repeat motivated us to determine the distance between the 241 nt and the upstream and downstream genes (indicated by "d_{up}" and "d_{down}" on Figure 2B). Comparing the distances from the 241 nt repeat to the upstream and downstream genes, it was observed that the 241 nt repeat was found to be significantly closer to upstream genes than to downstream genes in all of the *T. cruzi* strains analyzed (Figure 3A and Supplementary Files S8 and S9) including *T. cruzi marinkellei* (Supplementary Files S8 and S9), with the exception of Sylvio X10/1 strain (Supplementary Files S8 and S9).

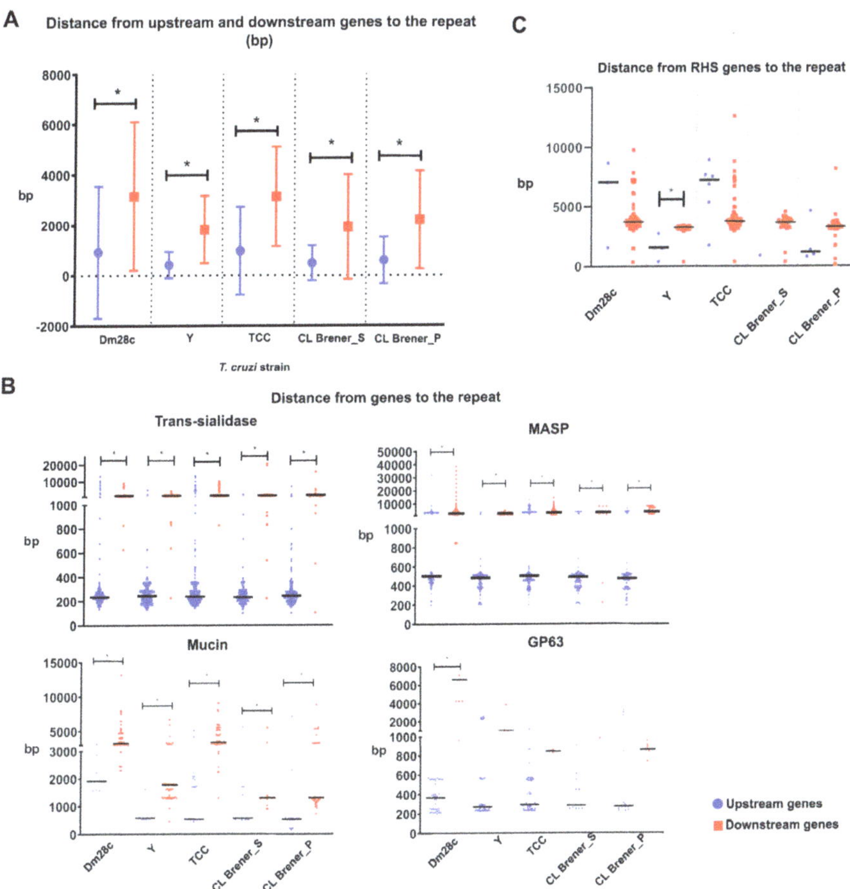

Figure 3. Distance from 241 nt repeats of upstream and downstream genes. The distances of each gene upstream (blue symbols) and downstream (red symbols) of the 241 nt repeat are plotted on the graph. Horizontal bars indicate the mean, and * indicates a p value < 0.001 from the Student's t-test. (**A**) Distance from the 241 nt repeat to upstream genes and downstream genes on Dm28c, Y, TCC and CL Brener Esmeraldo-Like haplotype and non-Esmeraldo-Like haplotype. (**B**) Distance from 241 nt to the four main multigenic families of genes among upstream and downstream genes on Dm28c, Y, TCC and CL Brener genome sequences. (**C**) Distance from the repeat to hypothetical proteins.

When the distance between the repeat and each multigenic family was analyzed, the 241 nt repeat was found to be significantly closer to some upstream multigenic families, including trans-sialidase, MASP and mucin, on all strains analyzed. The GP63 from upstream genes was closer to the repeat than from downstream genes; however, it was significant only in the Dm28c strain (Figure 3B). In contrast, the distance from RHS to the repeat was different among strains, while in Y and CL Brener P strains, the repeat was closer to upstream RHS genes (significant only in the Y strain), and in Dm28c and TCC, the repeat is closer to RHS from downstream genes (Figure 3C).

The proximity of the 241 nt repeat to upstream genes raised the question whether this repeat could be transcribed as part of the 3'UTRs mRNA of upstream genes. Since the UTR length of mRNA from *T. cruzi* varies in size, ranging from 17 to 2800 nucleotides, and is generally limited to 56.65% of the final mRNA [31], we used these two pieces of information to infer the possible presence of the 241 nt repeat in the 3'UTR of trans-sialidase, MASP, mucin and GP63 final mRNA. To this end, we first calculated the distance from the first nucleotide after the stop codon (of upstream gene) to the last nucleotide of the 241 nt sequence, as shown in Figure 4A (dB), and then analyzed the proportion of genes where dB was lower than 2800 bp. Second, we calculated the distance from the first nucleotide of upstream genes to the last nucleotide of the 241 nt repeat (dA in Figure 4A) and calculated the ratio of dB/dA. Then, the proportion of genes where dB/dA was lower than 56.65% was investigated. Therefore, when dB is lower than 2800 bp and the dB/dA ratio is lower than 56.65%, it is possible that the repeat is enclosed into the 3'UTR of the final mRNA.

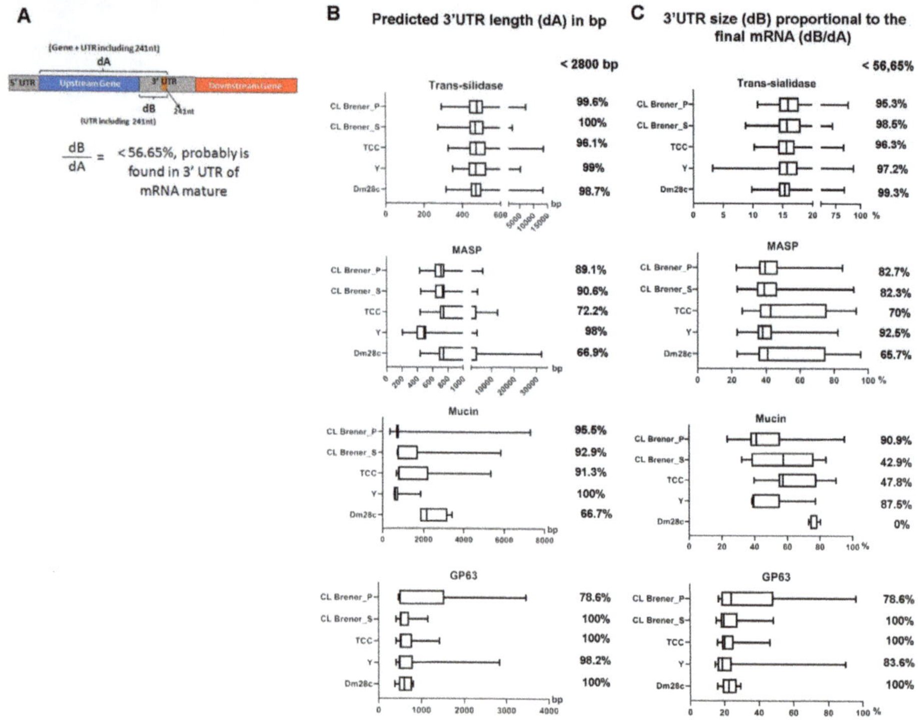

Figure 4. The 241 nt repeat is found in the 3'UTR of upstream genes. (**A**) Schematic representation of upstream and downstream genes to the 241 nt sequence and distances used to predict the 3'UTR length (dA). (**B**) Predicted 3'UTR including the 241 nt sequence (distance dA is represented on "b"). The percentage of genes that predicted 3'UTR represents less than 2800 bp of the mRNA and is listed on the left of the graphs. (**C**) The predicted 3'UTR size (distance dB is represented on "b", which includes the 241 nt sequence) in proportion to the final full mRNA was plotted. The percentage of genes that predicted 3'UTR represents less than 56.65% of the mRNA is listed on the left of the graphs. In (**B**) and (**C**), the four major representative genes among upstream and downstream genes are represented. Abbreviations: MASP-mucin-associated surface protein and GP-glycoprotein.

Over 96% of the trans-sialidase genes from DM28c, Y, TCC, and CL Brener showed a dB lower than 2800 bp and a dB/dA ratio lower than 56.65% (Figure 4B,C). For MASP genes, there was variation among the strains: In Dm28c, 66.9% of the MASP genes showed a dB lower than 2800 bp (Figure 4B), and 65.7% of the MASP genes showed a dB/dA ratio lower than 56.65% (Figure 4C). In Y strains, 98% of the MASP genes showed a dB lower than 2800 bp (Figure 4B), and 92.5% of the MASP genes showed a dB/dA lower than 56.65% (Figure 4C). In TCC, 72.2% of the MASP genes showed a dB lower than 2800 bp (Figure 4B), and 70% of the MASP genes showed a dB/dA lower than 56.65% (Figure 4C). In the CL Brener strain, 90.6% (S) and 89.1% (P) of the MASP genes showed a dA lower than 2800 bp (Figure 4B), and 82.3% (S) and 86.7% (P) of the MASP genes showed a dB/dA ratio lower than 56.65%

(Figure 4C). The analysis of GP63 genes showed that 100% of the GP63 genes from Dm28c, TCC and CL Brener S had a dB lower than 2800 bp (Figure 4B) and a dB/dA lower than 55.65% (Figure 4C). In the Y strain, 98.2% of GP63 genes showed a dB lower than 2800 bp (Figure 4B), and 83.6% of the GP63 genes showed a dB/dA ratio lower than 56.65% (Figure 4C). In CL Brener P, 78.6% of the GP63 genes showed a dB lower than 2800 bp (Figure 4B) and a dB/dA ratio lower than 56.65% (Figure 4C).

As can be observed from the results described above, trans-sialidase, MASP and GP63 genes (from the four strains) had similar proportions of genes with a dB lower than 2800 bp and a dB/dA ratio lower than 56.65%. However, the analyzed mucin genes presented greater differences between the dB and dB/dA ratio analyses. In Dm28c, 66.7% of mucin genes showed a dB lower than 2800 bp (Figure 4B), and 0% of mucin genes showed a dB/dA lower than 56.65% (Figure 4C). In the Y strain, 92.5% of mucin genes showed a dB lower than 2800 bp (Figure 4B), and 83.6% of mucin genes showed a dB/dA lower than 56.65% (Figure 4C). In TCC, 91.3% of mucin genes showed a dB lower than 2800 bp (Figure 4B), and 47.8% of mucin genes showed a dB/dA ratio lower than 56.65% (Figure 4C). In CL Brener, 92.9% (S) and 95.5% (P) of mucin genes showed a dB lower than 2800 bp (Figure 4B), and 42.9% (S) and 90.9% (P) of mucin genes showed a dB/dA ratio lower than 56.65% (Figure 4C).

These data suggest that the 241 nt repeat can be part of the 3′UTR of the final mRNA of most trans-sialidase, MASP and GP63 genes. The mucin genes analyzed here showed a lower proportion where the 241 nt repeat is located in the 3′UTR. Additionally, mucin is the multigenic family with the fewest genes in the genome associated with the repeat (approximately 5%; Supplementary File S8); thus, any function of this repeat may have a minor role on mucin genes.

3.5. The 241 nt Repeats Are Found Significantly Expressed in Transcriptomes and Highly Correlated to the mRNA 3′UTR Sequence

To answer the question whether the 241 nt repeat is indeed expressed in the 3′ UTR of the final RNA and the possible role of this repeat in gene expression, transcriptome datasets available in GenBank [32] were analyzed. First, we selected all RNA-seqs of epimastigote and trypomastigote forms with at least two replicates each. Only the Dm28c, Y and CL Brener strains were available, and only the Y and CL Brener transcriptome data could be analyzed due to the percentage of aligned reads (Supplementary File S10). Therefore, RNAseq data of the *T. cruzi* CL Brener strain (Franco G.R., unpublished data) and Y strain [33,34] from epimastigote and trypomastigote forms were aligned against a reference genome sequence, coverage analysis was performed, and the expression profiles of the 241 nt repeat and surrounding regions were obtained.

To determine the presence of reads covering the 241 nt repeat sequence, the counts per million reads mapped (CPM) parameter was used. The cut-off value of 2 was established, and CPMs over 2 were considered as a significant expression of the analyzed region. Figure 5A shows the CPM values from epimastigote and trypomastigote forms of Y and CL Brener strains, and the great majority of the repeats are expressed in epimastigote forms (96.9% in Y and 70.6% and 76.5% in CL Brener S and P, respectively) and trypomastigote forms (66.4% in Y and 100% in CL Brener S and P). In fact, the CPM mean is over 60 in both strains (in the epimastigote form of the Y strain and in the trypomastigote form of the CL Brener strain). Thus, the 241 nt repeats are not just a repetitive element on the genome but also are expressed as constituent of the final RNA.

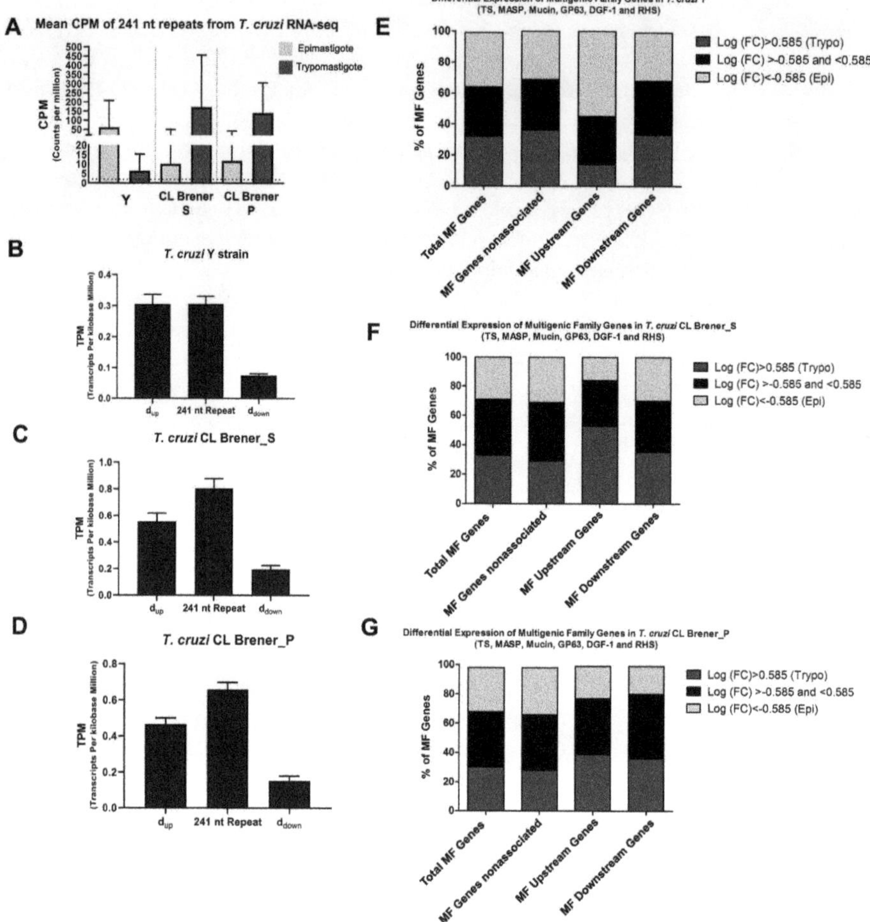

Figure 5. RNA-seq analysis from Y and CL Brener strains of *T. cruzi*. After RNA-seq alignment with reference genomes (Y and CL Brener strains), the coverage and expression profile were obtained from 241 nt repeats and surrounding regions. (**A**) CPM values from 241 nt repeats in epimastigotes (light gray) and trypomastigotes (dark gray) (CL Brener and Y strain). The dashed line indicates the cut-off (2). (**B–D**) TPM values from transcripts aligned in the 241 nt repeat, d_{up} and d_{down} segments of CL Brener_S (**B**), CL Brener_P (**C**) and Y strain (**D**). (**E–G**) Gene expression profile of multigenic families associated and nonassociated with the 241 nt repeat. The Log2(FC) values of the trypomastigote/epimastigote ratios were calculated, and a FC of 1.5 x was chosen as the cut-off. A Log2(FC) > 0.585 indicates genes upregulated in trypomastigotes (dark gray), a log2(FC) < −0.585 indicates genes upregulated in epimastigotes (light gray) and a log2(FC) between −0.585 and 0.585 indicates nondifferential expression (black). Abbreviations: CPM-counts per million; TPM-transcripts per million, FC-fold change, TS-trans-sialidase, MASP-mucin-associated surface protein, DGF-1-dispersed gene family, RHS-retrotransposon hot spot and MF-multigenic family.

Once the presence of the 241 nt repeat is confirmed in mRNAs, we then analyzed if the 241 nt repeat was indeed in the 3′ UTR of multigenic family genes (trans-sialidase, MASP, mucin and GP63), as predicted by the genomic analysis. For that, the TPM of three regions were considered: the 241 nt repeat, the region between the repeat and upstream gene (indicated by d_{up} in Figure 2B) and the region between the repeat and downstream gene (indicated by d_{down} in Figure 2B). Additionally, only the

241 nt repeats flanked by one upstream gene and one downstream gene were considered (patterns ++ and −− of Figure 2B and Table 3).

Some background information is provided below:

(i)   The TPM parameter interprets the transcriptional abundances of determined regions, allowing comparison of the proportion of reads among mapped regions because the TPM normalizes the depth and length of the sequencing data.

(ii)   The d_{up} encloses the 3′ UTR, while the d_{down} corresponds to the entire region between the nucleotide just after the 241 nt repeat to the last nucleotide before downstream gene. Therefore, the d_{down} encloses the segment transcribed from the genome but lost after trans-splicing as well as the 5′ UTR of the downstream gene. Thus, d_{up} will have higher amounts of transcripts (higher TPM), and d_{down} will have lower amounts of transcripts (lower TPM).

Therefore, the TPM analysis rationale was that if the 241 nt repeat is part of the mRNA 3′ UTR, the TPM from the repeat and d_{up} would be more similar, but if the 241 nt repeat is not part of the 3′ UTR, the repeat TPM would be similar to the d_{down} TPM. Figure 5B–D show the mean TPM values from the Y strain (Figure 5B), CL Brener_S (Figure 5C) and CL Brener_P (Figure 5D). The TPMs of the 241 nt repeat regions are higher than the TPMs of d_{down}, and repeat TPM values are closer to the d_{up} TPMs. To assess whether there is an association of the TPM value from repeat and d_{up}, a statistical test was applied.

In the CL Brener (S and P) and Y strains, the 241 nt repeats were significantly associated with d_{up} according to the one-tailed sign test (p-value < 0.01 for both strains). There was a strong association between the repeat and the d_{up} corresponding to the genes of multigenic families (trans-sialidase, MASP, mucin and GP63). The remaining genes showed no significant association with the repeats in CL Brener-S (p = 0.82), borderline significance for CL Brener-P (p = 0.046), and a significant association for the Y genome (p = 0.002), although in all cases, the association was not as strong as that in the set of the four gene families (trans-sialidase, MASP, mucin and GP63).

Additionally, the regions corresponding to d_{up} and the 241 nt repeat were analyzed in terms of their coverage, and most of these regions on CL Brener and Y strains were completely covered (Supplementary File S11). Taken together, these data strongly indicate that the 241 nt repeat is indeed expressed in the 3′ UTR of the genes of multigenic families such as trans-sialidase, MASP, mucin and GP63.

3.6. Distinct Expression Profile Between the Epimastigote and Trypomastigote of Genes Is Associated with the 241 nt Repeat

The RNA-seq data also allowed us to quantify the expression profile of epimastigotes and trypomastigotes from CL Brener and Y strains. In that manner, six multigenic family (MF) genes were selected, and four were found to be enriched among upstream genes (trans-sialidase, MASP, mucin and GP63) plus DGF-1 and RHS (enriched among downstream genes). For this analysis, only upstream genes and downstream genes from patterns ++ and −− (Figure 2B) were considered, and all genes annotated as "pseudogene" were excluded.

The log2(fold change, FC) ratio of trypomastigote/epimastigote from the six multigenic family genes were calculated, and the results are summarized in Figure 5E–G. A FC (fold change) of 1.5 was established, and thus, genes were considered upregulated in trypomastigotes when the log2(FC) was higher than 0.585 and upregulated in epimastigotes when the log2(FC) was lower than −0.585. Figure 5E–G shows the percentage of genes differentially expressed in epimastigotes or trypomastigotes as well as the percentage of genes not differentially expressed. The MF genes were organized into four different groups so that the FC could be compared among them: 1. total MF genes from the genome; 2. MF genes that are not flanking the 241 nt repeat (nonassociated); 3. MF genes located upstream to the 241 nt repeat; and 4. MF genes located downstream to the 241 nt repeat. Analyzing the FCs of the genes of the six multigenic families of *T. cruzi* Y strain, a similar distribution among the three

FC ranges was observed (Figure 5E): 32% of the total MF genes were upregulated in trypomastigotes, 35% were upregulated in epimastigotes, and 35% had no differential expression.

MF genes not associated with the 241 nt repeat and MF genes among downstream genes had similar results to the total MF genes; however, the group of MF genes among the upstream genes showed a decrease in the percentage of genes upregulated in trypomastigotes (~14%).

In *T. cruzi* CL Brener, the total MF genes showed that ~38% in S and ~39% in P of the genes are not upregulated in epimastigotes and trypomastigotes; meanwhile, ~29% in S and ~30% in P are upregulated in epimastigotes, and ~33% in S and ~31% in P are upregulated in trypomastigotes. Slight changes were observed among MF nonassociated genes; however, MF among upstream genes showed an increase in the percentage of genes upregulated in trypomastigotes (~53% in S and ~39% in P) together with a decrease in the percentage of upregulated genes in epimastigotes (~16% in S and ~22% in P). Additionally, MF among the downstream genes of CL Brener P showed a decrease in epimastigote upregulated genes (~20%), while in CL Brener S the percentage of MF among downstream genes was similar to that of the MF nonassociated genes.

These changes in differential expressed genes (DEG) of MF genes containing the 241 nt repeat in 3′ UTR strongly point to a relevant role of the 241 nt repeat in gene expression regulation among different life cycles of *T. cruzi*.

4. Discussion

The *T. cruzi* genome presents a highly repetitive DNA fraction that comprises at least 50% of its genome [14]. Apart from multigenic families, which encode mostly surface proteins and have some of their functions established, most of the repetitive elements in the *T. cruzi* genome do not yet have a defined function. Through a new approach of genome screening using a sliding window of 150 nucleotides, sequential filtering steps, and alignment of the resulting sequences, a resulting repetitive sequence of 241 nts was identified and mapped in each chromosome of clone CL Brener through Blast-n search in TriTrypDB. Further analysis showed that the 241 nt repeat is found in all strains of *T. cruzi* as well as in the ancestral strain *T. cruzi marinkellei*. This sequence was found to be distributed on almost all chromosomes of CL Brener and Y strains as an interspersed repetitive element and enriched in chromosomes with high concentrations of multigenic families, such as chromosomes 18, 28, 38, and 48 from CL Brener S [35]. However, the 241 nt repeat seems to not be randomly distributed along the genome, as it has a close relationship with its upstream genes (defined according to its transcription orientation). Analyzing the distance between the 241 nt and the upstream and downstream genes, a significantly shorter distance was observed from this element to upstream genes. Furthermore, the repertoire of genes found upstream from the 241 nt repeat was mostly composed of surface protein genes in all analyzed strains (Dm28c, Y, TCC, CL Brener, Brazil A4) and *T. cruzi marinkellei*. Surprisingly, in the Sylvio strain, we did not observe the proximity of the 241 nt repeat to upstream genes, and genes flanking the repeat were not the same as those found in all other strains. Since the other strains from TcI, TcII, and TcVI as well as the ancestral *T. cruzi marinkellei* presented the same repertoire of genes flanking the 241 nt repeat, the differences in the repertoire found in the Sylvio strain could be from genome assembly that is fragmented in repetitive regions [30].

In the *T. cruzi* genome, repeated elements (micro- and minisatellite repetitive DNA, retroelements) and multigenic families encoding surface proteins (TS, GP63, MASP, and mucins), DGF-1 and RHS are located in large nonsyntenic regions [16,36,37], also named disruptive compartments [37]. Since the 241 nt repeats are primarily associated with the surface protein genes RHS and DGF-1, they have been mapped on the nonsyntenic regions of the genome. Interestingly, even the hypothetical protein genes carrying the 241 nt repeat were mapped to this region. Taken together, these results suggest that the duplication of 241 nt repeats occurred together with the expansion of multigenic families in *T. cruzi*.

Transcriptome data analysis showed that the 241 nt repeat is indeed expressed in epimastigote and trypomastigote forms of *T. cruzi* (strains Y and CL Brener). Moreover, the 241 nt repeat is transcribed as part of 3′ UTR of trans-sialidases, MASP, mucin and GP63 and its presence seems to be involved in gene expression regulation. Gene expression analysis of trypomastigotes and epimastigotes (Y and CL Brener) indicated that MF genes associated with the 241 nt repeat are differentially expressed when compared to the MF genes nonassociated with the repeat. In CL Brener strain, a higher percentage of MF genes among upstream genes are upregulated in trypomastigotes, while in the Y strain, the MF genes among upstream genes are downregulated in trypomastigotes. However, the molecular bases that contribute to different expression patterns of genes harboring the 241 nt repeat in each of the two strains (Y and CL Brener) remain unknown.

Three findings reinforce the possibility of a biological function for this repeat: (i) the presence of the 241 nt repeat in the genomes of all analyzed DTUs and in the ancestral *T. cruzi marinkellei*; (ii) the conserved repertoire of genes flanking the 241 nt repeat in different strains and in the ancestral subspecies; and (iii) the presence of the 241 nt in the 3′ UTR region of MF genes whose expression changes in different forms of *T. cruzi* life cycle. Therefore, we propose that the 241 nt repeat could serve as a cis-regulatory element on mRNA, playing a role in the posttranscriptional regulation of surface proteins of *T. cruzi*. UTR segments are involved in gene expression regulation, regulating mRNA transcription to mRNA decay [38] and in the interaction of mRNA with other RNA molecules [39]. Diverse elements in the 3′ UTR region have been described to have cis-regulatory functions in gene expression [40,41] not only in later divergent eukaryotes but also in trypanosomatids [10,20,42–44]. In *T. cruzi*, for example, mRNAs harboring a 43-nt U-rich element in its 3′UTR are upregulated in amastigote forms. This U-rich sequence is subject to TcUBP1 (a RNA binding protein) binding, which leads to mRNA destabilization in epimastigotes and mRNA expression in amastigotes [45]. Moreover, in *Leishmania*, a 450-nt sequence was identified and showed the cis-regulatory function of mRNAs, causing an amastigote stage-specific expression of mRNA harboring it on its 3′ UTR [10]. Further experiments are necessary to investigate the proposed biological function for this repeat that, if confirmed, will contribute to the understanding of the controlled expression of genes in *T. cruzi*, a medically important organism that presents a unique system of gene expression among eukaryotes.

5. Conclusions

Through a new approach of genome screening that involves nucleotide window sliding, filtering steps, and sequence alignment, a novel repetitive sequence of 241 nts was identified. The 241 nt element (named 241 nt repeat) is not found on the *T. brucei* and *Leishmania sp* genomes, and it is interspersed on almost all chromosomes of *T. cruzi* (Y and CL Brener strain). The repertoire of genes found upstream from the 241 nt repeat was mostly composed of surface protein genes encoding trans-sialidases, MASP, mucins and GP63 protease. Since (i) this new repeat was found to be transcribed as part of the 3′ UTR of mRNAs of these multigenic families and (ii) MF harboring the 241 nt repeat presents a gene expression profile different from those not harboring the repeat, the involvement of the 241 nt repeat in the control of gene expression in *T. cruzi* is strongly suggested.

Supplementary Materials: The following are available online at http://www.mdpi.com/2073-4425/11/10/1235/s1; Supplementary File S1. DNA sequence of the 241 nt repeat; Supplementary File S2. Number of retrieved sequences from the 241 nt Blast-n search on available genome sequences of *Leishmania*, *Trypanosoma brucei* and *T. cruzi* from TritrypDB; Supplementary File S3. Graphs representing the frequency and location of 241 nt repeats on each chromosome of the Esmeraldo-like haplotype from the *T. cruzi* CL Brener genome; Supplementary File S4. Graphs representing the frequency and location of 241 nt repeats on each chromosome of the non-Esmeraldo-like haplotype from the *T. cruzi* CL Brener genome; Supplementary File S5. Graphs representing the frequency and location of 241 nt repeats on each chromosome of the *T. cruzi* Y strain (YC6 from TritrypDB). Supplementary File S6. List of 241 nt repeats found in intergenic region of *T. cruzi* genome sequences from the following strains: Dm28c, Y, TCC and CL Brener; Supplementary File S7. List of 241 nt repeats found inside coding regions of *T. cruzi* genome sequences from the following strains: Dm28c, Y, TCC and CL Brener; Supplementary File S8. Tables containing (i) the total number of genes found upstream and downstream to the 241 nt repeat and their representation in percentage, (ii) List of genes and their percentages among the total genes from the genome sequence and (iii) genes from the genome found upstream and downstream to the 241 nt repeat (%); Supplementary File S9. Repertoire of genes flanking the 241 nt repeat in *T. cruzi* Brazil A4, Sylvio X10/1 and *T. cruzi* marinkellei and a graph showing the distance from the 241 nt repeat to upstream and downstream genes of *T. cruzi* Brazil A4, Sylvio X10/1 and *T. cruzi* marinkellei; Supplementary File S10. Table containing information of the RNAseq alignment of *T. cruzi* CL

Brener, Y and Dm28c strains. Supplementary File S11. Table containing the percentage of d_{up} region covered by RNAseq reads of *T. cruzi* CL Brener and Y strains.

Author Contributions: Conceptualization: M.C.E.; methodology: S.G.C., M.d.S.R., J.S.L.P., M.Y.N.J., and M.C.E.; formal analysis: S.G.C., M.M., M.d.S.R., J.S.L.P., M.Y.N.J., and M.C.E.; investigation: S.G.C., M.M., M.Y.N.J., J.P.C.d.C., and M.C.E.; resources: M.C.E.; data curation: S.G.C., M.M., N.d.O.N., and M.Y.N.J.; writing—original draft preparation: S.G.C., M.Y.N.J.; writing—review and editing: J.F.d.S., J.P.C.d.C., and M.C.E.; supervision, J.F.d.S.; project administration: M.C.E.; funding acquisition: M.C.E. All authors have read and agreed to the published version of the manuscript.

Funding: This work was supported by FAPESP under grants CeTICS 2013/07467-1 and 2016/50050-2. M.C.E. is a fellow from CNPq (grant 306199/2018-1).

Conflicts of Interest: The authors declare no conflict of interest. The funders had no role in the design of the study; in the collection, analyses, or interpretation of data; in the writing of the manuscript; or in the decision to publish the results.

References

1.	Stanaway, J.D.; Roth, G. The burden of Chagas disease: Estimates and challenges. *Glob. Heart* **2015**, *10*, 139–144. [CrossRef] [PubMed]

2.	Queiroz, R.M.; Charneau, S.; Bastos, I.M.; Santana, J.M.; Sousa, M.V.; Roepstorff, P.; Ricart, C.A. Cell surface proteome analysis of human-hosted *Trypanosoma cruzi* life stages. *J. Proteome Res.* **2014**, *13*, 3530–3541. [CrossRef] [PubMed]

3.	Atwood, J.; Weatherly, D.B.; Minning, T.A.; Bundy, B.; Cavola, C.; Opperdoes, F.R.; Orlando, R.; Tarleton, R.L. The *Trypanosoma* cruzi proteome. *Science* **2005**, *309*, 473–476. [CrossRef] [PubMed]

4.	Minning, T.A.; Weatherly, D.B.; Atwood, J.; Orlando, R.; Tarleton, R.L. The steady-state transcriptome of the four major life-cycle stages of *Trypanosoma cruzi*. *BMC Genom.* **2009**, *10*, 1–15. [CrossRef] [PubMed]

5.	Martinez-Calvillo, S.; Vizuet-de-Rueda, J.C.; Florencio-Martinez, L.E.; Manning-Cela, R.G.; Figueroa-Angulo, E.E. Gene expression in trypanosomatid parasites. *J. Biomed. Biotechnol.* **2010**, *2010*, 525241. [CrossRef]

6.	Jager, A.V.; De Gaudenzi, J.G.; Cassola, A.; D'Orso, I.; Frasch, A.C. mRNA maturation by two-step trans-splicing/polyadenylation processing in trypanosomes. *Proc. Natl. Acad. Sci. USA* **2007**, *104*, 2035–2042. [CrossRef] [PubMed]

7.	Nardelli, S.C.; Ávila, A.R.; Freund, A.; Motta, M.C.; Manhães, L.; De Jesus, T.C.L.; Schenkman, S.; Fragoso, S.P.; Krieger, M.A.; Goldenberg, S.; et al. Small-subunit rRNA processome proteins are translationally regulated during differentiation of *Trypanosoma cruzi*. *Eukaryot. Cell* **2007**, *6*, 337–345. [CrossRef]

8.	Holetz, F.B.; Correa, A.; Avila, A.R.; Nakamura, C.V.; Krieger, M.A.; Goldenberg, S. Evidence of P-body-like structures in *Trypanosoma cruzi*. *Biochem. Biophys. Res. Commun.* **2007**, *356*, 1062–1067. [CrossRef] [PubMed]

9.	Cassola, A.; De Gaudenzi, J.G.; Frasch, A.C. Recruitment of mRNAs to cytoplasmic ribonucleoprotein granules in trypanosomes. *Mol. Microbiol.* **2007**, *65*, 655–670. [CrossRef]

10.	McNicoll, F.; Müller, M.; Cloutier, S.; Boilard, N.; Rochette, A.; Dubé, M.; Papadopoulou, B. Distinct 3′-untranslated region elements regulate stage-specific mRNA accumulation and translation in Leishmania. *J. Biol. Chem.* **2005**, *280*, 35238–35246. [CrossRef]

11.	Jeacock, L.; Faria, J.; Horn, D. Codon usage bias controls mRNA and protein abundance in trypanosomatids. *Elife* **2018**, *7*, e32496. [CrossRef]

12.	Horn, D. Codon usage suggests that translational selection has a major impact on protein expression in trypanosomatids. *BMC Genom.* **2008**, *9*, 2. [CrossRef]

13.	Michaeli, S. Trans-splicing in trypanosomes: Machinery and its impact on the parasite transcriptome. *Future Microbiol.* **2011**, *6*, 459–474. [CrossRef] [PubMed]

14.	El-Sayed, N.M.; Myler, P.J.; Bartholomeu, D.C.; Nilsson, D.; Aggarwal, G.; Tran, A.N.; Ghedin, E.; Worthey, E.A.; Delcher, A.L.; Blandin, G.; et al. The genome sequence of *Trypanosoma cruzi*, etiologic agent of Chagas disease. *Science* **2005**, *309*, 409–415. [CrossRef] [PubMed]

15.	Pita, S.; Diaz-Viraque, F.; Iraola, G.; Robello, C. The Tritryps comparative repeatome: Insights on repetitive element evolution in Trypanosomatid pathogens. *Genome Biol. Evol.* **2019**, *11*, 546–551. [CrossRef] [PubMed]

16.	Callejas-Hernández, F.; Rastrojo, A.; Poveda, C.; Gironès, N.; Fresno, M. Genomic assemblies of newly sequenced *Trypanosoma cruzi* strains reveal new genomic expansion and greater complexity. *Sci. Rep.* **2018**, *8*, 14631. [CrossRef] [PubMed]

17. Ramírez, J.D.; Guhl, F.; Umezawa, E.S.; Morillo, C.A.; Rosas, F.; Marin-Neto, J.A.; Restrepo, S. Evaluation of adult chronic Chagas' heart disease diagnosis by molecular and serological methods. *J. Clin. Microbiol.* **2009**, *47*, 3945–3951. [CrossRef] [PubMed]

18. Pech-Canul, A.C.; Monteon, V.; Solis-Oviedo, R.L. A brief view of the surface membrane proteins from *Trypanosoma cruzi*. *J. Parasitol. Res.* **2017**, *2017*, 3751403. [CrossRef]

19. Di Noia, J.M.; D'Orso, I.; Sanchez, D.O.; Frasch, A.C. AU-rich elements in the 3′-untranslated region of a new mucin-type gene family of *Trypanosoma cruzi* confers mRNA instability and modulates translation efficiency. *J. Biol. Chem.* **2000**, *275*, 10218–10227. [CrossRef]

20. D'Orso, I.; Frasch, A.C. Functionally different AU-and G-rich cis-elements confer developmentally regulated mRNA stability in *Trypanosoma cruzi* by interaction with specific RNA-binding proteins. *J. Biol. Chem.* **2001**, *276*, 15783–15793. [CrossRef]

21. Clayton, C.; Shapira, M. Post-transcriptional regulation of gene expression in trypanosomes and leishmanias. *Mol. Biochem. Parasitol.* **2007**, *156*, 93–101. [CrossRef] [PubMed]

22. Smit, A.F.A.H.R.; Green, P. Repeat Masker Open-4.0 2013–2015. Available online: www.repeatmasker.org (accessed on 22 June 2016).

23. Aslett, M.; Aurrecoechea, C.; Berriman, M.; Brestelli, J.; Brunk, B.P.; Carrington, M.; Depledge, D.P.; Fischer, S.; Gajria, B.; Gao, X.; et al. TriTrypDB: A functional genomic resource for the Trypanosomatidae. *Nucleic Acids Res.* **2010**, *38*, D457–D462. [CrossRef] [PubMed]

24. Kroese, D.P.; Brereton, T.; Taimre, T.; Botev, Z.I. Why the Monte Carlo method is so important today. *WIREs Comput. Stat.* **2014**, *6*, 386–392. [CrossRef]

25. Kim, D.; Langmead, B.; Salzberg, S.L. HISAT: A fast spliced aligner with low memory requirements. *Nat. Methods* **2015**, *12*, 357–360. [CrossRef] [PubMed]

26. Quinlan, A.R.; Hall, I.M. BEDTools: A flexible suite of utilities for comparing genomic features. *Bioinformatics* **2010**, *26*, 841–842. [CrossRef] [PubMed]

27. Robinson, M.D.; McCarthy, D.J.; Smyth, G.K. edgeR: A Bioconductor package for differential expression analysis of digital gene expression data. *Bioinformatics* **2010**, *26*, 139–140. [CrossRef]

28. R Core Team. A Language and Environment for Statistical Computing. R Found. *Stat. Comput.* **2020**. Available online: https://www.R-project.org/ (accessed on 11 May 2020).

29. Martin, C.; Baptista, C.S.; Ienne, S.; Cerqueira, G.C.; Bartholomeu, D.C.; Zingales, B. Genomic organization and transcription analysis of the 195-bp satellite DNA in *Trypanosoma cruzi*. *Mol. Biochem. Parasitol.* **2008**, *160*, 60–64. [CrossRef]

30. Franzén, O.; Ochaya, S.; Sherwood, E.; Lewis, M.D.; Llewellyn, M.S.; Miles, M.A.; Andersson, B. Shotgun sequencing analysis of *Trypanosoma cruzi* I Sylvio X10/1 and comparison with *T. cruzi* VI CL Brener. *PLoS Negl. Trop. Dis.* **2011**, *5*, e984. [CrossRef] [PubMed]

31. Brandao, A.; Jiang, T. The composition of untranslated regions in *Trypanosoma cruzi* genes. *Parasitol. Int.* **2009**, *58*, 215–219. [CrossRef]

32. Database resources of the National Center for Biotechnology Information. *Nucleic Acids Res.* **2018**, *46*, D8–D13. [CrossRef]

33. Li, Y.; Shah-Simpson, S.; Okrah, K.; Belew, A.T.; Choi, J.; Caradonna, K.L.; Padmanabhan, P.; Ndegwa, D.M.; Temanni, M.R.; Corrada Bravo, H.; et al. Transcriptome Remodeling in *Trypanosoma cruzi* and Human Cells during Intracellular Infection. *PLoS Pathog.* **2016**, *12*, e1005511. [CrossRef]

34. Houston-Ludlam, G.A.; Belew, A.T.; El-Sayed, N.M. Comparative transcriptome profiling of human foreskin fibroblasts infected with the Sylvio and y strains of *Trypanosoma cruzi*. *PLoS ONE* **2016**, *11*, e0159197. [CrossRef]

35. Reis-Cunha, J.L.; Rodrigues-Luiz, G.F.; Valdivia, H.O.; Baptista, R.P.; Mendes, T.A.; De Morais, G.L.; Guedes, R.; Macedo, A.M.; Bern, C.; Gilman, R.H.; et al. Chromosomal copy number variation reveals differential levels of genomic plasticity in distinct *Trypanosoma cruzi* strains. *BMC Genom.* **2015**, *16*. [CrossRef]

36. Ghedin, E.; Bringaud, F.; Peterson, J.; Myler, P.; Berriman, M.; Ivens, A.; Andersson, B.; Bontempi, E.; Eisen, J.; Angiuoli, S.; et al. Gene synteny and evolution of genome architecture in trypanosomatids. *Mol. Biochem. Parasitol.* **2004**, *134*, 183–191. [CrossRef] [PubMed]

37. Berná, L.; Rodríguez, M.; Chiribao, M.L.; Parodi-Talice, A.; Pita, S.; Rijo, G.; Alvarez-Valin, F.; Robello, C. Expanding an expanded genome: Long-read sequencing of *Trypanosoma cruzi*. *Microb. Genom.* **2018**, *4*. [CrossRef] [PubMed]

38. Romagnoli, B.A.A.; Holetz, F.B.; Alves, L.R.; Goldenberg, S. RNA binding proteins and gene expression regulation in *Trypanosoma cruzi*. *Front. Cell Infect. Microbiol.* **2020**, *10*. [CrossRef] [PubMed]

39. Hughes, T.A. Regulation of gene expression by alternative untranslated regions. *Trends Genet.* **2006**, *22*, 119–122. [CrossRef] [PubMed]

40. Chen, C.Y.; Shyu, A.B. AU-rich elements: Characterization and importance in mRNA degradation. *Trends Biochem. Sci.* **1995**, *20*, 465–470. [CrossRef]

41. Bakheet, T.; Frevel, M.; Williams, B.R.; Greer, W.; Khabar, K.S. ARED: Human AU-rich element-containing mRNA database reveals an unexpectedly diverse functional repertoire of encoded proteins. *Nucleic Acids Res.* **2001**, *29*, 246–254. [CrossRef]

42. D'Orso, I.; Frasch, A.C. TcUBP-1, a developmentally regulated U-rich RNA-binding protein involved in selective mRNA destabilization in trypanosomes. *J. Biol. Chem.* **2001**, *276*, 34801–34809. [CrossRef] [PubMed]

43. D'Orso, I.; De Gaudenzi, J.G.; Frasch, A.C. RNA-binding proteins and mRNA turnover in trypanosomes. *Trends Parasitol.* **2003**, *19*, 151–155. [CrossRef]

44. De Gaudenzi, J.G.; Carmona, S.J.; Aguero, F.; Frasch, A.C. Genome-wide analysis of 3'-untranslated regions supports the existence of post-transcriptional regulons controlling gene expression in trypanosomes. *PeerJ* **2013**, *1*, e118. [CrossRef] [PubMed]

45. Li, Z.H.; De Gaudenzi, J.G.; Alvarez, V.E.; Mendiondo, N.; Wang, H.; Kissinger, J.C.; Frasch, A.C.; Docampo, R. A 43-nucleotide U-rich element in 3'-untranslated region of large number of *Trypanosoma cruzi* transcripts is important for mRNA abundance in intracellular amastigotes. *J. Biol. Chem.* **2012**, *287*, 19058–19069. [CrossRef] [PubMed]

Publisher's Note: MDPI stays neutral with regard to jurisdictional claims in published maps and institutional affiliations.

 *genes*

Article

Comparative Analysis of the Minimum Number of Replication Origins in Trypanosomatids and Yeasts

Marcelo S. da Silva *, **Marcela O. Vitarelli, Bruno F. Souza and Maria Carolina Elias** *

Laboratório de Ciclo Celular, Center of Toxins, Immune Response and Cell Signaling (CeTICS), Instituto Butantan, São Paulo 05503-900, Brazil; marcela.vitarelli@butantan.gov.br (M.O.V.); fsouza.bruno@gmail.com (B.F.S.)
* Correspondence: mamasantos2003@yahoo.com.br (M.S.d.S.); carolina.eliassabbaga@butantan.gov.br (M.C.E.)

Received: 27 March 2020; Accepted: 5 May 2020; Published: 8 May 2020

Abstract: Single-celled eukaryote genomes predominantly replicate through multiple origins. Although origin usage during the S-phase has been elucidated in some of these organisms, few studies have comparatively approached this dynamic. Here, we developed a user-friendly website able to calculate the length of the cell cycle phases for any organism. Next, using a formula developed by our group, we showed a comparative analysis among the minimum number of replication origins (MO) required to duplicate an entire chromosome within the S-phase duration in trypanosomatids (*Trypanosoma cruzi*, *Leishmania major*, and *Trypanosoma brucei*) and yeasts (*Saccharomyces cerevisiae* and *Schizosaccharomyces pombe*). Using the data obtained by our analysis, it was possible to predict the MO required in a situation of replication stress. Also, our findings allow establishing a threshold for the number of origins, which serves as a parameter for genome approaches that map origins. Moreover, our data suggest that when compared to yeasts, trypanosomatids use much more origins than the minimum needed. This is the first time a comparative analysis of the minimum number of origins has been successfully applied. These data may provide new insight into the understanding of the replication mechanism and a new methodological framework for studying single-celled eukaryote genomes.

Keywords: trypanosomatids; yeasts; trypanosomatids genome; cell cycle phases; S-phase duration; DNA replication; replication origins

1. Introduction

In cellular organisms, DNA replication is a crucial and highly regulated process that follows specific steps, which vary slightly between prokaryotes and eukaryotes. In general, the earliest step in DNA replication is the establishment of replication origins, i.e., the genomic loci where DNA synthesis begins [1]. The start of replication is preceded by the binding of an initiator at the replication origins, which recruits and activates the replisome in a process called origin firing. Each origin fired produces two replication forks in opposite directions (bidirectional movement), which are responsible for synthesizing DNA at a rate that varies according to the organism and cell type [1–3]. The replication time required for all chromosomes determines the S-phase duration. Although the S-phase length is referred to as a way of regulating the cell cycle progression [4,5], recent studies have suggested that it is extremely robust [6–8].

Studies indicated that bacteria [1,9] and some protozoan parasites, such as *Leishmania* spp. [10], typically have one single origin per chromosome. On the other hand, most other eukaryotes, such as *S. cerevisiae* and *S. pombe*, generally have multiple origins per chromosome [9,11–13]. The exact number of origins per chromosome can vary according to cell type and the cellular environment [14]. However, the minimum number of origins (MO) required to duplicate an entire chromosome within a specific S-phase duration must show minimal variation because it depends on two very constant factors: average replication rate and chromosome size [8].

In trypanosomatids, single-celled eukaryotes that encompass human pathogens are of great medical importance, and the question about how many origins are needed to replicate an entire chromosome during the S-phase is totally open [8,10,15,16]. Even for the widely studied domain Bacteria and the model eukaryote *S. cerevisiae*, this discussion is not yet a closed subject [1,17–19].

Here, we developed a website that is able to determine the duration of each cell cycle phase—G1, S, G2, mitosis (M), and cytokinesis (C)—in any organism. After using this website to obtain the S-phase duration for the organisms analyzed, we applied a formula developed by our group [8] and showed a comparative analysis between the minimum number of origins (MO) in trypanosomatids (*T. cruzi*, *L. major*, and *T. brucei*) and yeasts (*S. cerevisiae* and *S. pombe*). In addition to contributing to a discussion of why some organisms use far more origins than the minimum required, this study provides a clue about the dynamic of replication during the S-phase, raising questions about the possible phenomena involved in this process.

2. Materials and Methods

2.1. Trypanosomatids Culture, Growth Curves, and Morphological Patterns

Epimastigote forms of *T. cruzi* (CL Brener strain) were cultured at 28 °C in liver infusion tryptose (LIT) medium supplemented with 10% (*v/v*) fetal bovine serum and 1% (*v/v*) antibiotic/antimycotic solution. Promastigote forms of *L. major* (strain Friedlin) were cultured at 26 °C in an M199 medium supplemented with 10% (*v/v*) heat-inactivated fetal calf serum, 25 mM HEPES, and 1% (*v/v*) antibiotic/antimycotic solution.

For the growth curves, each parasite culture was initiated with 1×10^6 cells.mL^{-1}. Each growth curve was harvested and counted daily until it reached the stationary phase. For the establishment of the morphological patterns, formaldehyde-fixed and DAPI-stained exponentially growing parasites (*T. cruzi* and *L. major*) were examined under an Olympus BX51 fluorescent microscope (Olympus, Tokyo, Japan) (100× oil objective) to observe the profile of organelles that contain DNA (nucleus and kinetoplast).

2.2. EdU Incorporation Assays and 'Click' Chemistry Reaction

Exponentially growing parasites were incubated with 100 µM 5-ethynyl-2′-deoxyuridine (EdU) (ThermoFisher Scientific, Waltham, MA, USA) for the time required according to each assay at species-specific temperatures (28 °C for *T. cruzi* and 26 °C for *L. major*). The parasites were then harvested by centrifugation at 2500 *g* for 5 min, washed three times in 1× PBS (137 mM NaCl, 2.7 mM KCl, 10 mM Na_2HPO_4, and 2 mM KH_2PO_4, pH 7.4), and the pellet was resuspended in 200 µL of the same buffer solution. Afterward, 100 µL of the cell suspension was loaded onto poly-L-lysine-pretreated microscope slides (Tekdon, Myakka, FL, USA), fixed for 20 min using 4% sterile paraformaldehyde (Merck, Darmstadt, Germany) diluted in 1× PBS, washed three times with 1× PBS, and then washed three times with 3% BSA (Sigma-Aldrich, Saint Louis, MO, USA) diluted in 1× PBS. Then, parasites were permeabilized for 10 min with 0.1% sterile Triton X-100 (Sigma Aldrich, Saint Louis, MO, USA) diluted in 1× PBS, washed three times with 1× PBS, and then washed three times with 3% BSA in 1× PBS. To detect incorporated EdU, we used the Click-iT EdU detection solution for 45 min protected from light. The Click-iT EdU detection mix solution consisted of 25 µL 500 mM ascorbic acid ($C_6H_8O_6$), 5 µL 100 mM copper sulfate ($CuSO_4$), 2.5 µL Alexa fluor azide 488 (ThermoFisher Scientific, Waltham, MA, USA), and 467.5 µL distilled water (for details about EdU procedure, see ref. [20]). Finally, the parasites were washed five times with 1× PBS. Vectashield Mounting Medium (Vector, Burlingame, CA, USA) containing 4′,6-diamidino-2-phenylindole dihydrochloride (DAPI) was used as an antifade mounting solution and to stain nuclear and kinetoplast DNA. Images were acquired using an Olympus Bx51 fluorescent microscope (100× oil objective) attached to an EXFO Xcite series 120Q lamp and a digital

Olympus XM10 camera with camera controller software Olympus Cell F (Olympus, Tokyo, Japan). Images were further analyzed using ImageJ software (National Institutes of Health, USA) to count the numbers of EdU-positive parasites, and the percentage of proliferating parasites was calculated for each sample relative to the total number of DAPI-positive parasites.

2.3. Development of the CeCyD Website and Analysis of the Cell Cycle

The website CeCyD (Cell Cycle Duration estimator) was developed using the Python programming language plus the Django v.1.8 framework. CeCyD is available at the following address http://cecyd.vital.butantan.gov.br/, and its source code is released under the GNU GPL-3 license at https://github.com/bruno-fs/CeCyD.

To estimate the duration of mitosis (M) and cytokinesis (C), the CeCyD uses the Williams (1971) equation [21]:

$$x = \frac{\ln(1 - y/2)}{-\alpha} \tag{1}$$

where x is the cumulative time within the cell cycle necessary to reach the start of the phase in question, i.e., the difference between the doubling time and x will give the time of the remaining phase(s); y is the cumulative proportion of cells up the phase in question (expressed as a fraction of one unit), i.e., the difference between the total cells (1% or 100%) and the percentage of cells in C or M+C, will provide the y value for C and M, respectively. Finally, α is the specific growth rate.

To estimate the G2 phase, the CeCyD must receive from the user the value corresponding to the period required for a cell to pass through G2 and M phases. For this, the user must apply an EdU pulse (e.g., 15 min) and then collected parasites every 15 min until a single cell containing two EdU-labeled nuclei (2N2K in case of trypanosomatids) is observed. The difference between this value and the duration of mitosis previously calculated corresponds to the G2-phase duration.

The S-phase duration is estimated by the CeCyD according to the Stanners and Till (1960) equation [22]:

$$S = \frac{1}{\alpha}\ln[L + e^{\alpha(Z)}] - (Z + t) \tag{2}$$

where L is the proportion of cells exhibiting EdU-labeled nuclei, α = ln 2/T (T = doubling time expressed in hours), Z = G2 + M + C, and t is the duration of the EdU labeling period in hours. Finally, the G1-phase duration is estimated by the difference between the doubling time and the sum of the remaining phases.

2.4. Estimation of the Minimum Number of Replication Origins (MO)

To estimate the MO needed to replicate an entire chromosome within the S-phase duration, we developed a mathematical inequation [8]. This formula uses as argument the S-phase duration (S) (which can be estimated by the CeCyD website), the size of the chromosome in question (N), and the replication rate (v). The lower bound MO for the number of origins required to replicate an entire chromosome is given by:

$$mo \geq \lceil \frac{N}{2, v, S} \rceil, \tag{3}$$

Of note, if the right-hand side of this inequation results in a fraction of a unit, then the next higher integer unit must be taken as the result of the inequation, which is represented by the ceiling function ($\lceil \ \rceil$).

For each organism analyzed, we used as parameters for the formula up-to-date data available in the TriTrypDB database (www.tritrypdb.org), NCBI database (www.ncbi.nlm.nih.gov), and data reported in the studies related [3,8,11–13,23–26] (see Table 1 for more details).

2.5. Origins Estimated by DNA Combing

To estimate how many origins are activated (on average) during the S-phase in any organism, we develop a simple mathematical equation. This equation uses a ratio between the size of the chromosome in question (N), and the inter-origin distance (IOD) obtained by DNA combing to estimate, on average, the total number of origins fired during the S-phase. The equation is given by:

$$Oc = \left\lceil \frac{N}{IOD} \right\rceil, \tag{4}$$

If the right-hand side of this equation results in a fraction of a unit, then the next higher integer unit must be taken as the result of the inequation, which is represented by the ceiling function ($\lceil \ \rceil$).

3. Results and Discussion

3.1. The CeCyD Website Allows a Quick Estimation of the Cell Cycle Phases Duration

Many studies have been using the two formulas developed by Williams (1971) [21] and Stanners and Till (1960) [22] to estimate the length of the cell cycle phases [8,27–32]. However, these estimations demand time and attention due to a large number of calculations involved. Also, they are subject to errors during the calculations. To facilitate the calculations and optimize the time consumed of these estimations, we developed a website called CeCyD (Cell Cycle Duration estimator), as shown in Figure 1A. CeCyD is available at the address http://cecyd.vital.butantan.gov.br/.

Briefly, CeCyD is a user-friendly website able to calculate the values of cytokinesis (C), mitosis (M), G2, S, and G1 phases of the cell cycle, for any organism. For this, the user needs the following experimental parameters: doubling time, percentage of cells in cytokinesis, percentage of cells in mitosis, minimum time to detect two EdU-labeled nuclei in the same cell, percentage of cells EdU-labeled after EdU pulse, and the duration of this EdU pulse.

To test and evaluate the efficiency of the CeCyD, we first obtained the parameters required for *L. major* and *T. cruzi* (CL Brener strain) from experimental analyses, as displayed in Figure S1. Then, we withdrew the same parameters for *T. brucei* from our previous study [8]. Next, we imputed the parameters in the CeCyD and estimated the duration of the cell cycle phases for each of these organisms, as shown in Figure 1B. *T. cruzi* presented G1 = 5.91 h (0.246 ccu), S = 9.86 h (0.411 ccu), and G2 + M + C = 8.23 h (0.343 cuu); *L. major* presented G1 = 5.52 h (0.53 ccu), S = 3.2 h (0.31 ccu), and G2 + M + C = 1.78 h (0.16 ccu); and *T. brucei* presented G1 = 3.37 h (0.397 ccu), S = 2.31 h (0.272 ccu), and G2 + M + C = 2.82 h (0.331 ccu). Of note, ccu means cell cycle unit, where one unit corresponds to the doubling time of each organism.

As expected, for *T. brucei*, the values provided by CeCyD were the same as those obtained in our previous work [8]. For both *L. major* and *T. cruzi*, when we compare the values provided by CeCyD with those obtained from other studies [30,32,33], we can observe similarities among the length of the cell cycle phases when EdU is used to monitor DNA replication [30]. However, when 5-bromo-2′-deoxyuridine (BrdU) is used to monitor DNA replication instead of EdU, the values obtained shown pronounced differences [33]. As already reported by our group [30], this discrepancy can be explained by the fact that there are differences in the detection of BrdU/EdU incorporation assays, i.e., EdU is more sensitive in monitoring DNA replication than the halogenated thymidine analogs (e.g., BrdU).

For our analyses, we used the S-phase duration from *T. cruzi*, *L. major*, *T. brucei*, *S. cerevisiae*, and *S. pombe*. Of note, for *T. brucei*, *S. cerevisiae*, and *S. pombe*, we did not use the CeCyD because the cell cycle phases duration for these organisms were already available [8,25,26,34,35], as shown in Figure 1B. Also, the cell cycle parameters used here were obtained from epimastigote cells of *T. cruzi*, promastigote cells of *L. major*, procyclic cells of *T. brucei*, mother cells of *S. cerevisiae*, and mitotic cells of *S. pombe*.

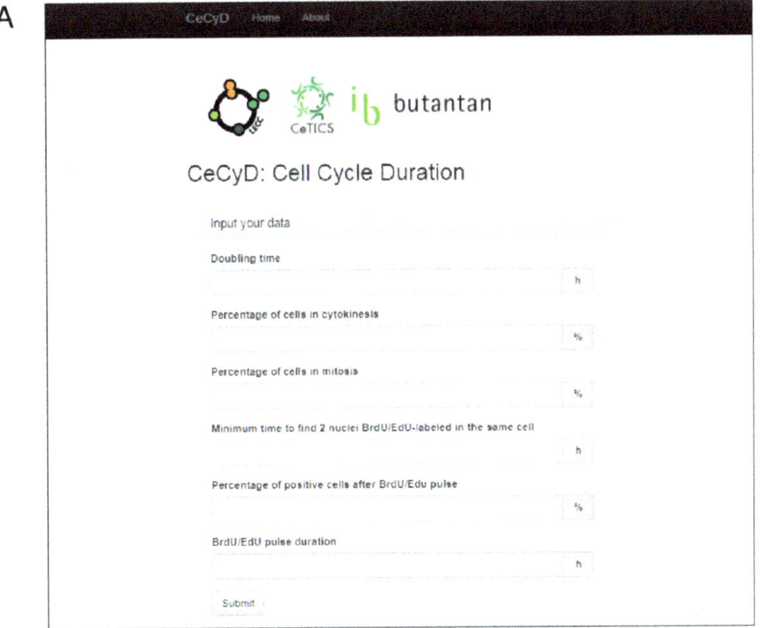

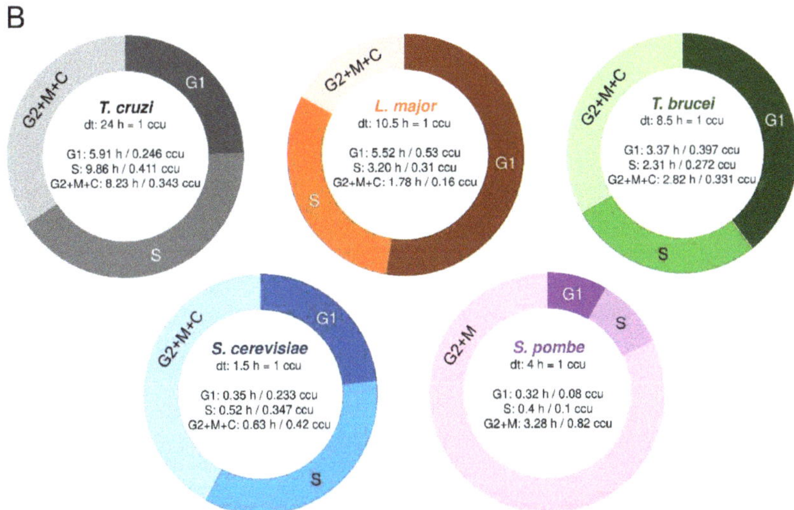

Figure 1. The user-friendly website CeCyD allows a quick estimation of the cell cycle phases duration for any cell type. (**A**) Screenshot of the CeCyD website showing the parameters to be inserted. This website is available at the address http://cecyd.vital.butantan.gov.br/. (**B**) Estimation of the cell cycle phases lengths (G1, S, and G2+M+C/G2+M) for *T. cruzi*, *L. major*, *T. brucei*, *S. cerevisiae*, and *S. pombe*. For *T. cruzi*, *L. major*, and *T. brucei* from calculations made using the CeCyD website. For *S. cerevisiae* and *S. pombe*, the values were obtained from other studies [25,26,34,35].

3.2. The Parameters Chromosome Size, S-Phase Duration, and Replication Rate Allow Estimating the MO per Chromosome in Any Organism

In a recent study, our group developed a formula able to estimate the MO required to duplicate an entire chromosome within the S-phase duration [8]. The development of this formula was based on the bidirectional movement of the replication forks, replication rate, S-phase duration, and the chromosome size in question. Although used only in *T. brucei* so far, this formula can be applied in any cell type. To demonstrate this, we estimated the MO in *T. cruzi*, *L. major*, *T. brucei* (using updated parameters), *S. cerevisiae*, and *S. pombe*, as shown in Table 1.

Table 1. Calculation of the minimum number of origins (MO) per chromosome in trypanosomatids (*T. cruzi*, *L. major*, and *T. brucei*) and yeasts (*S. cerevisiae* and *S. pombe*).

Chrom.	*T. cruzi* [1]		*L. major* [2]		*T. brucei* [3]		*S. cerevisiae* [4]		*S. pombe* [5]	
	Size (bp)	MO	Size (bp)	MO	Size (bp)	MO	Size (bp)	MO	Size (bp)	MO
I	77,958	1	268,988	1	1,064,672	2	230,19	3	5,598,923	129
II	151,740	1	355,712	1	1,193,948	2	813,14	9	4,397,795	101
III	196,644	1	384,502	1	1,653,225	2	315,34	4	2,465,919	57
IV	200,401	1	472,852	1	1,590,432	2	1,522,19	16	-	-
V	227,319	1	465,823	1	1,802,303	2	574,86	7	-	-
VI	389,024	1	516,869	1	1,618,915	2	270,15	3	-	-
VII	391,095	1	596,352	1	2,205,233	3	1,090,94	12	-	-
VIII	393,423	1	574,960	1	2,481,190	3	562,64	6	-	-
IX	509,634	1	573,434	1	3,542,885	4	439,88	5	-	-
X	518,846	1	570,865	1	4,144,375	5	745,44	8	-	-
XI	526,141	1	582,573	1	5,223,313	6	666,45	7	-	-
XII	533,093	1	675,346	1	-	-	1,078,17	12	-	-
XIII	558,364	1	654,595	1	-	-	924,43	10	-	-
XIV	598,625	1	622,644	1	-	-	784,33	9	-	-
XV	612,853	1	629,517	1	-	-	1,091,28	12	-	-
XVI	646,207	1	714,651	1	-	-	948,06	10	-	-
XVII	648,584	1	684,829	1	-	-	-	-	-	-
XVIII	655,081	1	739,748	1	-	-	-	-	-	-
XIX	671,453	1	702,208	1	-	-	-	-	-	-
XX	656,799	1	742,537	1	-	-	-	-	-	-
XXI	704,149	1	772,972	1	-	-	-	-	-	-
XXII	710,778	1	716,602	1	-	-	-	-	-	-
XXIII	655,477	1	772,565	1	-	-	-	-	-	-
XXIV	779,922	1	840,950	1	-	-	-	-	-	-
XXV	822,374	2	912,845	1	-	-	-	-	-	-
XXVI	801,422	1	1,091,540	2	-	-	-	-	-	-
XXVII	850,241	2	1,130,424	2	-	-	-	-	-	-
XXVIII	853,233	1	1,160,104	2	-	-	-	-	-	-
XXIX	870,934	1	1,212,663	2	-	-	-	-	-	-
XXX	863,882	1	1,403,434	2	-	-	-	-	-	-
XXXI	947,473	1	1,484,328	2	-	-	-	-	-	-
XXXII	968,069	1	1,604,637	2	-	-	-	-	-	-
XXXIII	1,041,172	1	1,583,653	2	-	-	-	-	-	-
XXXIV	1,065,764	1	1,866,748	2	-	-	-	-	-	-
XXXV	1,186,946	1	2,090,474	3	-	-	-	-	-	-
XXXVI	1,180,744	1	2,682,151	3	-	-	-	-	-	-
XXXVII	1,355,803	1	-	-	-	-	-	-	-	-
XXXVIII	1,444,805	1	-	-	-	-	-	-	-	-
XXXIX	1,854,104	1	-	-	-	-	-	-	-	-
XL	2,036,760	1	-	-	-	-	-	-	-	-
XLI	2,371,736	1	-	-	-	-	-	-	-	-

[1] *T. cruzi*: S-phase duration = 591.6 min (current study), replication rate = 2.05 kb·min^{-1} [23]; [2] *L. major*: S-phase duration = 192 min (current study), replication rate = 2.44 kb·min^{-1} [3]; [3] *T. brucei*: S-phase duration = 138.6 min [8], replication rate = 3.06 kb·min^{-1} [8]; [4] *S. cerevisiae*: S-phase duration = 30 min [34,35], replication rate = 1.6 kb·min^{-1} [11]; [5] *S. pombe*: mitotic S-phase duration = 24 min [25,26], mitotic replication rate = 0.91 kb·min^{-1} [24].

Among the single-celled eukaryotes analyzed here, *T. cruzi* draws attention because it is the only organism that requires only one origin per chromosome (MO = 1) to replicate its nuclear genome within the S-phase duration, as displayed in Table 1. *L. major*, on the other hand, requires more than one origin per chromosome to replicate its larger chromosomes (>1000 kb), while *T. brucei*, *S. cerevisiae*, and *S. pombe* requires more than one origin per chromosome to replicate their nuclear genomes, even for small chromosomes (<1000 kb), as shown in Table 1. As the formula to estimate the MO (Equation (3)) depends on the chromosome size, S-phase duration, and replication rate, the explanation for these organisms possess different MOs is related to these variables. For instance, *T. cruzi* has a long S-phase duration (9.86 h or 0.411 ccu) relative to other organisms analyzed here shown in Figure 1B, which justifies its MO per chromosome equaling 1, as presented in Table 1. Figure 1B and Table 1 show that *T. brucei* has an S-phase duration and replication rate similar to *L. major*; however, its chromosomes are larger than 1,000,000 bp (called megabase chromosomes [36]), which justifies the use of more than one origin per chromosome. *S. cerevisiae*, on the other hand, has a short S-phase duration (0.52 h or 0.347 ccu) and a low replication rate (1.6 kb·min^{-1} [11]), which imply, according to our formula, high MO values, as shown in Figure 1B and Table 1. *S. pombe*, in turn, has longer chromosomes, a short S-phase duration (0.4 h or 0.1 ccu) and a low replication rate (0.91 kb·min^{-1} [24]), which also imply high MO values. These data are also shown in Figure 1B and Table 1.

It is difficult to establish a reason why some organisms need a different number of origins during the S-phase. However, we can speculate that the number of origins needed to replicate all chromosomes during the S-phase is closely related to the S-phase duration itself. The question that remains is as follows: does the number of fired origins determine the S-phase duration, or is the S-phase duration robust, and a different number of origins is required to maintain this robustness? Although some studies point to robustness in S-phase duration [6–8], further studies are necessary to confirm which of these questions is the correct one.

It is worth to mention that among the parameters used to determine the MO, the replication rate is the most prone to alterations. Many factors can change the replication rate, such as decreased nucleotide pool [37,38], replication-transcription conflicts [39,40], DNA damage [41,42], among others, all of which leads mostly to some replication stress [43]. In other words, replication stress can be defined, in general, as the slowing of replication rate [44]. Thus, cells under replication stress probably would show different MO values relative to those estimated using the average replication rate from a wild type population.

3.3. In the Presence of Hypothetical Replication Stress, the MO Increase to Maintain Robustness in S-Phase Duration

To predict the behavior of MO in the presence of hypothetical replication stress, we simulated two conditions considering that the S-phase duration is robust [6–8]. The first one was mild replication stress, which was characterized here by a replication rate at 2/3 of the average value from the wild type population. The second situation was harsh replication stress, with a replication rate at a 1/3 of the average value. After applying these hypothetical values in Equation (3), we estimated the MO for the first (MO^{MR}) and second (MO^{HR}) conditions in trypanosomatids and *S. cerevisiae*, as shown in Table 2. It is worth mentioning that we did not perform this prediction for *S. pombe* because the peculiar behavior of its cell cycle seems to contribute to a flexible (non-robust) S phase duration [25,26].

According to our prediction, the only way a cell can maintain certain robustness in the S-phase duration in the presence of mild or harsh replication stress is to increase origin activation, which was evidenced by the increase in the MO values shown in Table 2. In other words, the demand for a higher number of activated origins in the presence of replication stress characterized by a slowing of replication rate can suggest that the cell tries to maintain robustness over the S-phase duration. This predicted behavior has already been evidenced by several cell types [45–48], including trypanosomatids [8] and *S. cerevisiae* [49]. However, *S. pombe* is an exception to this, because in addition to its S-phase not being robust [25,26], in the presence of replication stress, the origin firing is inhibited [24,50]. Moreover, *S. pombe* has other features that make its cell cycle unique when compared to other organisms: the S-phase is initiated before completion of the cytokinesis of the ongoing cell cycle [26,51], the cell mass influences the duration of the S-phase [25], and the main cell cycle control point is a size control in G2 phase [51]. Altogether, these peculiarities seem to contribute to more flexibility in the S-phase duration of this yeast. Nevertheless, further studies are still needed to better understand the dynamics of the replication stress response and origin usage through the S-phase in these non-metazoan organisms.

Table 2. Calculation of the minimum number of origins (MO) per chromosome in the presence of mild (MO^{MR}) and harsh (MO^{HR}) replication stress.

Chrom.	T. cruzi [1]			L. major [2]			T. brucei [3]			S. cerevisiae [4]		
	MO	MO^{MR}	MO^{HR}	MO	MO^{MR}	MO^{HR}	MO	MO^{MR}	MO^{HR}	MO	MO^{MR}	MO^{HR}
I	1	1	1	1	1	1	2	2	4	3	4	8
II	1	1	1	1	1	2	2	3	5	9	13	26
III	1	1	1	1	1	2	2	3	6	4	5	10
IV	1	1	1	1	1	2	2	3	6	16	24	48
V	1	1	1	1	1	2	2	4	7	7	10	19
VI	1	1	1	1	1	2	2	3	6	3	5	9
VII	1	1	1	1	1	2	3	4	8	12	18	35
VIII	1	1	1	1	1	2	3	5	9	6	9	18
IX	1	1	1	1	1	2	4	7	13	5	7	14
X	1	1	1	1	1	2	5	8	15	8	12	24
XI	1	1	1	1	1	2	6	10	19	7	11	21
XII	1	1	1	1	2	3	-	-	-	12	17	34
XIII	1	1	1	1	2	3	-	-	-	10	15	30
XIV	1	1	1	1	1	3	-	-	-	9	13	25
XV	1	1	1	1	2	3	-	-	-	12	18	35
XVI	1	1	1	1	2	3	-	-	-	10	15	30
XVII	1	1	1	1	2	3	-	-	-	-	-	-
XVIII	1	1	1	1	2	3	-	-	-	-	-	-
XIX	1	1	1	1	2	3	-	-	-	-	-	-
XX	1	1	1	1	2	3	-	-	-	-	-	-
XXI	1	1	1	1	2	3	-	-	-	-	-	-
XXII	1	1	1	1	2	3	-	-	-	-	-	-
XXIII	1	1	1	1	2	3	-	-	-	-	-	-
XXIV	1	1	1	1	2	3	-	-	-	-	-	-
XXV	1	1	2	1	2	3	-	-	-	-	-	-
XXVI	1	1	1	2	2	4	-	-	-	-	-	-
XXVII	1	1	2	2	2	4	-	-	-	-	-	-
XXVIII	1	1	2	2	2	4	-	-	-	-	-	-
XXIX	1	1	2	2	2	4	-	-	-	-	-	-
XXX	1	1	2	2	3	5	-	-	-	-	-	-
XXXI	1	1	2	2	3	5	-	-	-	-	-	-
XXXII	1	1	2	2	3	6	-	-	-	-	-	-
XXXIII	1	1	2	2	3	6	-	-	-	-	-	-
XXXIV	1	1	2	2	3	6	-	-	-	-	-	-
XXXV	1	1	2	3	4	7	-	-	-	-	-	-
XXXVI	1	1	2	3	5	9	-	-	-	-	-	-
XXXVII	1	1	2	-	-	-	-	-	-	-	-	-
XXXVIII	1	1	2	-	-	-	-	-	-	-	-	-
XXXIX	1	2	2	-	-	-	-	-	-	-	-	-
XL	1	2	2	-	-	-	-	-	-	-	-	-
XLI	1	2	2	-	-	-	-	-	-	-	-	-

[1] T. cruzi: S-phase duration = 591.6 min (current study), replication rate = 2.05 kb·min^{-1} [23]; [2] L. major: S-phase duration = 192 min (current study), replication rate = 2.44 kb·min^{-1} [3]; [3] T. brucei: S-phase duration = 138.6 min [8], replication rate = 3.06 kb·min^{-1} [8]; [4] S. cerevisiae: S-phase duration = 30 min [34,35], replication rate = 1.6 kb·min^{-1} [11].

3.4. The MO Allows the Establishment of a Threshold That Can Serve as a Parameter by Other Methods That Detect Origins

To compare the MO with the origins obtained by different experimental approaches, we set up graphs in order to show trend lines for each methodology analyzed, as shown in Figure 2A–E. We observed an expected positive correlation between the number of origins and the size of each chromosome, i.e., the larger the chromosome, the more origins are required to replicate it within the S-phase duration. As the MO is estimated from relatively constant parameters in a wild type population, the trend line of the MO, shown in Figure 2A–E in black lines, allows the establishment of a threshold that can serve as a parameter when estimating the number of origins by other methods.

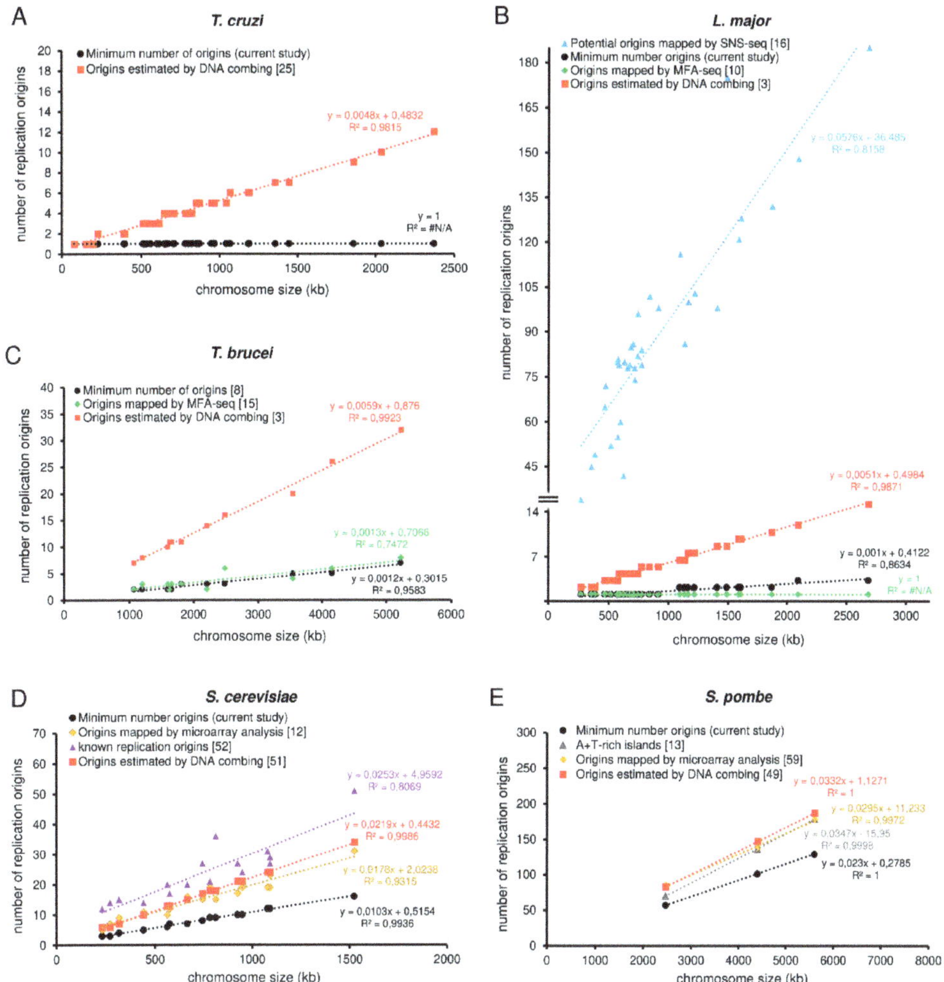

Figure 2. Comparative analysis among different approaches used to estimate replication origins in trypanosomatids (*T. cruzi*, *L. major*, *T. brucei*) and yeasts (*S. cerevisiae* and *S. pombe*) (a–e). Graphs showing positive correlations between chromosome length and the number of replication origins estimated by different approaches: minimum number of origins—MO (black), origins estimated by DNA combing (red), origins estimated by MFA-seq (green), potential origins mapped by SNS-seq (blue), origins estimated by microarray (yellow), known origins (purple), and A+T rich islands (gray). (**A**) *T. cruzi*, (**B**) *L. major*, (**C**) *T. brucei*, (**D**) *S. cerevisiae*, and (**E**) *S. pombe*. The trend lines for all approaches, as well as the equations, are shown. Studies are referenced in each graph.

Before we go on with our analysis, it is worth mentioning that according to their different usages, replication origins can be classified into three categories: constitutive, which are always activated in all cells of a given population; flexible, whose usage varies from cell to cell; or dormant, which are not fired during a normal cell cycle but are activated in the presence of replication stress [52]. However, due to the technical difficulty in distinguishing flexible and dormant origins, we refer to these only as non-constitutive origins.

Thus, when comparing the trend line of the origins estimated by DNA combing (the red lines in Figure 2A–E) with the trend lines of MO (the black lines in Figure 2A–E,), we observed that *T. cruzi*,

L. major, *T. brucei*, *S. cerevisiae*, and *S. pombe* use, on average, more origins per chromosome than the minimum required, i.e., the red lines are above from the black ones, as shown in Figure 2A–E). This makes sense, given that the DNA combing approach estimates the pool of all origins (constitutive + non-constitutive) fired in a population.

For *L. major*, in addition to the trend line for origins estimated by DNA combing (the red line in Figure 2B), we also plotted a trend line for potential origins mapped by small leading nascent strand purification coupled to next-generation sequencing (SNS-seq) [16] (the blue line in Figure 2B), and a trend line for origins mapped by marker frequency analysis (MFA-seq) [10] (the green line in Figure 2B). The trend line of the potential origins mapped by SNS-seq is far above from the others (the blue line in comparison with the others in Figure 2B). This makes sense because the SNS-seq approach has high accuracy and resolution in detecting small sites of replication, which can include DNA repair, potential origins, and other events that generate DNA synthesis in a population. However, the trend line of origins mapped by MFA-seq is below the threshold imposed by the minimal origins (MO trend line) (the black line in comparison with the green one in Figure 2B). This implies that only with origins mapped by MFA-seq [10], *L. major* is not able to replicate its nuclear genome within the S-phase duration. Although it seems meaningless, this can be easily explained by the fact that the MFA-seq analysis has low resolution and accuracy, probably being able to identify only the constitutive origins in a population and not the entire pool of fired origins, as occurs in the DNA combing approach for example [3,8]. Nevertheless, further studies are necessary to figure out how many origins are indeed used for a single cell of *L. major* during a standard cell cycle.

In *T. brucei*, we also plotted a trend line for the origins mapped by MFA-seq, and the situation is similar to that presented by *L. major* (the black dots in comparison with the green ones in Figure 2C), i.e., some chromosomes are not able to be duplicated only with the origins mapped by MFA-seq, as already explained in a recent study of our group [8]. Briefly, the reason is the same as previously explained: the low resolution of the MFA-seq analysis.

For *S. cerevisiae*, in addition to plotting a trend line for origins estimated by DNA combing [53] (the red line in Figure 2D) and a trend line for origins mapped by microarray analysis [12] (the green line in Figure 2D), we also plotted a trend line for the known positions of origins [54] (the purple line in Figure 2D). The known position of origins refers to a conserved DNA sequence (called autonomous replicating sequences—ARS) where the assembly of the pre-replication complex occurs [53,55,56]. The trend line of the known origins is above from the MO trend line (the purple line in comparison with the black one in Figure 2D), and above from the trend line of the origins estimated by DNA combing (the purple line in comparison with the red one in Figure 2D). This makes sense because the known positions of origins are potential sites for the establishment of origins. However, not all of these sites are activated during the S-phase, i.e., in *S. cerevisiae*, there are many more potential sites for the establishment of origins than those that are indeed used to complete replication within the S-phase duration [19,57–60]. The trend line of origins mapped by microarray analysis is above the threshold imposed by the minimum origins (MO trend line) (the green line in comparison with the black one in Figure 2D), but below the trend lines of both the known origins and the origins estimated by DNA combing (the comparison amongst the green, purple and red lines in Figure 2D). This was expected for the same reason raised before, i.e., just like MFA-seq, microarray analysis also has low accuracy in detecting the entire pool of origins activated in a population and maps mainly the constitutive origins.

Unlike *S. cerevisiae*, *S. pombe* lacks a consensus DNA sequence that determines origin sites. However, its origins coincide with chromosomal A+T-rich islands [13]. Thus, in addition to plotting a trend line for origins estimated by DNA combing [50] (the red line in Figure 2E) and a trend line for origins mapped by microarray analysis [61] (the yellow line in Figure 2E), we also plotted a trend line for the A+T rich islands [13] (the gray line in Figure 2E). All three of these trend lines (red, yellow, and gray) are above from the threshold imposed by the minimum origins (MO trend line) and practically overlap each other, although as the comparison amongst all the trend lines in Figure 2E shows, the trend line of the origins estimated by DNA combing is slightly above, as expected. This overlapping of the trend lines raises a question about the dynamics of origin usage during the S-phase in *S. pombe*, which seems to be relatively peculiar when compared to other single-celled eukaryotes [50].

Of note, so far, there is no data about MFA-seq or microarray analysis in *T. cruzi*, which prevents a deeper comparative analysis in this organism.

3.5. Trypanosomatids Can Use Around Fivefold More Origins than the Minimum Required to Complete Replication within the S-Phase Duration

To investigate how many times more origins than the minimum (MO) the organisms analyzed can use, we calculated the ratio between the angular coefficient (a value) of linear equations (y = ax + b) of maximum origins used and MO, shown as the red and black lines, respectively, in Figure 2. Here, we defined maximum origins used as the origins estimated by DNA combing, represented by the red lines in Figure 2; see material and methods.

Using this reasoning, we can estimate that *S. pombe* uses, on average, 1.44 times more origins than the MO, while in *S. cerevisiae*, this ratio is 2.12. Interestingly, in trypanosomatids, this ratio is higher. In *T. brucei* this ratio is 4.91, in *L. major* is 5.1, and in *T. cruzi* this ratio is so high that tends to infinity since *T. cruzi* needs only one origin per chromosome to replicate its nuclear genome (MO = 1 for all chromosomes), i.e., the MO linear equation is y = 1, as shown in Figure 3.

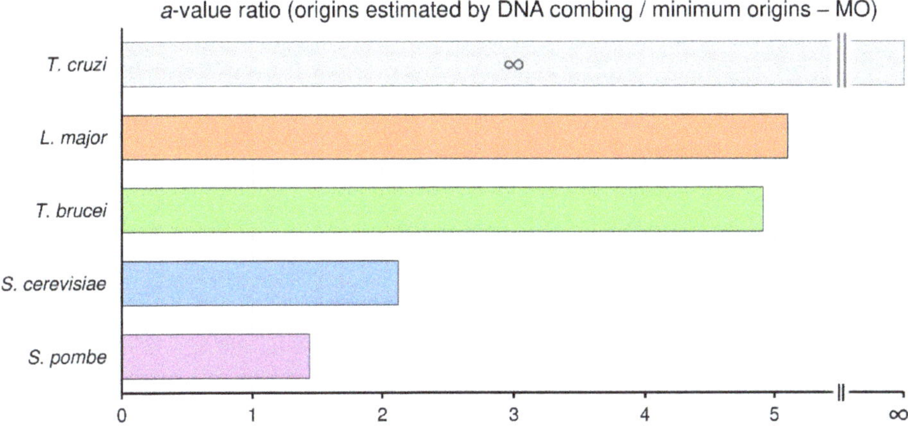

Figure 3. Trypanosomatids use around fivefold more origins than the minimum required. Angular coefficient (a-value) ratios between origins estimated by DNA combing and the minimum origins (MO) for *T. cruzi* (gray bar), *L. major* (orange bar), *T. brucei* (green bar), *S. cerevisiae* (blue bar), and *S. pombe* (purple bar).

Although we cannot classify the total origins used as constitutive or non-constitutive, one question can be raised: what makes trypanosomatids apparently use a pool of origins much higher than the MO when compared to the yeasts *S. cerevisiae* and *S. pombe*? One possible explanation is that in trypanosomatids, unlike other eukaryotes, the majority of their genes are organized into large polycistronic clusters, which could favor replication stress through replication–transcription conflicts [8]. Replication stress, as reported in some studies [48,62], is a potential contributor for the activation of replication origins. However, although proposed by our group [8], this hypothesis needs to support more experimental assays to gain credibility. Another possibility is that the replication rate of *S. cerevisiae* and *S. pombe* are lower than those in trypanosomatids (1.6 kb·min^{-1} in *S. cerevisiae*, 0.91 kb·min^{-1} in *S. pombe*, and 2–3 kb·min^{-1} in trypanosomatids), as shown in Table 1. *S. cerevisiae* has a chromosomes size and an S-phase duration similar to those found in trypanosomatids shown in Table 1 and Figure 1B. Thus, the only way to maintain robustness in the S-phase duration is by activating more origins. Apparently, *S. cerevisiae* does just that, but further studies are necessary to figure out its exact dynamics of origin usage during the S-phase. On the other hand, *S. pombe* has larger chromosomes and a relatively short S-phase duration when compared to trypanosomatids, as displayed in Table 1 and Figure 1B. Moreover, as already mentioned, *S. pombe* does not have a robust S-phase [25,26] and its origins fire stochastically [50], which precludes any speculation regarding its peculiar dynamics of origin usage. However, unlike trypanosomatids, *S. pombe* appears to use a number of origins very close to the minimum required.

This is the first time a comparative analysis of the minimum number of origins has been successfully applied. These data may provide new insight into the understanding of origin usage during the S-phase and a new methodological framework for studying single-celled eukaryotes genomes.

4. Conclusions

Here, we demonstrate that the minimum number of origins (MO) required to duplicate an entire chromosome within the S-phase duration can be easily obtained from the parameters chromosome size, S-phase duration, and replication rate. Predictions performed by us suggest that in the presence of replication stress, all the organisms analyzed here demands higher MO values. Moreover, we evidenced here that the MO allows the establishment of a threshold that can serve as a parameter by other methods that detect origins. Also, our data strongly suggest that trypanosomatids can use around fivefold more origins than the MO. This value is relatively higher than other single-celled organisms, such as the yeasts *S. cerevisiae* and *S. pombe*. However, further studies are required to figure out the dynamics of origin usage during the S-phase in these organisms, especially in trypanosomatids.

Supplementary Materials: The following are available online at http://www.mdpi.com/2073-4425/11/5/523/s1, Figure S1: Parameters required to use the CeCyD website.

Author Contributions: M.S.d.S. conceived the rationale of the experimental design and this manuscript, with fundamental insights from M.C.E. M.S.d.S., and M.O.V. carried out the experiments. B.F.S. and M.O.V. developed the CeCyD website with insights from M.S.d.S. M.S.d.S. wrote the manuscript with essential contribution from M.O.V., B.F.S., and M.C.E. All authors read and approved the final version of the manuscript. M.C.E. supervised the project.

Funding: This research was funded by São Paulo Research Foundation (FAPESP) and Center of Toxins, Immune Response, and Cell Signaling (CeTICS) under grants 2013/07467-1, 2016/50050-2, 2014/24170-5, 2017/18719-2, 2017/07693-2). MCE is also fellow from the National Council for Scientific and Technological Development (CNPq) under grant 306199/2018-1 and MOV was fellow from CNPq under grant 870219/1997-9.

Acknowledgments: The authors thank the São Paulo Research Foundation (FAPESP) and Center of Toxins, Immune Response, and Cell Signaling (CeTICS) for their support.

Conflicts of Interest: The authors declare no conflict of interest.

References

1. Leonard, A.C.; Méchali, M. DNA replication origins. *Cold Spring Harb. Perspect. Biol.* **2013**, *5*, a010116. [CrossRef] [PubMed]
2. Myllykallio, H.; Lopez, P.; López-García, P.; Heilig, R.; Saurin, W.; Zivanovic, Y.; Philippe, H.; Forterre, P. Bacterial mode of replication with eukaryotic-like machinery in a hyperthermophilic archaeon. *Science* **2000**, *288*, 2212–2215. [CrossRef] [PubMed]
3. Stanojcic, S.; Sollelis, L.; Kuk, N.; Crobu, L.; Balard, Y.; Schwob, E.; Bastien, P.; Pagès, M.; Sterkers, Y. Single-molecule analysis of DNA replication reveals novel features in the divergent eukaryotes Leishmania and Trypanosoma brucei versus mammalian cells. *Sci. Rep.* **2016**, *6*, 23142. [CrossRef] [PubMed]
4. Turrero García, M.; Chang, Y.; Arai, Y.; Huttner, W.B. S-phase duration is the main target of cell cycle regulation in neural progenitors of developing ferret neocortex. *J. Comp. Neurol.* **2016**, *524*, 456–470. [CrossRef]
5. Günesdogan, U.; Jäckle, H.; Herzig, A. Histone supply regulates S phase timing and cell cycle progression. *Elife* **2014**, *3*, e02443. [CrossRef]
6. Gindin, Y.; Valenzuela, M.S.; Aladjem, M.I.; Meltzer, P.S.; Bilke, S. A chromatin structure-based model accurately predicts DNA replication timing in human cells. *Mol. Syst. Biol.* **2014**, *10*, 722. [CrossRef]

7. Zhang, Q.; Bassetti, F.; Gherardi, M.; Lagomarsino, M.C. Cell-to-cell variability and robustness in S-phase duration from genome replication kinetics. *Nucleic Acids Res.* **2017**, *45*, 8190–8198. [CrossRef]

8. da Silva, M.S.; Cayres-Silva, G.R.; Vitarelli, M.O.; Marin, P.A.; Hiraiwa, P.M.; Araújo, C.B.; Scholl, B.B.; Ávila, A.R.; McCulloch, R.; Reis, M.S.; et al. Transcription activity contributes to the firing of non-constitutive origins in African trypanosomes helping to maintain robustness in S-phase duration. *Sci. Rep.* **2019**, *9*, 18512. [CrossRef]

9. Robinson, N.P.; Bell, S.D. Origins of DNA replication in the three domains of life. *FEBS J.* **2005**, *272*, 3757–3766. [CrossRef]

10. Marques, C.A.; Dickens, N.J.; Paape, D.; Campbell, S.J.; McCulloch, R. Genome-wide mapping reveals single-origin chromosome replication in Leishmania, a eukaryotic microbe. *Genome Biol.* **2015**, *16*, 230. [CrossRef]

11. Sekedat, M.D.; Fenyö, D.; Rogers, R.S.; Tackett, A.J.; Aitchison, J.D.; Chait, B.T. GINS motion reveals replication fork progression is remarkably uniform throughout the yeast genome. *Mol. Syst. Biol.* **2010**, *6*, 353. [CrossRef] [PubMed]

12. Yabuki, N.; Terashima, H.; Kitada, K. Mapping of early firing origins on a replication profile of budding yeast. *Genes Cells* **2002**, *7*, 781–789. [CrossRef] [PubMed]

13. Segurado, M.; de Luis, A.; Antequera, F. Genome-wide distribution of DNA replication origins at A+T-rich islands in Schizosaccharomyces pombe. *EMBO Rep.* **2003**, *4*, 1048–1053. [CrossRef] [PubMed]

14. Parker, M.W.; Botchan, M.R.; Berger, J.M. Mechanisms and regulation of DNA replication initiation in eukaryotes. *Crit. Rev. Biochem. Mol. Biol.* **2017**, *52*, 107–144. [CrossRef] [PubMed]

15. Tiengwe, C.; Marcello, L.; Farr, H.; Dickens, N.; Kelly, S.; Swiderski, M.; Vaughan, D.; Gull, K.; Barry, J.D.; Bell, S.D.; et al. Genome-wide Analysis Reveals Extensive Functional Interaction between DNA Replication Initiation and Transcription in the Genome of Trypanosoma brucei. *Cell Rep.* **2012**, *2*, 185–197. [CrossRef] [PubMed]

16. Lombraña, R.; Álvarez, A.; Fernández-Justel, J.M.; Almeida, R.; Poza-Carrión, C.; Gomes, F.; Calzada, A.; Requena, J.M.; Gómez, M. Transcriptionally Driven DNA Replication Program of the Human Parasite Leishmania major. *Cell Rep.* **2016**, *16*, 1774–1786. [CrossRef]

17. Wang, X.; Lesterlin, C.; Reyes-Lamothe, R.; Ball, G.; Sherratt, D.J. Replication and segregation of an Escherichia coli chromosome with two replication origins. *Proc. Natl. Acad. Sci. USA* **2011**, *108*, 243–250. [CrossRef]

18. Gao, F. Bacteria may have multiple replication origins. *Front. Microbiol.* **2015**, *6*, 324. [CrossRef]

19. Müller, C.A.; Hawkins, M.; Retkute, R.; Malla, S.; Wilson, R.; Blythe, M.J.; Nakato, R.; Komata, M.; Shirahige, K.; De Moura, A.P.S.; et al. The dynamics of genome replication using deep sequencing. *Nucleic Acids Res.* **2013**, *42*, e3. [CrossRef]

20. Salic, A.; Mitchison, T.J. A chemical method for fast and sensitive detection of DNA synthesis in vivo. *Proc. Natl. Acad. Sci. USA* **2008**, *105*, 2415–2420. [CrossRef]

21. Williams, F.M. Dynamics of microbial populations. In *Systems Analysis and Simulation in Ecology*; Patten, B.C., Ed.; Academic Press Inc.: New York, NY, USA, 1971; pp. 198–268.

22. Stanners, C.P.; Till, J.E. DNA synthesis in individual L-strain mouse cells. *Biochim. Biophys. Acta* **1960**, *37*, 406–419. [CrossRef]

23. de Araujo, C.B.; Calderano, S.G.; Elias, M.C. The Dynamics of Replication in Trypanosoma cruzi Parasites by Single-Molecule Analysis. *J. Eukaryot. Microbiol.* **2019**, *66*, 514–518. [CrossRef] [PubMed]

24. Iyer, D.R.; Rhind, N. The intra-S checkpoint responses to DNA damage. *Genes* **2017**, *8*, 74. [CrossRef] [PubMed]

25. Nasmyth, K.; Nurse, P.; Fraser, R.S.S. The effect of cell mass on the cell cycle timing and duration of S-phase in fission yeast. *J. Cell Sci.* **1979**, *39*, 215–233.

26. Carlson, C.R.; Grallert, B.; Stokke, T.; Boye, E. Regulation of the start of DNA replication in Schizosaccharomyces pombe. *J. Cell Sci.* **1999**, *112*, 939–946.

27. Woodward, R.; Gull, K. Timing of nuclear and kinetoplast DNA replication and early morphological events in the cell cycle of Trypanosoma brucei. *J. Cell Sci.* **1990**, *95 Pt 1*, 49–57.

28. Ploubidou, A.; Robinson, D.R.; Docherty, R.C.; Ogbadoyi, E.O.; Gull, K. Evidence for novel cell cycle checkpoints in trypanosomes: Kinetoplast segregation and cytokinesis in the absence of mitosis. *J. Cell Sci.* **1999**, *112*, 4641–4650.

29. da Silva, M.S.; Monteiro, J.P.; Nunes, V.S.; Vasconcelos, E.J.; Perez, A.M.; Freitas-Júnior, L.d.H.; Elias, M.C.; Cano, M.I.N. Leishmania amazonensis Promastigotes Present Two Distinct Modes of Nucleus and Kinetoplast Segregation during Cell Cycle. *PLoS ONE* **2013**, *8*, e81397. [CrossRef]

30. da Silva, M.S.; Muñoz, P.A.M.; Armelin, H.A.; Elias, M.C. Differences in the Detection of BrdU/EdU Incorporation Assays Alter the Calculation for G1, S, and G2 Phases of the Cell Cycle in Trypanosomatids. *J. Eukaryot. Microbiol.* **2017**, *64*, 756–770. [CrossRef]

31. Tavernelli, L.E.; Motta, M.C.M.; Gonçalves, C.S.; da Silva, M.S.; Elias, M.C.; Alonso, V.L.; Serra, E.; Cribb, P. Overexpression of Trypanosoma cruzi High Mobility Group B protein (TcHMGB) alters the nuclear structure, impairs cytokinesis and reduces the parasite infectivity. *Sci. Rep.* **2019**, *9*, 192. [CrossRef]

32. Ambit, A.; Woods, K.L.; Cull, B.; Coombs, G.H.; Mottram, J.C. Morphological events during the cell cycle of leishmania major. *Eukaryot. Cell* **2011**, *10*, 1429–1438. [CrossRef] [PubMed]

33. Elias, M.C.; da Cunha, J.P.C.; de Faria, F.P.; Mortara, R.A.; Freymüller, E.; Schenkman, S. Morphological events during the Trypanosoma cruzi cell cycle. *Protist* **2007**, *158*, 147–157. [CrossRef] [PubMed]

34. Brewer, B.J.; Chlebowicz-Sledziewska, E.; Fangman, W.L. Cell cycle phases in the unequal mother/daughter cell cycles of Saccharomyces cerevisiae. *Mol. Cell. Biol.* **1984**, *4*, 2529–2531. [CrossRef] [PubMed]

35. Ivanova, T.; Maier, M.; Missarova, A.; Ziegler-Birling, C.; Carey, L.B.; Mendoza, M. Budding yeast complete DNA replication after chromosome segregation begins. *bioRxiv* **2018**, 407957.

36. Berriman, M.; Ghedin, E.; Hertz-Fowler, C.; Blandin, G.; Renauld, H.; Bartholomeu, D.C.; Lennard, N.J.; Caler, E.; Hamlin, N.E.; Haas, B.; et al. The Genome of the African Trypanosome Trypanosoma brucei. *Science* **2005**, *309*, 416–422. [CrossRef]

37. Saxena, S.; Somyajit, K.; Nagaraju, G. XRCC2 Regulates Replication Fork Progression during dNTP Alterations. *Cell Rep.* **2018**, *25*, 3273–3282.e6. [CrossRef]

38. Wilhelm, T.; Magdalou, I.; Barascu, A.; Techer, H.; Debatisse, M.; Lopez, B.S. Spontaneous slow replication fork progression elicits mitosis alterations in homologous recombination-deficient mammalian cells. *Proc. Natl. Acad. Sci. USA* **2014**, *111*, 763–768. [CrossRef]

39. Azvolinsky, A.; Giresi, P.G.; Lieb, J.D.; Zakian, V.A. Highly Transcribed RNA Polymerase II Genes Are Impediments to Replication Fork Progression in Saccharomyces cerevisiae. *Mol. Cell* **2009**, *34*, 722–734. [CrossRef]

40. Deshpande, A.M.; Newlon, C.S. DNA replication fork pause sites dependent on transcription. *Science* **1996**, *272*, 1030–1033. [CrossRef]

41. Dulev, S.; De Renty, C.; Mehta, R.; Minkov, I.; Schwob, E.; Strunnikov, A. Essential global role of CDC14 in DNA synthesis revealed by chromosome underreplication unrecognized by checkpoints in cdc14 mutants. *Proc. Natl. Acad. Sci. USA* **2009**, *106*, 14466–14471. [CrossRef]

42. Branzei, D.; Foiani, M. Regulation of DNA repair throughout the cell cycle. *Nat. Rev. Mol. Cell Biol.* **2008**, *9*, 297–308. [CrossRef] [PubMed]

43. Rodriguez-Acebes, S.; Mourón, S.; Méndez, J. Uncoupling fork speed and origin activity to identify the primary cause of replicative stress phenotypes. *J. Biol. Chem.* **2018**, *293*, 12855–12861. [CrossRef] [PubMed]

44. Zeman, M.K.; Cimprich, K.A. Causes and consequences of replication stress. *Nat. Cell Biol.* **2014**, *16*, 2–9. [CrossRef] [PubMed]

45. Ge, X.Q.; Blow, J.J. Chk1 inhibits replication factory activation but allows dormant origin firing in existing factories. *J. Cell Biol.* **2010**, *191*, 1285–1297. [CrossRef] [PubMed]

46. McIntosh, D.; Blow, J.J. Dormant Origins, the Licensing Checkpoint, and the Response to Replicative Stresses. *Cold Spring Harb. Perspect. Biol.* **2012**, *4*, a012955. [CrossRef]

47. Chen, Y.-H.; Jones, M.J.K.; Yin, Y.; Crist, S.B.; Colnaghi, L.; Sims, R.J.; Rothenberg, E.; Jallepalli, P.V.; Huang, T.T. ATR-mediated phosphorylation of FANCI regulates dormant origin firing in response to replication stress. *Mol. Cell* **2015**, *58*, 323–338. [CrossRef]

48. Courtot, L.; Hoffmann, J.S.; Bergoglio, V. The Protective Role of Dormant Origins in Response to Replicative Stress. *Int. J. Mol. Sci.* **2018**, *19*, 3569. [CrossRef]

49. Poli, J.; Tsaponina, O.; Crabbé, L.; Keszthelyi, A.; Pantesco, V.; Chabes, A.; Lengronne, A.; Pasero, P. dNTP pools determine fork progression and origin usage under replication stress. *EMBO J.* **2012**, *31*, 883–894. [CrossRef]

50. Patel, P.K.; Arcangioli, B.; Baker, S.P.; Bensimon, A.; Rhind, N. DNA replication origins fire stochastically in fission yeast. *Mol. Biol. Cell* **2006**, *17*, 308–316. [CrossRef]

51. Oliva, A.; Rosebrock, A.; Ferrezuelo, F.; Pyne, S.; Chen, H.; Skiena, S.; Futcher, B.; Leatherwood, J. The cell cycle-regulated genes of Schizosaccharomyces pombe. *PLoS Biol.* **2005**, *3*, 1239–1260. [CrossRef]

52. Fragkos, M.; Ganier, O.; Coulombe, P.; Méchali, M. DNA replication origin activation in space and time. *Nat. Rev. Mol. Cell Biol.* **2015**, *16*, 360–374. [CrossRef] [PubMed]

53. Lengronne, A. Monitoring S phase progression globally and locally using BrdU incorporation in TK+ yeast strains. *Nucleic Acids Res.* **2001**, *29*, 1433–1442. [CrossRef] [PubMed]

54. Nieduszynski, C.A.; Knox, Y.; Donaldson, A.D. Genome-wide identification of replication origins in yeast by comparative genomics. *Genes Dev.* **2006**, *20*, 1874–1879. [CrossRef]

55. Broach, J.R.; Li, Y.Y.; Feldman, J.; Jayaram, M.; Abraham, J.; Nasmyth, K.A.; Hicks, J.B. Localization and sequence analysis of yeast origins of DNA replication. *Cold Spring Harb. Symp. Quant. Biol.* **1982**, *47*, 1165–1173. [CrossRef] [PubMed]

56. Stinchcomb, D.T.; Struhl, K.; Davis, R.W. Isolation and characterisation of a yeast chromosomal replicator. *Nature* **1979**, *282*, 39–43. [CrossRef] [PubMed]

57. Dershowitz, A.; Newlon, C.S. The effect on chromosome stability of deleting replication origins. *Mol. Cell. Biol.* **1993**, *13*, 391–398. [CrossRef] [PubMed]

58. Raghuraman, M.K.; Winzeler, E.A.; Collingwood, D.; Hunt, S.; Wodicka, L.; Conway, A.; Lockhart, D.J.; Davis, R.W.; Brewer, B.J.; Fangman, W.L. Replication dynamics of the yeast genome. *Science* **2001**, *294*, 115–121. [CrossRef]

59. Wyrick, J.J.; Aparicio, J.G.; Chen, T.; Barnett, J.D.; Jennings, E.G.; Young, R.A.; Bell, S.P.; Aparicio, O.M. Genome-wide distribution of ORC and MCM proteins in S. cerevisae: High-resolution mapping of replication origins. *Science* **2001**, *294*, 2357–2360. [CrossRef]

60. Pasero, P.; Bensimon, A.; Schwob, E. Single-molecule analysis reveals clustering and epigenetic regulation of replication origins at the yeast rDNA locus. *Genes Dev.* **2002**, *16*, 2479–2484. [CrossRef]

61. Heichinger, C.; Penkett, C.J.; Bähler, J.; Nurse, P. Genome-wide characterization of fission yeast DNA replication origins. *EMBO J.* **2006**, *25*, 5171–5179. [CrossRef]

62. Yekezare, M.; Gómez-González, B.; Diffley, J.F.X. Controlling DNA replication origins in response to DNA damage—Inhibit globally, activate locally. *J. Cell Sci.* **2013**, *126*, 1297–1306. [CrossRef] [PubMed]

Article

The Experimental Proteome of *Leishmania infantum* Promastigote and Its Usefulness for Improving Gene Annotations

África Sanchiz, Esperanza Morato, Alberto Rastrojo, Esther Camacho,
Sandra González-de la Fuente, Anabel Marina, Begoña Aguado and Jose M. Requena *

Centro de Biología Molecular "Severo Ochoa" (CBMSO, CSIC-UAM) Campus de Excelencia Internacional (CEI) UAM+CSIC, Universidad Autónoma de Madrid, 28049 Madrid, Spain; africa.sanchiz@gmail.com (Á.S.); emorato@cbm.csic.es (E.M.); arastrojo@cbm.csic.es (A.R.); ecamacho@cbm.csic.es (E.C.); sandra.g@cbm.csic.es (S.G.-d.l.F.); amarina@cbm.csic.es (A.M.); baguado@cbm.csic.es (B.A.)
* Correspondence: jmrequena@cbm.csic.es

Received: 27 July 2020; Accepted: 28 August 2020; Published: 2 September 2020

Abstract: *Leishmania infantum* causes visceral leishmaniasis (kala-azar), the most severe form of leishmaniasis, which is lethal if untreated. A few years ago, the re-sequencing and de novo assembling of the *L. infantum* (JPCM5 strain) genome was accomplished, and now we aimed to describe and characterize the experimental proteome of this species. In this work, we performed a proteomic analysis from axenic cultured promastigotes and carried out a detailed comparison with other *Leishmania* experimental proteomes published to date. We identified 2352 proteins based on a search of mass spectrometry data against a database built from the six-frame translated genome sequence of *L. infantum*. We detected many proteins belonging to organelles such as glycosomes, mitochondria, or flagellum, as well as many metabolic enzymes and many putative RNA binding proteins and molecular chaperones. Moreover, we listed some proteins presenting post-translational modifications, such as phosphorylations, acetylations, and methylations. On the other hand, the identification of peptides mapping to genomic regions previously annotated as non-coding allowed for the correction of annotations, leading to the N-terminal extension of protein sequences and the uncovering of eight novel protein-coding genes. The alliance of proteomics, genomics, and transcriptomics has resulted in a powerful combination for improving the annotation of the *L. infantum* reference genome.

Keywords: *Leishmania infantum*; proteome; post-translational modifications (PTMs); proteogenomics; mass spectrometry

---

1. Introduction

The genus *Leishmania* belongs to the order Trypanosomatida and includes protozoan parasites that are responsible for a complex of diseases named leishmaniasis, which is the second most common cause of mortality among tropical infectious diseases, after malaria [1]. Some species of *Leishmania* are human pathogens that cause different clinical manifestations as cutaneous (CL), mucocutaneous (MCL), or visceral (VL) leishmaniasis. Old World species *Leishmania infantum* and *Leishmania donovani* cause VL or kala-azar, which is often lethal if untreated, whereas *Leishmania major* causes CL and *Leishmania braziliensis* is associated with MCL. The VL-causative species are genetically almost identical, although they differ in geographic distribution: *L. donovani* is found in the Indian subcontinent and East Africa, while *L. infantum* is endemic in the countries around the Mediterranean basin, Latin America, and China [2]. Two stages, promastigote and amastigote, alternate in the *Leishmania* life cycle. Promastigotes are flagellated and motile forms develop extracellularly in the gut of their sand-fly vector. The infection of the mammal host takes place during the sand-fly blood meal; afterwards, parasites are phagocytized by macrophages, and amastigote forms develop inside these host cells.

Within the *Leishmania* genus, the *L. major* genome was the first to be sequenced [3], followed by the *L. infantum* and *L. braziliensis* ones [4]. During the last decade, the extraordinary progress in DNA sequencing methodologies has allowed for the drafting of the genomes for many other *Leishmania* species [5–12] and for the improvement of the assemblies of the first sequenced genomes [13–15].

The availability of a complete and well-annotated genome provides the ultimate resource for genome-wide scale approaches, such as transcriptome and proteome analyses [16]. In parallel to the advances in sequencing technologies, proteomics methodologies are achieving unprecedented levels of sensitivity, and novel MS-based experimental approaches have become the method of choice for the analysis of complex protein mixtures such as cells, tissues, and even whole organisms. In particular, several proteomic technologies are being used to study diverse aspects of *Leishmania* biology such as parasite development, virulence, and drug resistance [17]. Thus, proteomics approaches have been used to determine differential patterns of protein expression between the promastigote and amastigote stages in *Leishmania mexicana* [18], *L. infantum* [19], and *L. donovani* [20], among others. Other studies have been aimed to ascertain specific proteomes by means of organelle fractionation to obtain enriched fractions of mitochondria, flagella, or glycosomes [21,22]. The identification and mapping of protein post-translational modifications (PTMs) provide additional information about the activation of specific pathways in a given growing condition, thus improving the knowledge on protein interactions in complex networks.

Here, we present a wide and detailed proteome of the *L. infantum* JPCM5 strain, based on axenically grown promastigotes in the logarithmic growth phase. A careful comparative analysis with other published proteomes from different Old and New World *Leishmania* species has also been carried out. Additionally, the MS data allowed for the identification of PTMs (phosphorylation, acetylation, methylation, formylation, and glycosylation) at specific protein sites that might have regulatory functions. Furthermore, we showed how the integration of proteomics with genomic and transcriptomic data represents a powerful and complementary strategy for gene annotation, as demonstrated before in a plethora of species [23]. Hence, by applying this proteogenomic approach, it was possible to improve the annotations for several *L. infantum* genes, as well as the identification of eight novel genes.

2. Materials and Methods

2.1. Leishmania Infantum Culture and Protein Extraction

L. infantum JPCM5 strain parasites were grown at 26 °C in Roswell Park Memorial Institute (RPMI) 1640 medium supplemented with 15% of heat inactivated fetal calf serum (Biowest SAS, Nuaillé, France). Promastigote cultures were initiated at 1×10^6 parasites/mL and harvested at the mid-logarithmic phase ($1–2 \times 10^7$ parasites/mL). Around $1–2 \times 10^8$ parasites were collected and washed twice with phosphate-buffered saline (PBS); finally, parasites were suspended by pipetting in 300 µL of a RIPA (RadioImmunoPrecipitation Assay) lysis buffer (Thermo Fisher Scientific, Rockford, IL, USA) in the presence of EDTA-free Easy Pack Protease inhibitor (Roche, Diagnostics, Mannheim, Germany). After 6 cycles (30 s pulse/30 s pause) of sonication in a bath at 4 °C, samples were incubated for 90 min at 4 °C, and, afterwards, protein lysates were centrifuged at 14,000 g for 30 min. The supernatant was collected and used for proteomics analyses.

2.2. In-Gel and In-Solution Digestion of Samples by Trypsin and Chymotrypsin

For the in-gel digestion of proteins, samples were mixed with an equal volume of a 2× Laemmli buffer and loaded onto 1.2-cm wide wells of a conventional SDS-PAGE gel (0.75 mm-thick, 4% polyacrylamide stacking-gel, and 10% polyacrylamide resolving-gel). The electrophoresis was stopped as soon as the electrophoretic front entered 3 mm into the resolving gel, so the proteins became concentrated in the stacking/resolving gel interface. After Coomassie staining, the protein-containing gel was cut into small pieces (2 × 2 mm cubes) and placed into a microcentrifuge tube, as described elsewhere [24]. The gel pieces were destained in acetonitrile:water (ACN:H$_2$O, 1:1), reduced and alkylated (disulfide bonds from cysteinyl residues were reduced with 10 mM dithiothreitol (DTT) for 1 h at 56 °C, and then thiol groups were alkylated with 10 mM iodoacetamide for 1 h at room temperature in darkness), and digested in situ with sequencing grade trypsin (Promega, Madison, WI) or chymotrypsin (Roche Diagnostics), as described by Shevchenko et al. [25], with minor modifications. The gel pieces were shrunk by removing all liquid using sufficient ACN. Acetonitrile was pipetted out, and the gel pieces were dried in a speedvac. The dried gel pieces were re-swollen in 100 mM Tris-HCl and 10 mM CaCl$_2$ at pH 8 with 60 ng/μL trypsin or chymotrypsin at a 5:1 protein:enzyme (*w*/*w*) ratio. The tubes were kept on ice for 2 h and incubated at 37 °C (trypsin) or 25 °C (chymotrypsin) for 12 h. Digestion was stopped by the addition of 1% trifluoroacetic acid (TFA). Whole supernatants were dried down and then desalted onto OMIX C18 pipette tips (Agilent Technologies, Santa Clara, CA, USA) before the MS analysis.

Additionally, in-solution digestion was performed as described elsewhere [26]. After the denaturation of proteins with an 8 M urea, the protein sample was reduced and alkylated: disulfide bonds from cysteinyl residues were reduced with 10 mM DTT for 1 h at 37 °C, and then thiol groups were alkylated with 50 mM iodoacetamide for 1 h at room temperature in darkness. The sample was diluted to reduce urea concentration below 1.4 M and digested using sequencing-grade trypsin (Promega, Madison, WI, USA) or chymotrypsin (Roche Diagnostics) overnight at 37 °C (trypsin) or 25 °C (chymotrypsin) using a 1:20 (*w*/*w*) enzyme/protein ratio. Digestion was stopped by the addition of 1% TFA. Whole supernatants were dried down and then desalted onto OMIX C18 pipette tips (Agilent Technologies) before the MS analysis.

2.3. Reverse Phase-Liquid Chromatography Mass Spectrometry Analysis (RP-LC-MS/MS)

The digested protein samples (above) were resuspended in 10 μL of 0.1% formic acid and analyzed by RP-LC-MS/MS in an Easy-nLC II system coupled to an ion trap LTQ-Orbitrap-Velos-Pro hybrid mass spectrometer (Thermo Fisher Scientific). The peptides were concentrated (on-line) by reverse phase chromatography using a 0.1 × 20 mm C18 RP precolumn (Thermo Fisher Scientific) and then separated using a 0.075 × 250 mm C18 RP column (Thermo Fisher Scientific) operating at 0.3 μL/min. Peptides were eluted using a 180-min dual gradient. The gradient profile was set as follows: 5–25% solvent B for 135 min, 25–40% solvent B for 45 min, 40–100% solvent B for 2 min, and 100% solvent B for 18 min (Solvent A: 0.1% formic acid in water; solvent B: 0.1% formic acid and 80% acetonitrile in water). ElectroSpray ionization (ESI) was done using a nano-bore emitter stainless steel ID 30 μm (Proxeon) interface. The Orbitrap resolution was set at 30,000. Peptides were detected in survey scans from 400 to 1600 amu (1 μscan), followed by twenty data-dependent MS/MS scans (Top 20) using an isolation width of 2 u (in mass-to-charge ratio units), a normalized collision energy of 35%, and a dynamic exclusion that was applied during 60 s periods.

2.4. Data Analysis

Peptide identification from raw data was carried out using the PEAKS Studio X search engine (Bioinformatics Solutions Inc, Waterloo, ON, Canada). A custom Python script was used to create a database comprising all possible open reading frames (ORF) coding for protein sequences of ≥20 amino acids existing in any of the six-frames in the *L. infantum* JPCM5 strain genome sequence [13]. This database (named LINF-all-ORFs) consisted of 294,654 entries. In parallel, a fusion-database, created by merging the *L. infantum* protein sequences annotated in UniProt and the LINF-all-ORFs entries, was also used by the search engine. Finally, a search against a decoy database (decoy fusion-database) was also performed. The following constraints were used for the searches: tryptic or chymotryptic cleavage (semispecific), up to two missed cleavage sites, tolerances of 20 ppm for precursor ions and 0.6 Da for MS/MS fragment ions, and optional Met oxidation and Cys carbamidomethylation were allowed. The false discovery rates (FDRs) for peptide spectrum matches

(PSMs) were limited to 0.01 or lower. Those proteins that were identified with at least two distinct peptides were considered for further analysis [27–29].

The LINF-all-ORFs entries with mapped peptides were compared with the annotated *L. infantum* proteins (UniProt database) using in-house Python scripts in order to identify both the misannotations and novel proteins. Additionally, Python scripts were used to ascribe post-translational modifications to particular protein entries.

Functional categories and enzymatic pathways using the DAVID program (Functional Annotation Tool, DAVID Bioinformatics Resources 6.8) and the KEGG Pathway (Kyoto Encyclopedia of Genes and Genomes) were used for the classification of the proteins identified by MS.

3. Results and Discussion

3.1. Protein Identification from the LC–MS/MS Peptide Spectra

The main objective of this study was to obtain the experimental proteome of the *L. infantum* promastigote stage with the additional aim of improving current genomic annotations. For this purpose, a recently published re-sequenced genome [13] was used. However, we did not restrict the search of peptide spectra on currently annotated protein-coding genes; instead, a database consisting of all possible polypeptides (equal or larger than 20 amino acids) was used (see Materials and Methods for further details). The workflow used for sample preparation and proteomics is shown in Figure 1. From the MS/MS data, it could be seen that only those associated with peptides longer than seven amino acids were considered for protein identification. Among the identified proteins, 2344 proteins matched with previously annotated proteins [13]. Moreover, eight novel proteins were uncovered, thus legitimating the search strategy. In addition, some ORFs had to be extended to accommodate the MS-identified peptides (see below for further details about these findings). Most of the proteins—70.5% (1659 out of 2352)—were identified by three or more unique peptides per protein, 14.5% (341) of the protein identifications were supported by two unique peptides, and only 15% (352) of the identifications were done by a single unique peptide. Currently, 3482 out of 8590 annotated proteins (around 40%) in the *L. infantum* genome have the status of hypothetical proteins; the MS spectra obtained in this work provided experimental evidence of the real existence for 456 of those hypothetical proteins.

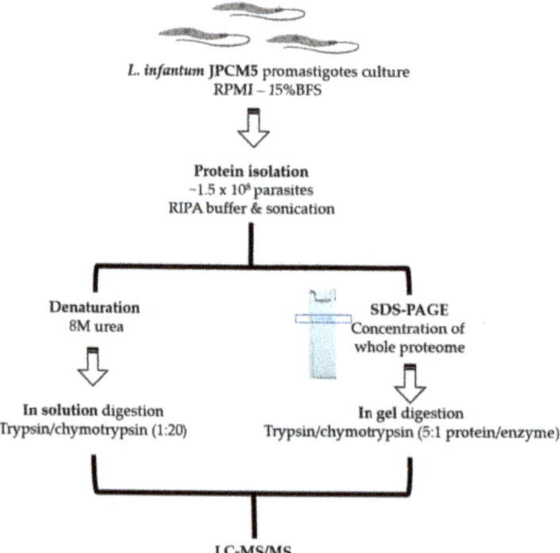

Figure 1. Workflow for protein extraction and proteomic analyses of *Leishmania infantum* promastigotes. The experimental MS data were searched against the UniProt protein database and a database consisting of all possible polypeptides encoded in the six-frames of the *L. infantum* genome (based on v2/2018; www.leish-esp.cbm.uam.es; [13]).

The first comprehensive study aimed to characterize the *L. infantum* proteome was carried out by the Papadopoulou's group [30]. Using two-dimensional (2D) gel electrophoresis, these authors visualized 2261 protein spots in promastigote samples and 2273 spots in amastigote ones. However, after MS analysis, only 168 protein spots, derived from 71 different genes, could be identified [30]. A better proteome resolution was attained after a fractionation step including digitonin extraction; hence, 153 *L. infantum* proteins were identified by MS analysis of selected spots [31]. The combination of two-dimensional liquid chromatography (2DLC), electrospray ionization mass spectrometry (2DLC-ESI-MS), and 2DLC-matrix-assisted laser desorption/ionization mass spectrometry (2DLC-MALDI-MS) allowed Leifso and co-workers to identify 91 *L. infantum* proteins [19]. An enrichment for basic proteins using the technique of free-flow electrophoresis prior to separation by 2D gel electrophoresis led to the identification of around 200 *L. infantum* proteins [32]. Alcolea and coworkers [33] identified 28 proteins in a proteomic study aimed to uncover differentially expressed proteins between the early-logarithmic and the stationary phases during the culturing of *L. infantum* promastigotes. In two different studies using MS analysis of the exoproteome derived from *L. infantum* promastigote cultures, a total of 102 [34] and 494 [35] proteins were identified. Therefore, our work provides the most complete, to date, experimentally evidenced proteome for *L. infantum*.

Outstanding studies on proteome identification have been performed in both *L. donovani* and *L. major*. In 2008, Rosenzweig and collaborators reported the identification of 1713 proteins in *L. donovani* [20]. A comparison between the proteins identified in our work (*L. infantum* JPCM5) and those identified in *L. donovani* showed that 1218 proteins were common (orthologs) in both studies (Figure 2). We failed to identify 207 proteins of those reported in *L. donovani*, whereas we found 1130 proteins that are absent from the *L. donovani* proteome reported by Rosenzweig et al. [20]

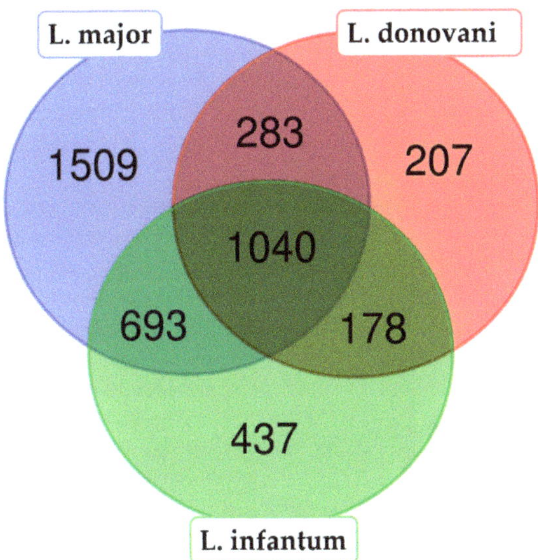

Figure 2. Comparison (Venn diagram) between the identified proteins in this work (*L. infantum*, in green) and those identified in two previous studies [20,36] performing proteomic analysis in *Leishmania donovani* (in red) and *Leishmania major* (in blue). The Venn diagram was created by the tool available at bioinformatics.psb.ugent.be/webtools/. Note—the discrepancy between the number of proteins identified by Rosenzweig et al. [20] (see text) and those represented in the Venn diagram (1713 vs. 1708) was due to 5 gene duplications that were corrected after re-assembling of the *L. infantum* genome [13].

More recently, Pandey and coworkers reported the identification of 3386 different proteins in *L. donovani* promastigote and amastigote stages [37,38]. After comparing their data and the proteins identified in this study, 1650 of the proteins observed in *L. infantum* promastigotes were found to be present (their orthologues) in the *L. donovani* promastigote proteome. However, among the 613 proteins that Nirujogi et al. [37] reported to be exclusively expressed in *L. donovani* amastigotes, 126 proteins were also identified in our proteomics study, thus indicating that these proteins are also being expressed in the promastigotes stage, at least in *L. infantum* (see Supplementary File, Table S1). Most of them were annotated as hypothetical proteins or with unknown function, but there are also metabolic enzymes, translation machinery components (ribosomal proteins and eukaryotic initiation factors), and RNA binding proteins.

In 2014, Pawar et al. [36] reported a quite wide proteome of the *L. major* promastigote stage, in which 3613 proteins were identified. These authors followed a proteogenomic approach, as we did in this study, consisting of searching the mass spectra against a six-frame translated database generated from a complete genome sequence. An orthology-based comparison indicated that the *L. major* promastigote proteome and the *L. infantum* proteome of this study shared 1733 proteins (Figure 2). Moreover, considering the 1792 proteins identified in the *L. major* proteome, though not in our study, and the 615 proteins exclusively identified by us in the *L. infantum* proteome, the total number of identified proteins presumably expressed in the promastigote stage is 4140 (roughly half of the predicted proteins to be encoded in the *Leishmania* genome).

3.2. Representativeness of the Translational Machinery and RNA Binding Proteins in the L. infantum Experimental Proteome

Around 200 proteins from the *L. infantum* promastigote proteome were categorized as components of the translational machinery: 122 ribosomal proteins, 51 translation regulatory factors, and 24 tRNA synthases (Supplementary File, Table S2). As expected for a highly proliferative stage (the promastigotes were growing in the logarithmic phase when harvested for analysis), in which protein synthesis needs to be very active, all the ribosome components, tRNA-loading enzymes, and regulatory factors were abundant and easily detected by mass spectrometry. Nevertheless, in contrast, very few of the annotated mitoribosomal proteins [39] were identified in the *L. infantum* promastigote proteome. This observation may indicate that mitochondrial ribosomes are in relatively low amounts when promastigotes are grown in nutrient-rich culture media.

Proteins with RNA binding properties deserve special attention, since gene expression in *Leishmania* and related trypanosomatids is essentially being controlled at the post-transcriptional level [40,41]. In this scenario, RNA-binding proteins are key players in controlling RNA metabolism [42–44]. In the *L. infantum* promastigote proteome, a large number of known RNA binding proteins were detected (Supplementary File, Table S3). Apart from the mentioned ribosomal proteins and translation factors, 15 RNA helicases were detected, as well as many of the RNA-binding domain-containing proteins reported in a recent study aimed to the capture and identification of RNA-bound proteins in *L. donovani* [45]. The RNA-binding proteins of the Pumilio family (aka PUF proteins) are especially numerous (11 members) in *Leishmania* [46]. In this study, we identified 6 out of 11 PUF proteins that are being expressed in the promastigote stage of *L. infantum*; these are PUF 1, PUF 4, PUF 6, PUF 7, PUF 8, and PUF 10 (see Table S3 to see their gene IDs).

3.3. Metabolic Enzymes and Pathways

Going deeper, we performed an in-silico pathway reconstruction using the detected proteins in the *L. infantum* promastigote proteome. By using the KEGG database resource (http://www.genome.jp/kegg/) accessed via the DAVID package, a total of 578 (27.6%) of the detected proteins could be classified into pathways representing classical cellular processes. In particular, 236 proteins were identified as metabolic enzymes; 31% of these enzymes belong to glycolysis (Tables 1 and 2), the tricarboxylic acid (TCA) cycle, and the pentose phosphate cycle (Supplementary File, Table S4), which are three metabolic pathways playing essential maintenance functions in the cell [47]. Remarkably, the complete set of 29 enzymes that make up the TCA cycle were identified in the promastigote proteome (Figure 3).

Table 1. List of glycosomal enzymes related to gluconeogenesis and glycolysis identified in *L. infantum* (according to the Kyoto Encyclopedia of Genes and Genomes (KEGG) pathway database).

Gene ID	Unique Peptides	Description
LINF_040016700	12	Fructose-1-6-bisphosphatase
LINF_120010600	43	Glucose-6-phosphate isomerase
LINF_200006000	41	Phosphoglycerate kinase C-glycosomal
LINF_210007800	47	Hexokinase
LINF_210012000	13	Phosphoglucomutase
LINF_230009500	10	Aldose 1-epimerase-like protein
LINF_240013700	32	Triosephosphate isomerase
LINF_250017300	46	Aldehyde dehydrogenase—mitochondrial precursor
LINF_270024900	56	Glycosomal phosphoenolpyruvate carboxykinase
LINF_290032900	23	ATP-dependent phosphofructokinase
LINF_300035000	49	Glyceraldehyde 3-phosphate dehydrogenase—glycosomal
LINF_300039500	13	PAS-domain containing phosphoglycerate kinase
LINF_340040800	13	Aldose 1-epimerase-like protein

Table 2. List of cytosolic enzymes related to gluconeogenesis and glycolysis identified in the *L. infantum* proteome (according to the KEGG pathway database).

Gene ID	Unique Peptides	Description
LINF_140018000	55	Enolase
LINF_180019200	23	Pyruvate dehydrogenase e1 component α subunit
LINF_210011100	10	Dihydrolipoamide acetyltransferase
LINF_210012000	13	Phosphoglucomutase
LINF_230009000	43	NADP-dependent alcohol dehydrogenase
LINF_230014800	47	Acetyl-CoA synthetase
LINF_250023800	29	Pyruvate dehydrogenase e1 β subunit
LINF_290025700	4	Dihydrolipoamide dehydrogenase
LINF_310034500	2	Dihydrolipoamide dehydrogenase
LINF_320040600	37	Dihydrolipoamide dehydrogenase
LINF_350005300 LINF_350005400	43	Pyruvate kinase
LINF_360030600	43	Glyceraldehyde 3-phosphate dehydrogenase—cytosolic
LINF_360034400	19	Dihydrolipoamide acetyltransferase precursor

The glycosome is a trypanosomatid-specific, membrane-enclosed organelle that contains glycolytic enzymes, among others. Thus, glycolysis in *Leishmania* takes place in these organelles for the steps between glucose and 3-phosphoglycerate [48], as well as in the cytosol for those late steps leading to the formation of pyruvate [20]. The identified enzymes involved in these two stages are listed in Tables 1 and 2. Among them, there are 32 enzymes belonging to the glycosomal/cytosolic glycolysis (and gluconeogenesis) pathway until the formation of pyruvate by pyruvate kinases (IDs LINF_350005400 and LINF_350005300). Some enzymes involved in the mitochondrial electron transport respiratory chain were detected (Supplementary File, Table S5). Similar findings were found by Rosenzweig and collaborators in the *L. donovani* promastigote proteome [20]. Several proteins of the electron transport chain are encoded by the kinetoplast DNA maxicircles [49] such as cytochrome oxidase subunits and NADH dehydrogenase, but they were not searched in this study.

An active energy metabolism requires enzymes to be involved in redox homeostasis. Several of these enzymes have been identified in the *L. infantum* proteome, and they are likely abundant, as judged by the large number of unique peptides that were mapped to them. The detected proteins were tryparedoxin (LINF_150019000, with 31 unique peptides), peroxidoxin (LINF_230005400, 31 peptides), cyclophilin (LINF_060006300, 20 peptides), iron superoxide dismutases (LINF_080007900 and LINF_320024000, with 14 and 18 peptides, respectively), and several elongation factors 1β (LINF_340014200 and LINF_340014000 with 16 peptides each and LINF_360020500 with 19 peptides).

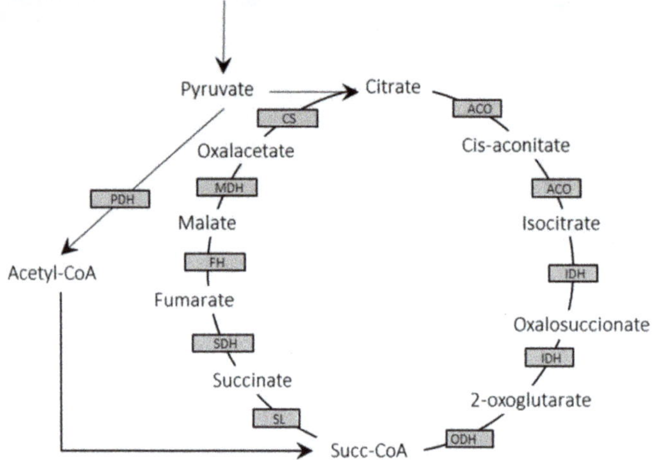

Figure 3. Detected enzymes in the *L. infantum* JPCM5 proteome composing the complete tricarboxylic acid (TCA) cycle. PDH: pyruvate dehydrogenase; ACO: aconitase; IDH: isocitrate dehydrogenase; ODH: 2-oxoglutarate dehydrogenase; SL: succinyl-CoA ligase; SDH: succinyl dehydrogenase; FH: fumarate hydratase; MDH: malate dehydrogenase; and CS: citrate synthase.

3.4. Components of the Proteostasis Network

The proteasome is a complex of multi-subunit proteases, associated with protein degradation, but in protozoan parasites, it has been also involved in cell differentiation and replication processes [50]. In fact, proteasomal inhibitors have been described as promising therapeutic targets for leishmaniasis and trypanosomiasis [51,52]. According to the KEGG database, the complete compendium of proteasomal proteins were identified in this study (Supplementary File, Table S6).

Protein degradation and protein folding cooperate to maintain protein homeostasis or proteostasis [53]. Multiple and drastic environmental changes (pH variation, sudden temperature up-/down-shifts, and oxidative stress) occur along the *Leishmania* life cycle. Most often, these environmental insults promote protein unfolding and aggregation; to counteract these effects, cells possess specialized molecular chaperones (or heat shock proteins: HSPs) that serve as central integrators of protein homeostasis. Not surprisingly, *Leishmania* parasites possess a large number and variety of molecular chaperones [54]. In this study, we identified proteins belonging to the different HSP families: HSP100, HSP83/90, HSP70, HSP60, HSP40/DnaJ, and HSP20 (listed in Supplementary File, Table S7). Mitochondrion is a cellular organelle in which molecular chaperones are of particular relevance because they are involved in protein transport across membranes and protein refolding inside the mitochondria. Recently, the mitochondrial proteome was analyzed in *L. tropica* [22]. Taking advantage of that study, in Table 3, we list those HSPs identified in the *L. infantum* proteome that are potentially mitochondrial proteins.

Table 3. Identified molecular chaperones of *L. infantum* promastigotes, with putative mitochondrial location (according to Tasbihi et al. [22]).

Gene ID	Unique Peptides	Molecular Chaperone
LINF_260011400	9	Heat shock protein 10 (HSP10)
LINF_320040500	5	HSP40/JDP45
LINF_350035100	7	HSP40/JDP50
LINF_360027100	19	HSP60/cpn60.2
LINF_360027200	26	HSP60/cpn60.3
LINF_240010000	5	HSP40/JDP8
LINF_260017400	42	HSP70.4
LINF_280017800	51	Grp78/BiP
LINF_330033000	48	HSP75/TRAP-1
LINF_020012400	23	HSP78
LINF_330009000	22	HSP83/90

3.5. Glycosomal Proteins Represent a Substantial Fraction of the Experimentally Detected Proteins in the L. infantum Promastigote

As mentioned above, glycosomes are specialized peroxisomes that contain key enzymes involved on energy metabolism and purine salvage [55]. Moreover, as occurs in peroxisomes, glycosomal proteins are targeted for import to and location in glycosomes by the presence of the peroxisomal targeting signals (PTSs) PTS1 and PTS2. Two essential proteins for targeting newly synthesized proteins, with a PTS2 import signal, into the glycosome are peroxins (PEXs) 5 and 7 [56]. Remarkably, both proteins, PEX5 (LINF_350019100) and PEX7 (LINF_290012400), were identified in the experimental proteome of *L. infantum* promastigotes. Moreover, we made a direct comparison between the *L. infantum* proteome reported here and two studies focused on glycosomal proteomes in *Leishmania tarentolae* and *L. donovani* [48,57]. Colasante and collaborators [48] identified 464 proteins in a glycosomal membrane preparation from *L. tarentolae*, and they concluded that 258 would be glycosomal proteins, including 40 enzymes. Interestingly, the orthologs of 165 (64%) of these proteins were experimentally identified in the *L. infantum* promastigote proteome. In particular, complete enzymatic complements involved in glycosomal glycolysis and gluconeogenesis steps were identified in both studies. Jamdhade and coworkers [57] reported the proteome analysis of an enriched glycosome fraction from *L. donovani* promastigotes, identifying 1355 proteins. In our study, orthologs to 853 of those putative *L. donovani* glycosomal proteins were found; these are listed in Supplementary File, Table S8.

The purine salvage pathway, essential for trypanosomes, also takes place in glycosomes [58]. Notably, the 13 enzymes composing the purine salvage pathway (Figure 4) [58] were identified in this *L. infantum* experimental proteome, which was in agreement with the relevance of this metabolic route for parasite survival. In this regard, it is somewhat unexpected that only two enzymes (adenylosuccinate synthetase (ADSS) and inosine monophosphate dehydrogenase (IMPDH)) were identified in the *L. tarentolae* glycosomal proteome, and five of these enzymes were identified in the glycosomal fractions of *L. donovani*. Similarly, we identified most of the enzymes constituting the de novo pyrimidine biosynthesis pathway (Table 4) [59] in the *L. infantum* proteome, whereas Jamdhade et al. [57] only found one enzyme of this pathway in the *L. donovani* glycosomal proteome—the orotate phosphoribosyltransferase (LDBPK_160560).

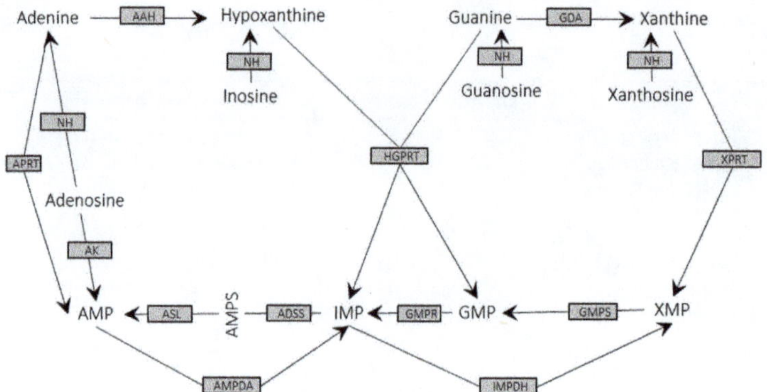

Figure 4. Enzymes from the purine salvage cycles identified in the *L. infantum* experimental proteome. All the enzymes (grey squares) that are required to complete the pathway were identified in this study. APRT: adenine phosphoribosyltransferase (LINF_130016900); NH: nucleoside hydrolase (LINF_180021400); AK: adenosine kinase (LINF_300014400); AAH: adenine aminohydrolase (LINF 350026800); ASL: adenylsuccinate lyase (LINF 040009600); AMPDA: AMP deaminase (LINF 130014700); ADSS: adenylosuccinate synthetase (LINF 130016900); GMPR: GMP reductase (LINF 170014800); GMPS: GMP synthase (LINF 220006100); HGPRT: hypoxanthine-guanine phosphoribosyltransferase (LINF 210014900); GDA: guanine deaminase (LINF 290014000); IMPDH: inosine monophosphate dehydrogenase (LINF 190022000); and XPRT: xanthine phosphoribosyltransferase (LINF 210015000). The cycle was depicted according to Boitz et al. [58].

Table 4. Enzymes involved in de novo pyrimidine biosynthesis identified in *L. infantum* promastigote proteome.

Gene ID	Unique Peptides	Description
LINF_060011200	7	Deoxyuridine triphosphatase
LINF_160010400	5	Dihydroorotate dehydrogenase (fumarate)
LINF_160010500	15	Aspartate carbamoyltransferase
LINF_160010700	28	Orotate phosphoribosyltransferase
LINF_160011200	6	Carbamoyl-phosphate synthase
LINF_180021400	17	Nonspecific nucleoside hydrolase
LINF_340016700	8	Uracil phosphoribosyltransferase

3.6. Exoproteome Components Identified in the L. infantum Experimental Proteome

Leishmania-secreted molecules and exosomes are particularly relevant for infection establishment because parasites and exosomes are co-egested during the insect blood meal [60]. A detailed analysis of the *L. infantum* secreted proteins (exoproteome) was carried out by Santarem et al. [35]. These authors found that the proteome profiles were distinct depending on the metabolic stage of the parasites (logarithmic or stationary phase promastigotes). The number of distinct proteins identified in that study was 297, and around 90% of them were also identified in the proteome reported here. In another outstanding study, Atayde et al. [61] analyzed the proteomic composition of *L. infantum* exosomes and extracellular vesicles that were directly isolated from the sand fly midgut. Table 5 lists proteins commonly present in exosome preparations; all of them were identified in the *L. infantum* experimental proteome reported here.

Table 5. Common components of *Leishmania* exosomes identified in this *L. infantum* proteome study.

Gene ID	Description	Features [Ref.]
LINF 050017500; LINF 040007000	Surface antigens	Virulence factor [61]
LINF_090013900	Oligopeptidase b	Virulence factor [35]
LINF_100010100	GP63-leishmanolysin	Virulence factor [60,61]
LINF_120014700	Surface antigen protein 2	Virulence factor [61]
LINF_140018000	Enolase	Virulence factor [35,62]
LINF_150019000	Tryparedoxin peroxidase	Virulence factor [61]
LINF_170005900	Elongation factor 1-α	Exosome marker [61]
LINF_190020600	Cysteine peptidase A (CPA)	Virulence factor [61]
LINF_200018000	Calpain-like cysteine peptidase	Virulence factor [61]
LINF_280035000 LINF_280036000	HSP70	Exosome marker [61]
LINF_280034700	Receptor for activated C kinase 1	Immunomodulator [35]
LINF_320036700	Nucleoside diphosphate kinase b	Immunomodulator [35]
LINF_330009000	HSP83/90	Exosome marker [61]
LINF_350027300 LINF_350027500	Kinetoplastid membrane protein 11 (KMP11)	Immunomodulator [35]
LINF_360018400	Fructose-1-6-bisphosphate aldolase	Immunomodulator [60]

3.7. Other Relevant Proteins Identified in the L. infantum Promastigote Proteome

Proteins with a high molecular weight (HMW) represent a challenge for mass spectrometry-based assays, as they are usually underrepresented in protein extracts used for proteomic analysis. To overcome this limitation, Brotherton et al. [63] optimized extraction protocols to enrich HMW proteins and membrane proteins in *L. infantum* promastigotes and amastigotes. In our study, we confirmed the presence of tryptic and/or chymotryptic peptides from 35 HMW proteins with a molecular weight (MW) higher than 200 kDa (Supplementary File, Table S9). Among them, the identification of a calpain-like cysteine peptidase was remarkable, as it had an estimated MW of around 700 kDa (LINF_270010200) and was identified by 124 unique peptides, thus covering 24% of the amino acid sequence.

The flagellum is a characteristic organelle of *Leishmania* that confers motility to the parasite in the promastigote stage, during which this structure is particularly prominent. In a recent publication, an exhaustive structural and functional characterization of the *L. mexicana* promastigote flagellum was reported [64]. In that study, flagella preparations were analyzed by proteomics, and this allowed for the identification of 701 unique proteins for this organelle. Orthologues to around 400 flagellum-specific proteins were identified in the *L. infantum* proteome described here. More importantly, most of the proteins relevant for flagellum assembly and motility in *L. mexicana* promastigotes [64] were identified in the experimental proteome of *L. infantum* promastigotes (Supplementary File, Table S10).

3.8. Detection of Post-Translational Modifications

The PTMs of proteins influence their activity, structure, turnover, localization, and capacity to interact with other proteins. In *Leishmania*, PTMs, together with mRNA stability and translation processes, are the essential regulators of gene expression. In this study, based on MS/MS spectra, a significant number of phosphorylated, methylated, acetylated, glycosylated, and/or formylated proteins were identified in the *L. infantum* proteome. Thus, even though specific enrichments of modified proteins were not performed, we identified modified peptides that accounted for 10 phosphorylated, 144 methylated, 192 acetylated, 28 formylated, and 3 glycosylated proteins.

The phosphorylation of serine (S), threonine (T), and tyrosine (Y) amino acids implies an increase of 79.97 Da in their molecular weights (unimod.org). The phosphorylated proteins identified in this study and the modified residues are listed in Table 6. Apart from two unknown phosphoproteins (LINF_040005600 and LINF_220013200), the ribosomal protein S10, α tubulins, an rRNA biogenesis protein-like protein, the 3-ketoacyl CoA-thiolase, the flagellar protein KHARON1, the glycogen synthase kinase 3 (GSK-3), and the prototypical HSP70 might be regulated by phosphorylation (Table 6). In fact, the phosphorylation of HSP70 has been reported to occur during the stress response in both promastigotes and amastigotes of *L. donovani* [65].

Table 6. Phosphoproteins in the *L. infantum* proteome.

Gene ID	Description	Position
LINF_040005600	Hypothetical protein-conserved	T187
LINF_130007700; LINF_130007800; LINF_130008000; LINF_130008200; LINF_130008300; LINF_130008400; LINF_130008600; LINF_130008700	α tubulin	T334; Y357
LINF_180007700	Glycogen synthase kinase 3 (GSK-3)	Y186
LINF_190017000	Hypothetical protein-conserved	S120
LINF_200011800	rRNA biogenesis protein-like protein	Y571
LINF_220013200	Hypothetical protein-conserved	S45
LINF_230014600	3-ketoacyl-CoA thiolase	S229
LINF_280035400	HSP70	T159; T164
LINF_360015600; LINF_360015700	40S ribosomal protein S10	S157
LINF_360068400	Flagellum targeting protein KHARON1	S158

Most of the observed phosphorylation events occurred on S and T, but in some proteins, phosphorylation on Y residues was also detected. Rosenzweig et al. [66] identified 16 phosphorylated proteins in *L. donovani* in either promastigotes or amastigotes; in that study, all of phosphorylations occurred at S or/and T residues. No coincidences exist between the phosphoproteins identified by these authors and those identified in this study (Table 6); however, this was not unexpected because phosphorylation is a dynamic modification and the numbers of phosphorylated proteins identified in both studies were low. Though the occurrence of phosphorylation events has been proven to be much lower in tyrosine residues [67], it is remarkable that phosphorylated tyrosines were found in α tubulins, the rRNA biogenesis protein, and in GSK-3 (Table 6). Some of these phosphorylated proteins have also been identified in previous studies. Thus, for instance, the 3-ketoacyl-CoA thiolase was found to be phosphorylated (at serine 229) in *L. donovani* promastigotes [68]. Kinases and phosphatases are enzymes implicated in the regulation of phosphorylation/dephosphorylation processes, and, accordingly, several serine, threonine, and tyrosine kinases and phosphatases have been identified in the *L. infantum* proteome (Supplementary File, Table S11).

Methylation (+14 Da) is a physiological PTM that occurs at the C- and N-terminal ends of proteins, and on the side chain nitrogen of arginine (R) and lysine (K); this modification is critical for regulating several cellular processes. Apart from those amino acids, methylations have been found to occur in other amino acids like aspartic acid (D), glutamic acid (E), histidine (H), glutamine (Q), and asparagine (N) [69]. In the *L. infantum* proteome, 139 proteins were predicted to be methylated, 76 of them showed methylation at K or R residues, 123 of the modified proteins showed D and/or E methylated residues, and a methylated-H was found in β-tubulin. All the methylated proteins detected in this study are listed in Supplementary File, Table S12. In summary, our findings pointed out that methylation at D and E residues would be relatively frequent in *Leishmania*; as suggested by Sprung et al. [69], methylations at E and D residues would increase the hydrophobicity of the modified proteins. Some examples of highly methylated proteins identified in this study are α and β tubulins, heat shock proteins HSP70 and HSP83/90, and the elongation factor 2 (eEF2). Many orthologs to the methylated proteins detected in this study were also identified as methylated in *L. donovani* [66].

Acetylation (+42.02 Da) is a PTM considered as relevant as phosphorylation in metabolic and signaling pathways. K acetylation has been described as a reversible enzymatic reaction that regulates protein function, as it is particularly relevant in chromatin compaction by the acetylation of histones [70]. Interestingly, the accumulation of acetylated histones has been observed at the polycistronic transcription initiation sites in *L. major* and *Trypanosoma cruzi* [71,72]. However, in the *L. infantum* proteomics data, peptides bearing acetylated K belonging to histones were not identified, as was the case in the *L. donovani* proteome [66], thus suggesting a relatively low proportion of acetylated histones in the bulk of total cellular histones. Some examples of proteins detected as acetylated in this study are β tubulins (LINF_330015100, LINF_330015200, and LINF_210028500; modified at K297), guanylate kinase (LINF_330018400, K3), the subunit β of ATP synthase (LINF_250018000; K511) and a calpain-like cysteine peptidase (LINF_140014400; K74).

In addition, acetylation at the N-terminal ends of proteins may occur either co- or post-translationally, as it is a frequent modification in eukaryotic proteins even though their physiological consequences remain poorly understood [73]. This irreversible modification affects protein fate in the cell and is carried out by N-terminal acetyltransferases (Nat). In the *L. infantum* promastigote proteome, we identified three of these enzymes: Nat-1, Nat-B, and Nat-C (Supplementary File, Table S11). On the other hand, among the 144 N-terminally acetylated proteins identified in this study (Supplementary File, Table S13), 40 proteins showed acetylation at their initial methionine (iM), and 104 were acetylated at the second amino acid, suggesting a cleavage of the iM during protein maturation [74]. In the cases in which acetylation takes place at the iM, we detected a bias in the amino acids located behind the iM. Thus, in 40% of the acetylated proteins, the second amino acid was the polar non-charged N or Q residues (Figure 5A also shows the other more frequent amino acids located at the second position). An acetylation reaction after iM removal was mainly found on S (55% of the cases) and alanine (A) (in 31%) residues (Figure 5B). These two amino acids, as well as threonine (found in 7.7% of the detected acetylated residues), have small side chains, a feature previously noted to favor a more efficient iM cleavage in the course of protein maturation [66,75].

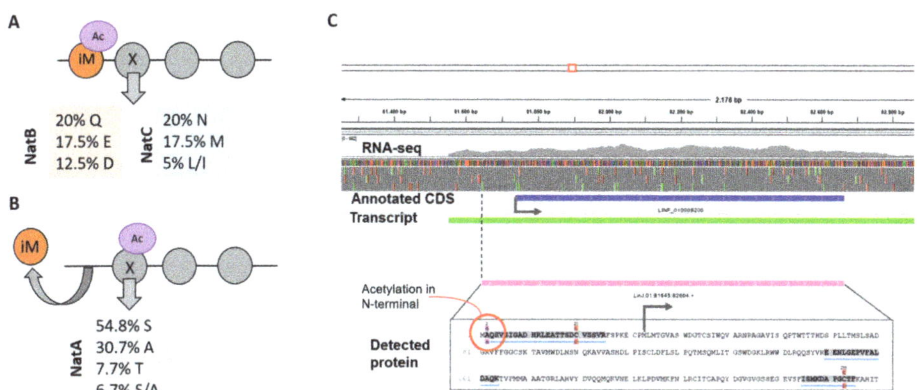

Figure 5. Features of the N-terminal acetylated proteins identified in this study. (**A**) Frequencies (percentages) of amino acids found next to the acetylated initial methionine (iM) and the putative enzymes responsible for the acetylation. (**B**) Percentages of amino acids found to be acetylated after cleavage of the iM and the putative enzyme involved in the reaction. (**C**) An example illustrating the usefulness of proteomics data for improving gene annotations. An acetylated peptide (red circle on the grey shaded sequence) was mapped to sequences located upstream of the currently annotated coding sequence for gene LINF_010008200 (blue box). The corrected gene (pink box) fit well into the transcript (green box). The image in (**C**) was generated using the Integrative Genome Viewer (IGV). CDS: coding sequence.

On the other hand, the analysis of acetylated peptides allowed us to correct the initiator AUG codon (and, therefore, the predicted amino acid sequence) of four previously misannotated genes (whose new coordinates are indicated in Supplementary File, Table S13). One example is illustrated in Figure 5C; the coding sequence of the LINF_010008200 gene (coding for a poly (A) export protein) should be extended upstream 36 triplets based on the existence of an acetylated peptide encoded in that region.

Glycosylation also plays a relevant role in protein maturation, as well as in signal transduction mechanisms [76]. In this *L. infantum* proteomic study, we detected hexosylations or N-acetylhexosamine addition in three proteins; these modifications were consistent with N-linked glycosylation events at N residues. These PTM-modified proteins are cysteine peptidase A (CPA; LINF_190020600) modified at N345, 3-hydroxy-3 methylglutaryl CoA synthase (at N340; LINF_240027300), and PUF6 (at N483; LINF_330019100). The characterization of glycoproteins and the nature of their modifications remain challenging due to the complexity and variety of glycan moieties. In this study, two single modifications (hexosylation and N-acetylhexosamine addition) were searched, and this explained the extremely low number of detected glycosylated proteins. Rosenzweig et al. [66] found 13 glycosylated proteins—only one in asparagine and the rest in serine and threonine residues.

Formylation has not been previously described in trypanosomatids, but N-terminal methionine formylation in eukaryotes has been linked to cellular stress and protein degradation processes [77]. In particular, formyl-lysine residues has been found in histones and other nuclear proteins [78]. In our study, eight proteins were detected as formylated, mainly at leucine (in β-tubulin and typaredoxin), glycine (in a hypothetical protein; LINF_260024300), serine (in an RNA-guanylyltransferase) and lysine (in HSP83/90, a paraflagellar rod protein and a transaldolase) residues (see Supplementary File, Table S14). Future research on protein formylations in *Leishmania* and other organisms will provide insight into the physiological significance of this kind of PTM.

3.9. Proteogenomics

After assembling a genome, dedicated programs conduct the automatic annotation of ORFs. However, this annotation is not definitive, and a continuous effort of curation is needed. Transcriptomic analysis enables the obtainment of complete gene model annotation, including untranslated regions that are key to understand post-transcriptional regulation mechanisms. In addition, a proteogenomic analysis, such as that reported here, represents a powerful and useful approach for the identification of non-annotated genes, the correction of misannotations, or the validation of gene annotations [23]. For this purpose, in this work, the experimentally obtained peptide spectra were searched against all the polypeptides longer than 20 amino acids predicted from the ORFs found in the six possible translation frames of the recently re-sequenced genome for *L. infantum* JPCM5 strain [13]. The majority of the identified peptides fit well in current gene annotation (available at TriTryDB.org and http://leish-esp.cbm.uam.es/). Nevertheless, some peptides mapped to non-annotated coding-regions in the *L. infantum* (JPCM5) genome, uncovering eight novel protein-coding genes (Supplementary File, Table S15). Figure 6 shows an example of a novel hypothetical protein found in chromosome 27, together with the MS spectra of the two peptides that allowed for its identification.

As mentioned above (and illustrated in Figure 5), some of the detected peptides were mapped to regions located upstream of annotated coding sequences (CDS). This led to the addition of N-terminal extensions to 34 annotated proteins and the establishment of new translation start codons for their corresponding genes (Supplementary File, Table S16). All the detected peptides were confirmed to be unique, and the accuracy of their MS/MS spectra was manually revised.

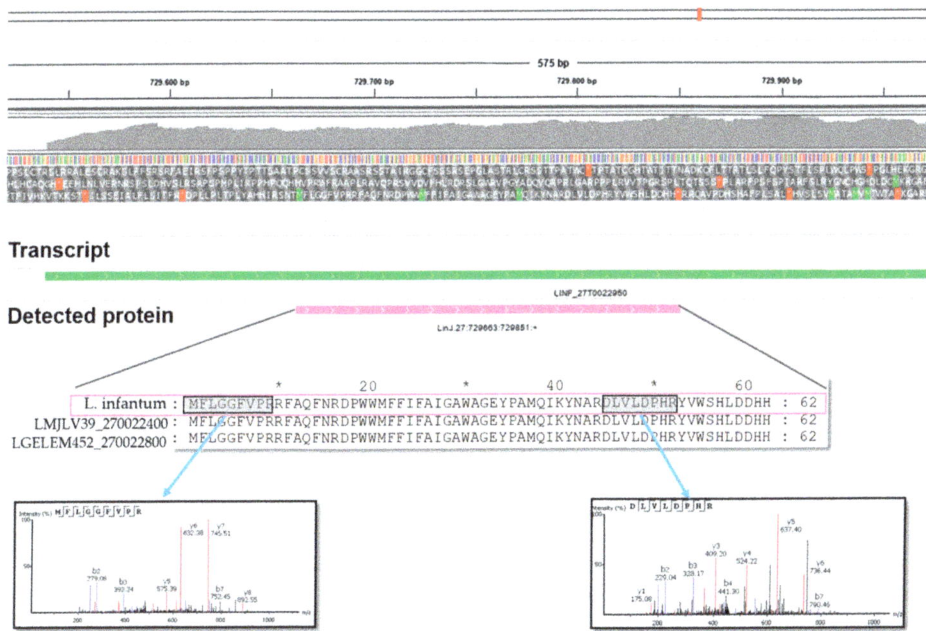

Figure 6. Identification of a novel protein based on the mapping of two experimentally detected-peptides in a region of *L. infantum* JPCM5 chromosome 27, which currently lacks an annotated ORF. The new CDS, named LINF_ 270,022,950 (pink box), fit well within a predicted transcript (LINF_27T0022950; green box). Interestingly, the predicted amino acid sequence was well-conserved when compared with proteins annotated in the genomic assemblies of *L. major* LV39 (ID: LmjVL39_270022400) and *Leishmania gerbilli* LEM452 (ID: LGELEM452_270022800).

4. Conclusions

In the last years, the characterization of trypanosomatid proteomes has become an active area of research. Here, we reported a proteomic analysis of *L. infantum*'s (JPCM5 strain) promastigote stage, and it was the first whole proteomic study in this species after the re-sequencing and de novo assembly of its genome in 2017 by González-De la Fuente et al. [13]. In addition, the search of the MS/MS spectra was performed against any possible ORFs larger than 20 triplets that existed in all-six frames from the *L. infantum* genome sequence. As a result, we identified 2352 proteins (Table S17), most of them corresponding to the predicted sequences in current gene annotations (TriTrypDB.org). Comparisons between the results of this study and previous proteomics data derived from promastigote stages in different *Leishmania* species showed a significant level of similarity regarding the type of detected proteins. Nevertheless, this proteomic study showed experimental evidence on the expression in this parasite stage of 123 proteins that were not detected in previous studies; these proteins are listed in Supplementary File, Table S18. In addition, this study allowed for the identification of several PTMs in proteins, such as phosphorylation, methylation, acetylation, glycosylation, and formylation. Finally, this study also allowed for the identification of eight new protein-coding genes and the extension of the ORFs for 34. In conclusion, whole proteomics and genomic studies are inextricable, the results of the former depend on an accurate genome annotation, and a genome cannot be only annotated in an automatics manner. Thus, the proteomics data obtained in this study have allowed for the correction of annotation mistakes, the discovery of new genes, and experimental evidence of the existence of a large number of proteins that had to date been annotated as hypothetical.

All this new information, at the level of individual genes, is already available at Wikidata.org (searchable by the ID gene) and is going to be incorporated in the TriTrypDB database and the Leish-ESP website (http://leish-esp.cbm.uam.es/).

Supplementary Materials: The following are available online at http://www.mdpi.com/2073-4425/11/9/1036/s1, Table S1: Proteins detected in *L. infantum* promastigotes whose orthologues in *L. donovani* were found to be only expressed in the amastigote stage, Table S2: Translational machinery components (ribosomal proteins, translation factors and tRNA synthetases), Table S3: Proteins annotated as RNA binding proteins, Table S4: Identified proteins belonging to the pentose phosphate pathway, Table S5: Proteins of the mithocondrial electron transport chain, Table S6: Identified proteins belonging to the proteasome, Table S7: Identified proteins in the categories of chaperones and heat shock proteins, Table S8: Putative glycosomal proteins, Table S9: Identified proteins with predicted molecular weight higher than 200 kDa, Table S10: Identified proteins in *L. infantum* putatively located in the flagellum and/or involved in parasite motility, Table S11: Enzymes putatively involved on post-translational modifications, Table S12: Methylated proteins and amino acids found to be modified, Table S13: N-terminal acetylated proteins and the modified position, Table S14: Formylated proteins and the modified position, Table S15: Identification of novel ORFs based on peptide MS spectra, Table S16: N-terminal extended proteins based on the identification of peptides mapping upstream of the annotated ORFs, Table S17: Experimentally identified proteins in promastigotes of *L. infantum* JPCM5 strain, Table S18; Proteins identified for the first time in experimental proteomes of *Leishmania* promastigotes.

Author Contributions: Conceptualization, Á.S., E.M., A.R., A.M., B.A., and J.M.R.; methodology, Á.S., E.M., A.R., E.C., A.M., and S.G.-d.l.F.; formal analysis, Á.S., E.M., A.R., E.C., and S.G.-d.l.F.; data curation, A.S., E.M., S.G.-d.l.F., and J.M.R.; writing—original draft preparation, Á.S.; writing—review and editing, Á.S., A.M., B.A., and J.M.R.; funding acquisition, A.M., B.A., and J.M.R. All authors have read and agreed to the published version of the manuscript.

Funding: This research was funded by grants (to B.A. and J.M.R.) from Proyecto del Ministerio de Economía, Industria y Competitividad SAF2017-86965-R, and by the Network of Tropical Diseases Research RICET (RD16/0027/0008); both grants are co-funded with FEDER funds. A.S. was funded by a postdoctoral contract from the "Programa de Empleo Juvenil" of the Community of Madrid, Spain, within the European Youth Employment Initiative (YEI). A.M. was funded by project PRB3-ISCIII (supported by grant PT17/0019) of the PE I+D+i 2013-2016, funded by ISCIII and ERDF. The CBMSO receives institutional grants from the Fundación Ramón Areces and from the Fundación Banco de Santander.

Acknowledgments: We thank the Genomics and NGS Core Facility at the Centro de Biología Molecular Severo Ochoa (CBMSO, CSIC-UAM) for helping with the bioinformatics analysis. The CBMSO Proteomics Facility is a member of Proteored.

Conflicts of Interest: The authors declare no conflict of interest.

References

1. World Health Organization (WHO). *Leishmaniasis: Disease, Epidemiology, Diagnosis, Detection and Surveillance, Vector Control, Access to Medicines and Information Resources*; WHO: Geneva, Switzerland, 2019.

2. Lukes, J.; Mauricio, I.L.; Schonian, G.; Dujardin, J.-C.; Soteriadou, K.; Dedet, J.-P.; Kuhls, K.; Tintaya, K.W.Q.; Jirku, M.; Chocholova, E.; et al. Evolutionary and geographical history of the Leishmania donovani complex with a revision of current taxonomy. *Proc. Natl. Acad. Sci. USA* **2007**, *104*, 9375–9380. [CrossRef] [PubMed]

3. Ivens, A.C.; Peacock, C.S.; Worthey, E.A.; Murphy, L.; Berriman, M.; Sisk, E.; Rajandream, M.; Adlem, E.; Anupama, A.; Apostolou, Z.; et al. The genome of the kinetoplastid parasite, Leishmania major. *Science* **2006**, *309*, 436–442. [CrossRef] [PubMed]

4. Peacock, C.S.; Seeger, K.; Harris, D.; Murphy, L.; Ruiz, J.C.; Quail, M.A.; Peters, N.; Adlem, E.; Tivey, A.; Aslett, M.; et al. Comparative genomic analysis of three Leishmania species that cause diverse human disease. *Nat. Genet.* **2007**, *39*, 839–847. [CrossRef] [PubMed]

5. Downing, T.; Imamura, H.; Decuypere, S.; Clark, T.G.; Coombs, G.H.; Cotton, J.A.; Hilley, J.D.; de Doncker, S.; Maes, I.; Mottram, J.C.; et al. Whole genome sequencing of multiple *Leishmania donovani* clinical isolates provides insights into population structure and mechanisms of drug resistance. *Genome Res.* **2011**, *21*, 2143–2156. [CrossRef] [PubMed]

6. Gupta, A.K.; Srivastava, S.; Singh, A.; Singh, S. De novo whole-genome sequence and annotation of a Leishmania strain isolated from a case of post-kala-azar dermal leishmaniasis. *Genome Announc.* **2015**, *3*, 4–5. [CrossRef]

7. Llanes, A.; Restrepo, C.M.; Del Vecchio, G.; Anguizola, F.J.; Lleonart, R. The genome of *Leishmania panamensis*: Insights into genomics of the L. (Viannia) subgenus. *Sci. Rep.* **2015**, *5*, 1–10. [CrossRef] [PubMed]

8. Raymond, F.; Boisvert, S.; Roy, G.; Ritt, J.F.; Légaré, D.; Isnard, A.; Stanke, M.; Olivier, M.; Tremblay, M.J.; Papadopoulou, B.; et al. Genome sequencing of the lizard parasite *Leishmania tarentolae* reveals loss of genes associated to the intracellular stage of human pathogenic species. *Nucleic Acids Res.* **2012**, *40*, 1131–1147. [CrossRef]

9. Real, F.; Vidal, R.O.; Carazzolle, M.F.; Mondego, J.M.C.; Costa, G.G.L.; Herai, R.H.; Würtele, M.; De Carvalho, L.M.E.; Ferreira, R.C.; Mortara, R.A.; et al. The genome sequence of leishmania (Leishmania) amazonensis: Functional annotation and extended analysis of gene models. *DNA Res.* **2013**, *20*, 567–581. [CrossRef]

10. Forsdyke, D.R. *Evolutionary Bioinformatics*; Springer: New York, NY, USA, 2006; pp. 1–424.

11. Imamura, H.; Downing, T.; Van den Broeck, F.; Sanders, M.J.; Rijal, S.; Sundar, S.; Mannaert, A.; Vanaerschot, M.; Berg, M.; De Muylder, G.; et al. Evolutionary genomics of epidemic visceral leishmaniasis in the Indian subcontinent. *Elife* **2016**, *5*, e12613. [CrossRef]

12. Coughlan, S.; Mulhair, P.; Sanders, M.; Schonian, G.; Cotton, J.A.; Downing, T. The genome of *Leishmania adleri* from a mammalian host highlights chromosome fission in *Sauroleishmania*. *Sci. Rep.* **2017**, *7*, 1–13. [CrossRef]

13. González-De La Fuente, S.; Peiró-Pastor, R.; Rastrojo, A.; Moreno, J.; Carrasco-Ramiro, F.; Requena, J.M.; Aguado, B. Resequencing of the *Leishmania infantum* (strain JPCM5) genome and de novo assembly into 36 contigs. *Sci. Rep.* **2017**, *7*, 18050.

14. Alonso, G.; Rastrojo, A.; López-Pérez, S.; Requena, J.M.; Aguado, B. Resequencing and assembly of seven complex loci to improve the Leishmania major (Friedlin strain) reference genome. *Parasites Vectors* **2016**, *9*, 74. [CrossRef] [PubMed]

15. González-De La Fuente, S.; Camacho, E.; Peiró-Pastor, R.; Rastrojo, A.; Carrasco-Ramiro, F.; Aguado, B.; Requena, J.M. Complete and de novo assembly of the *Leishmania braziliensis* (M2904) genome. *Mem. Inst. Oswaldo Cruz* **2019**, *114*, 1–6. [CrossRef] [PubMed]

16. Requena, J.M.; Alcolea, P.J.; Alonso, A.; Larraga, V. Omics approaches for understanding gene expression in Leishmania: Clues for tackling leishmaniasis. In *Protozoan Parasitism: From Omics to Prevention and Control*; Pablos-Torró, L.M., Lorenzo-Morales, J., Eds.; Caister Academic Press: Norfolk, UK, 2018; pp. 77–112.

17. Capelli-Peixoto, J.; Mule, S.N.; Tano, F.T.; Palmisano, G.; Stolf, B.S. Proteomics and Leishmaniasis: Potential clinical applications. *Proteom. Clin. Appl.* **2019**, *13*, 1800136. [CrossRef]

18. Nugent, P.G.; Karsani, S.A.; Wait, R.; Tempero, J.; Smith, D.F. Proteomic analysis of *Leishmania mexicana* differentiation. *Mol. Biochem. Parasitol.* **2004**, *136*, 51–62. [CrossRef]

19. Leifso, K.; Cohen-Freue, G.; Dogra, N.; Murray, A.; McMaster, W.R. Genomic and proteomic expression analysis of Leishmania promastigote and amastigote life stages: The Leishmania genome is constitutively expressed. *Mol. Biochem. Parasitol.* **2007**, *152*, 35–46. [CrossRef]

20. Rosenzweig, D.; Smith, D.; Opperdoes, F.; Stern, S.; Olafson, R.W.; Zilberstein, D. Retooling Leishmania metabolism: From sand fly gut to human macrophage. *FASEB J.* **2008**, *22*, 590–602. [CrossRef]

21. Jardim, A.; Hardie, D.B.; Boitz, J.; Borchers, C.H. Proteomic profiling of *Leishmania donovani* promastigote subcellular organelles. *J. Proteome Res.* **2018**, *17*, 1194–1215. [CrossRef]

22. Tasbihi, M.; Shekari, F.; Hajjaran, H.; Masoori, L.; Hadighi, R. Mitochondrial proteome profiling of *Leishmania tropica*. *Microb. Pathog.* **2019**, *133*, 103542. [CrossRef]

23. Armengaud, J. Proteogenomics and systems biology: Quest for the ultimate missing parts. *Expert Rev. Proteom.* **2010**, *7*, 65–77. [CrossRef]

24. Moreno, M.-L.; Escobar, J.; Izquierdo-Alvarez, A.; Gil, A.; Perez, S.; Pereda, J.; Zapico, I.; Vento, M.; Sabater, L.; Marina, A.; et al. Disulfide stress: A novel type of oxidative stress in acute pancreatitis. *Free Radic. Biol. Med.* **2014**, *70*, 265–277. [CrossRef] [PubMed]

25. Shevchenko, A.; Wilm, M.; Vorm, O.; Mann, M. Mass spectrometric sequencing of proteins from silver-stained polyacrylamide gels. *Anal. Chem.* **1996**, *68*, 850–858. [CrossRef] [PubMed]

26. Torres, L.L.; Cantero, A.; del Valle, M.; Marina, A.; Lopez-Gallego, F.; Guisan, J.M.; Berenguer, J.; Hidalgo, A. Engineering the substrate specificity of a thermophilic penicillin acylase from *Thermus thermophilus*. *Appl. Environ. Microbiol.* **2013**, *79*, 1555–1562. [CrossRef] [PubMed]

27. Zhang, J.; Xin, L.; Shan, B.; Chen, W.; Xie, M.; Yuen, D.; Zhang, W.; Zhang, Z.; Lajoie, G.A.; Ma, B. PEAKS DB: De novo sequencing assisted database search for sensitive and accurate peptide identification. *Mol. Cell. Proteom.* **2012**, *11*, M111.010587. [CrossRef] [PubMed]

28. Yonghua, H.; Bin, M.; Zaizhong, Z. SPIDER: Software for protein identification from sequence tags with de novo sequencing error. In Proceedings of the IEEE Computational Systems Bioinformatics Conference, Stanford, CA, USA, 19 August 2004.

29. Han, X.; He, L.; Xin, L.; Shan, B.; Ma, B. PeaksPTM: Mass spectrometry-based identification of peptides with unspecified modifications. *J. Proteome Res.* **2011**, *10*, 2930–2936. [CrossRef] [PubMed]

30. McNicoll, F.; Drummelsmith, J.; Müller, M.; Madore, É.; Boilard, N.; Ouellette, M.; Papadopoulou, B. A combined proteomic and transcriptomic approach to the study of stage differentiation in *Leishmania infantum*. *Proteomics* **2006**, *6*, 3567–3581. [CrossRef] [PubMed]

31. Foucher, A.L.; Papadopoulou, B.; Ouellette, M. Prefractionation by digitonin extraction increases representation of the cytosolic and intracellular proteome of *Leishmania infantum*. *J. Proteome Res.* **2006**, *5*, 1741–1750. [CrossRef]

32. Brotherton, M.-C.; Racine, G.; Foucher, A.L.; Drummelsmith, J.; Papadopoulou, B.; Ouellette, M. Analysis of stage-specific expression of basic proteins in *Leishmania infantum*. *J. Proteome Res.* **2010**, *9*, 3842–3853. [CrossRef]

33. Alcolea, P.J.; Alonso, A.; Larraga, V. Proteome profiling of *Leishmania infantum* promastigotes. *J. Eukaryot. Microbiol.* **2011**, *58*, 352–358. [CrossRef]

34. Braga, M.S.; Neves, L.X.; Campos, J.M.; Roatt, B.M.; de Oliveira Aguiar Soares, R.D.; Braga, S.L.; de Melo Resende, D.; Reis, A.B.; Castro-Borges, W. Shotgun proteomics to unravel the complexity of the *Leishmania infantum* exoproteome and the relative abundance of its constituents. *Mol. Biochem. Parasitol.* **2014**, *195*, 43–53. [CrossRef]

35. Santarém, N.; Racine, G.; Silvestre, R.; Cordeiro-da-Silva, A.; Ouellette, M. Exoproteome dynamics in *Leishmania infantum*. *J. Proteom.* **2013**, *84*, 106–118. [CrossRef] [PubMed]

36. Pawar, H.; Renuse, S.; Khobragade, S.N.; Chavan, S.; Sathe, G.; Kumar, P.; Mahale, K.N.; Gore, K.; Kulkarni, A.; Dixit, T.; et al. Neglected tropical diseases and omics science: Proteogenomics analysis of the promastigote stage of leishmania major parasite. *Omi. A J. Integr. Biol.* **2014**, *18*, 499–512. [CrossRef] [PubMed]

37. Nirujogi, R.S.; Pawar, H.; Renuse, S.; Kumar, P.; Chavan, S.; Sathe, G.; Sharma, J.; Khobragade, S.; Pande, J.; Modak, B.; et al. Moving from unsequenced to sequenced genome: Reanalysis of the proteome of *Leishmania donovani*. *J. Proteom.* **2014**, *97*, 48–61. [CrossRef] [PubMed]

38. Pawar, H.; Sahasrabuddhe, N.A.; Renuse, S.; Keerthikumar, S.; Sharma, J.; Kumar, G.S.S.; Venugopal, A.; Sekhar, N.R.; Kelkar, D.S.; Nemade, H.; et al. A proteogenomic approach to map the proteome of an unsequenced pathogen—*Leishmania donovani*. *Proteomics* **2012**, *12*, 832–844. [CrossRef]

39. Ramrath, D.J.F.; Niemann, M.; Leibundgut, M.; Bieri, P.; Prange, C.; Horn, E.K.; Leitner, A.; Boehringer, D.; Schneider, A.; Ban, N. Evolutionary shift toward protein-based architecture in trypanosomal mitochondrial ribosomes. *Science* **2018**, *362*, eaau7735. [CrossRef]

40. Clayton, C.; Shapira, M. Post-transcriptional regulation of gene expression in trypanosomes and leishmanias. *Mol. Biochem. Parasitol.* **2007**, *156*, 93–101. [CrossRef]

41. Requena, J.M. Lights and shadows on gene organization and regulation of gene expression in Leishmania. *Front. Biosci. Landmark Ed.* **2011**, *16*, 2069–2085. [CrossRef]

42. Beckmann, B.M.; Castello, A.; Medenbach, J. The expanding universe of ribonucleoproteins: Of novel RNA-binding proteins and unconventional interactions. *Pflugers Arch. Eur. J. Physiol.* **2016**, *468*, 1029–1040. [CrossRef]

43. Kramer, S.; Carrington, M. Trans-acting proteins regulating mRNA maturation, stability and translation in trypanosomatids. *Trends Parasitol.* **2011**, *27*, 23–30. [CrossRef]

44. De Gaudenzi, J.G.; Frasch, A.C.; Clayton, C. RNA-binding domain proteins in *Kinetoplastids*: A comparative analysis. *Eukaryot. Cell* **2014**, *4*, 2106–2114. [CrossRef]

45. Nandan, D.; Thomas, S.A.; Nguyen, A.; Moon, K.M.; Foster, L.J.; Reiner, N.E. Comprehensive identification of mRNABinding proteins of *Leishmania donovani* by interactome capture. *PLoS ONE* **2017**, *12*, e0170068. [CrossRef] [PubMed]

46. Folgueira, C.; Martínez-Bonet, M.; Requena, J.M. The *Leishmania infantum* PUF proteins are targets of the humoral response during visceral leishmaniasis. *BMC Res. Notes* **2010**, *3*, 13. [CrossRef] [PubMed]

47. Subramanian, A.; Jhawar, J.; Sarkar, R.R. Dissecting *Leishmania infantum* energy metabolism—A systems perspective. *PLoS ONE* **2015**, *10*, e0137976. [CrossRef] [PubMed]

48. Colasante, C.; Voncken, F.; Manful, T.; Ruppert, T.; Tielens, A.G.M.; van Hellemond, J.J.; Clayton, C. Proteins and lipids of glycosomal membranes from *Leishmania tarentolae* and *Trypanosoma brucei*. *F1000Research* **2013**, *2*, 1–15. [CrossRef]

49. Camacho, E.; Rastrojo, A.; Sanchiz, Á.; González-de la Fuente, S.; Aguado, B.; Requena, J.M. Leishmania mitochondrial genomes: Maxicircle structure and heterogeneity of minicircles. *Genes (Basel)* **2019**, *10*, 758. [CrossRef]

50. Paugam, A.; Bulteau, A.L.; Dupouy-Camet, J.; Creuzet, C.; Friguet, B. Characterization and role of protozoan parasite proteasomes. *Trends Parasitol.* **2003**, *19*, 55–59. [CrossRef]

51. Silva-Jardim, I.; Fátima Horta, M.; Ramalho-Pinto, F.J. The *Leishmania chagasi* proteasome: Role in promastigotes growth and amastigotes survival within murine macrophages. *Acta Trop.* **2004**, *91*, 121–130. [CrossRef]

52. Khare, S.; Nagle, A.S.; Biggart, A.; Lai, Y.H.; Liang, F.; Davis, L.C.; Barnes, S.W.; Mathison, C.J.N.; Myburgh, E.; Gao, M.-Y.; et al. Proteasome inhibition for treatment of leishmaniasis, Chagas disease and sleeping sickness. *Nature* **2016**, *537*, 229–233. [CrossRef]

53. Balchin, D.; Hayer-Hartl, M.; Hartl, F.U. In vivo aspects of protein folding and quality control. *Science* **2016**, *353*, aac4354. [CrossRef]

54. Requena, J.M.; Montalvo, A.M.; Fraga, J. Molecular chaperones of Leishmania: Central players in many stress-related and -unrelated physiological processes. *Biomed Res. Int.* **2015**, *2015*, 1–21. [CrossRef]

55. Bauer, S.; Morris, M.T. Glycosome biogenesis in trypanosomes and the de novo dilemma. *PLoS Negl. Trop. Dis.* **2017**, *11*, e0005333. [CrossRef] [PubMed]

56. Pilar, A.V.C.; Strasser, R.; McLean, J.; Quinn, E.; Cyr, N.; Hojjat, H.; Kottarampatel, A.H.; Jardim, A. Analysis of the Leishmania peroxin 7 interactions with peroxin 5, peroxin 14 and PTS2 ligands. *Biochem. J.* **2014**, *460*, 273–282. [CrossRef] [PubMed]

57. Jamdhade, M.D.; Pawar, H.; Chavan, S.; Sathe, G.; Umasankar, P.K.; Mahale, K.N.; Dixit, T.; Madugundu, A.K.; Prasad, T.S.K.; Gowda, H.; et al. Comprehensive Proteomics analysis of glycosomes from *Leishmania donovani*. *OMICS A J. Integr. Biol.* **2015**, *19*, 157–170. [CrossRef] [PubMed]

58. Boitz, J.M.; Ullman, B.; Jardim, A.; Carter, N.S. Purine salvage in Leishmania: Complex or simple by design? *Trends Parasitol.* **2012**, *28*, 345–352. [CrossRef] [PubMed]

59. Tiwari, K.; Dubey, V.K. Fresh insights into the pyrimidine metabolism in the trypanosomatids. *Parasites Vectors* **2018**, *11*, 1–15. [CrossRef]

60. Pérez-Cabezas, B.; Santarém, N.; Cecílio, P.; Silva, C.; Silvestre, R.; AM Catita, J.; Cordeiro da Silva, A. More than just exosomes: Distinct *Leishmania infantum* extracellular products potentiate the establishment of infection. *J. Extracell. Vesicles* **2019**, *8*, 1541708. [CrossRef]

61. Atayde, V.D.; Aslan, H.; Townsend, S.; Hassani, K.; Kamhawi, S.; Olivier, M. Exosome secretion by the parasitic protozoan Leishmania within the sand fly midgut. *Cell Rep.* **2015**, *13*, 957–967. [CrossRef]

62. Avilán, L.; Gualdrón-López, M.; Quiñones, W.; González-González, L.; Hannaert, V.; Michels, P.A.M.; Concepción, J.-L. Enolase: A key player in the metabolism and a probable virulence factor of *Trypanosomatid* parasites—Perspectives for its use as a therapeutic target. *Enzyme Res.* **2011**, *2011*, 932549. [CrossRef]

63. Brotherton, M.C.; Racine, G.; Ouameur, A.A.; Leprohon, P.; Papadopoulou, B.; Ouellette, M. Analysis of membrane-enriched and high molecular weight proteins in *Leishmania infantum* promastigotes and axenic amastigotes. *J. Proteome Res.* **2012**, *11*, 3974–3985. [CrossRef]

64. Beneke, T.; Demay, F.; Hookway, E.; Ashman, N.; Jeffery, H.; Smith, J.; Valli, J.; Becvar, T.; Myskova, J.; Lestinova, T.; et al. Genetic dissection of a *Leishmania flagellar* proteome demonstrates requirement for directional motility in sand fly infections. *PLoS Pathog.* **2019**, *15*, e1007828. [CrossRef]

65. Morales, M.A.; Watanabe, R.; Dacher, M.; Chafey, P.; Osorio, Y.; Fortéa, J.; Scott, D.A.; Beverley, S.M.; Ommen, G.; Clos, J.; et al. Phosphoproteome dynamics reveal heat-shock protein complexes specific to the *Leishmania donovani* infectious stage. *Proc. Natl. Acad. Sci. USA* **2010**, *107*, 8381–8386. [CrossRef] [PubMed]

66. Rosenzweig, D.; Smith, D.; Myler, P.J.; Olafson, R.W.; Zilberstein, D. Post-translational modification of cellular proteins during *Leishmania donovani* differentiation. *Proteomics* **2008**, *8*, 1843–1850. [CrossRef] [PubMed]

67. Santos, A.L.S.; Branquinha, M.H.; D'Avila-Levy, C.M.; Kneipp, L.; Sodré, C.L. *Proteins and Proteomics of Leishmania and Trypanosoma*; Springer: New York, NY, USA, 2014; ISBN 978-94-007-7304-2.

68. Tsigankov, P.; Gherardini, P.F.; Helmer-Citterich, M.; Späth, G.F.; Zilberstein, D. Phosphoproteomic analysis of differentiating Leishmania parasites reveals a unique stage-specific phosphorylation motif. *J. Proteome Res.* **2013**, *12*, 3405–3412. [CrossRef] [PubMed]

69. Sprung, R.; Chen, Y.; Zhang, K.; Cheng, D.; Zhang, T.; Peng, J.; Zhao, Y. Identification and validation of eukaryotic aspartate and glutamate methylation in proteins. *J. Proteome Res.* **2008**, *7*, 1001–1006. [CrossRef]

70. Alonso, V.L.; Serra, E.C. Lysine acetylation: Elucidating the components of an emerging global signaling pathway in trypanosomes. *J. Biomed. Biotechnol.* **2012**, *2012*. [CrossRef]

71. Thomas, S.; Green, A.; Sturm, N.R.; Campbell, D.A.; Myler, P.J. Histone acetylations mark origins of polycistronic transcription in Leishmania major. *BMC Genom.* **2009**, *10*, 152. [CrossRef]

72. Respuela, P.; Ferella, M.; Rada-Iglesias, A.; Aslund, L. Histone acetylation and methylation at sites initiating divergent polycistronic transcription in *Trypanosoma cruzi*. *J. Biol. Chem.* **2008**, *283*, 15884–15892. [CrossRef]

73. Cuervo, P.; Domont, G.B.; De Jesus, J.B. Proteomics of trypanosomatids of human medical importance. *J. Proteom.* **2010**, *73*, 845–867. [CrossRef]

74. Linster, E.; Wirtz, M. N-terminal acetylation: An essential protein modification emerges as an important regulator of stress responses. *J. Exp. Bot.* **2018**, *69*, 4555–4568. [CrossRef]

75. Gupta, N.; Tanner, S.; Jaitly, N.; Adkins, J.N.; Lipton, M.; Edwards, R.; Romine, M.; Osterman, A.; Bafna, V.; Smith, R.D.; et al. Whole proteome analysis of post-translational modifications: Applications of mass-spectrometry for proteogenomic annotation. *Genome Res.* **2007**, *17*, 1362–1377. [CrossRef]

76. Manzano-Román, R.; Fuentes, M. Relevance and proteomics challenge of functional posttranslational modifications in *Kinetoplastid* parasites. *J. Proteom.* **2020**, *220*, 103762. [CrossRef] [PubMed]

77. Eldeeb, M.A.; Fahlman, R.P.; Esmaili, M.; Fon, E.A. Formylation of eukaryotic cytoplasmic proteins: Linking stress to degradation. *Trends Biochem. Sci.* **2019**, *44*, 181–183. [CrossRef] [PubMed]

78. Wiśniewski, J.R.; Zougman, A.; Mann, M. Nε-Formylation of lysine is a widespread post-translational modification of nuclear proteins occurring at residues involved in regulation of chromatin function. *Nucleic Acids Res.* **2008**, *36*, 570–577. [CrossRef] [PubMed]

 *genes*

Article

The *Leishmania donovani* SENP Protease Is Required for SUMO Processing but Not for Viability

Annika Bea [1,†], Constanze Kröber-Boncardo [1,†], Manpreet Sandhu [1,2], Christine Brinker [1] and Joachim Clos [1,*] 

[1] Leishmaniasis Group, Bernhard Nocht Institute for Tropical Medicine, D-20359 Hamburg, Germany; annika.bea@bnitm.de (A.B.); kroeber@bnitm.de (C.K.-B.); manpreet.sandhu9898@gmail.com (M.S.); brinker@bnitm.de (C.B.)

[2] Boehringer Ingelheim RCV, A-1121 Vienna, Austria

* Correspondence: clos@bnitm.de; Tel.: +49-40-42818-481

† These authors contributed equally to this work.

Received: 10 September 2020; Accepted: 10 October 2020; Published: 14 October 2020

Abstract: The protozoan parasite *Leishmania donovani* is part of an early eukaryotic branch and depends on post-transcriptional mechanisms for gene expression regulation. This includes post-transcriptional protein modifications, such as protein phosphorylation. The presence of genes for protein SUMOylation, i.e., the covalent attachment of small ubiquitin-like modifier (SUMO) polypeptides, in the *Leishmania* genomes prompted us to investigate the importance of the sentrin-specific protease (SENP) and its putative client, SUMO, for the vitality and infectivity of *Leishmania donovani*. While SENP null mutants are viable with reduced vitality, viable SUMO null mutant lines could not be obtained. SUMO C-terminal processing is disrupted in SENP null mutants, preventing SUMO from covalent attachment to proteins and nuclear translocation. Infectivity *in vitro* is not affected by the loss of SENP-dependent SUMO processing. We conclude that SENP is required for SUMO processing, but that functions of unprocessed SUMO are critical for *Leishmania* viability.

Keywords: *Leishmania*; SENP; Ulp2; SUMO; CRISPR; protease

---

1. Introduction

Leishmania donovani is a protozoan parasite that causes the lethal visceral leishmaniasis, also known as *Kala azar*. It is a vector-borne pathogen, transmitted by female sandflies of the genus *Phlebotomus*, in particular *P. argentipes*. *Leishmania* exists in two main developmental stages. Promastigotes, elongated flagellates, proliferate rapidly in the sandfly gut. When transmitted to humans, the parasites are phagocytized by antigen-presenting cells and once inside the phagosomes, convert into ovoid, aflagellated amastigotes as which they may persist in the host for months or years.

The leishmaniae differ from their human host and from most other eukaryotes by their lack of gene-specific transcription regulation [1–3], relying on modulated RNA stability [4], inducible translation [5] and reversible gene amplification [6,7] instead.

In addition, *Leishmania* spp. have a full complement of protein kinases [8] and phosphatases [9] to modulate protein activity via phosphorylation and dephosphorylation. Heat shock proteins are important substrates for life cycle stage-dependent phosphorylation [10], but protein kinases also affect parasite morphology, infectivity and viability [8,11–13]. Methylation, acetylation and glycosylation of proteins, i.e., modifications of amino acid side chains, have also been described for *Leishmania* [14,15].

Another type of post-translational protein modifications (PTMs), the conjugation of modifying polypeptides to target proteins is not as well researched in *Leishmania*, but known to exist, e.g., the conjugation of a mitochondrial associated ubiquitin fold modifier (UFM) [16,17]. Conjugation of another modifier, small ubiquitin-like modifier (SUMO) was studied in *Trypanosoma* spp: SUMOylation of proteins was described for *Trypanosoma cruzi* [18] and *T. brucei* [19,20], where this PTM is involved in surface antigen expression and nuclear organization. A putative ortholog of *SUMO* is present in the *L. donovani* genome and expressed [5,21].

For SUMOylation to happen, the SUMO precursor must first undergo a proteolytic cleavage by a sentrin-specific protease (SENP), which removes the C-terminal amino acids and leaves an exposed, reactive, C-terminal di-glycine group [22,23]. A putative SENP ortholog is also encoded in the *L. donovani* genome and expressed [5,24]. In humans, the di-glycine is further activated by the E1 protein, transesterified to the E2 SUMO-conjugating enzyme and finally transferred to the substrate protein by the E3 SUMO ligase. DeSUMOylation is also facilitated by SENP [25], establishing SENP as a pivotal enzyme to control the SUMOylation state of substrate proteins.

SUMOylation of proteins may have different consequences and result in (i) interference with binding of partner proteins, (ii) additional interaction sites for other proteins, or (iii) SUMO-induced conformational changes of the modified protein [23]. SUMOylation may interfere or promote other PTMs, such as phosphorylation [26] or ubiquitination [27,28]. The SUMOylation status of proteins is highly dynamic, dependent on cell cycle phases, differentiation and stress exposure [23]. Incorrect or excessive SUMOylation is also associated with severe disease, such as cardiovascular or neurological dysfunctions, but also cancer [28]. It is therefore conceivable that in an organism such as *Leishmania*, which is highly dependent on post-transcriptional gene expression regulation, SUMOylation of proteins may play an important role in its adaption to vectors and hosts.

Here, we describe a reverse genetic analysis of SUMO and SENP in *L. donovani*. We test the SUMO-specific proteolytic activity of SENP *in vivo* and examine its impact on vitality and intracellular survival.

2. Materials and Methods

2.1. Leishmania Culture Conditions

Leishmania donovani strain 1S [29] promastigotes and derived mutants were cultured at 25 °C in M199+ medium [30] with the respective antibiotics: puromycin (25 µg/mL, AppliChem, Darmstadt, Germany), blasticidin (5 µg/mL), G418 (50 µg/mL) and hygromycin B (50 µg/mL, all Carl Roth, Karlsruhe, Germany). Cells were passaged every 3–4 days.

2.2. Electrotransfection of Leishmania Parasites

Electrotransfection and selection was performed as described [31]. Clonal parasite populations were obtained by limiting dilution in 96-well plates with an initial inoculum of 0.5 parasites/well in a final volume of 200 µL M199+ medium supplemented with the respective antibiotics and 1× penicillin/streptomycin (Sigma Aldrich, Munich, Germany).

2.3. In Vitro Infection of Murine Bone Marrow-Derived Macrophages

Isolation and *in vitro* infection of murine bone marrow derived macrophages was performed as described [30].

2.4. Construction and Preparation of Recombinant DNA

The SUMO (LdBPK_080480) and SENP (LdBPK_262070) coding sequences were amplified from *L. donovani* 1S genomic DNA using specific primer pairs (Table S1) that introduce restriction sites as indicated. PCR products were subsequently ligated into pCL2N [32], or derived plasmids pCL2N-3×HA (N-ter) and pCL2N-3×HA (C-ter), predigested with the cognate restriction enzymes. Plasmids were amplified in *Escherichia coli* DH5α and purified by CsCl density gradient ultracentrifugation as described previously [33].

2.5. PCR-Amplification of Targeting Constructs

For CRISPR/Cas9-mediated gene disruption, sgRNA templates and replacement constructs were PCR-amplified using the Expand High Fidelity PCR System (Roche, Mannheim, Germany) and PCR conditions essentially as described previously [34]. Oligonucleotides used are listed in Table S1.

2.6. RNA Extraction, cDNA Synthesis and Quantitative Real-Time PCR (qRT-PCR)

RNA extraction, cDNA detection and RT-qPCR were performed essentially as described [35,36]. Primer sequences are listed in Table S2.

2.7. Next Generation Sequencing

Isolation of genomic DNA, DNA library preparation and sequencing was performed following established protocols and carried out on a MiSeq sequencer (Illumina, San Diego, CA, USA) [36].

2.8. Western Blotting

Western blot was performed essentially as described [37,38]. Primary anti-HA IgG antibody (polyclonal, mouse, 1:5000; Invitrogen, Carlsbad, CA, USA) in blocking solution (5% milk/TBST solution) was used in conjunction with anti-mouse-AP IgG (polyclonal, goat, 1:1000; Dianova, Hamburg, Germany).

2.9. Immunofluorescence Assays

Immunofluorescence assays of log-phase promastigotes, heat-shocked promastigotes and axenic amastigotes were performed as described previously [38]. Briefly, 2×10^5 cells were washed with 1×PBS and applied on microscopic slides and fixed with ice-cold methanol. Following permeabilization and blocking, the cells were stained with primary anti-HA IgG antibody (polyclonal, mouse, 1:3000; Invitrogen, Carlsbad, CA, USA) and secondary anti-mouse Alexa Fluor® 594 IgG (polyclonal, goat, 1:1000; Thermo Fisher Scientific, Waltham, MA, USA) and DAPI (1:50; Sigma Aldrich, Munich, Germany). Fluorescence microscopy was carried out on an EVOS® FL Auto Cell Imaging System using a 64× magnification.

2.10. In Silico Procedures

In silico construction of plasmids, DNA and protein sequence analyses was performed using the MacVector software, version 17 (MacVector Inc., Cambridge, UK). Microscopy images were processed using Adobe Photoshop CS3 (Adobe Corp., San Jose, CA, USA) and juxtaposed using Intaglio (Version 3.9, Purgatory Design, Durango, CO, USA). Multi-panel figures were also assembled using the Intaglio software.

In silico design of *SUMO*- and *SENP*-specific sgRNAs and primers for the amplification of the donor repair cassettes was performed using the LeishGEdit online tool [39]. Oligonucleotides were purchased from Sigma-Aldrich (München, Germany).

Gene annotations and reference genomes (version 42) of *L. donovani* BPK were downloaded from the TriTrypDB server. Reads were aligned to the reference genomes using the MacVector software version 17 and Bowtie2 algorithm [40].

Statistical analyses were performed using Prism (version 8, GraphPad Software, San Diego, CA, USA). Ranking tests were performed using the U-test [41]. Differences were considered significant at a level of $p < 0.05$.

3. Results

3.1. Expression of SUMO and SENP Proteins in Leishmania spp.

When screening the *L. donovani* genome database using BLAST, we identified genes coding for SUMO (LdBPK_080480) and SENP (LdBPK_262070). A ClustalW amino acid sequence comparison of SUMO genes from five *Leishmania* species and two *Trypanosoma* species with four human paralogs and orthologs from *Drosophila* and yeast was performed and used to build a phylogenetic tree (Figure 1A). The SUMO orthologs from the lower eukaryotic clade are distinct from the metazoan SUMOs, but reasonably well conserved (Figure 1C). Notably, the di-glycine motif near the C terminus is present in all SUMO orthologs. The SENP/Ulp2 peptidases, too, were highly conserved among the *Leishmania* spp. and clearly related to the *Trypanosoma* orthologs (Figure 1B).

Both SUMO and SENP are constitutively expressed in *L. donovani*. RNA-seq and ribosome profiling data generated previously [5] show minor variations for SUMO protein synthesis and RNA abundance for *L. donovani* before and after radicicol-induced promastigote-to-amastigote differentiation (Figure 1D). SENP also shows a constitutive, stage-independent protein synthesis and RNA levels. The normalized [5] ribosome footprinting read densities for SUMO and SENP were slightly above those for ubiquitin fold modifier (UFM, LdBPK_161100), another PTM polypeptide [19,20], and lower than those recorded for polyubiquitin (LdBPK_090950), indicating a gene expression rate slightly above the median (1.0) for *L. donovani* genes. With expression of SUMO and SENP established, we decided to target both genes for replacement, using a CRISPR/Cas9 approach.

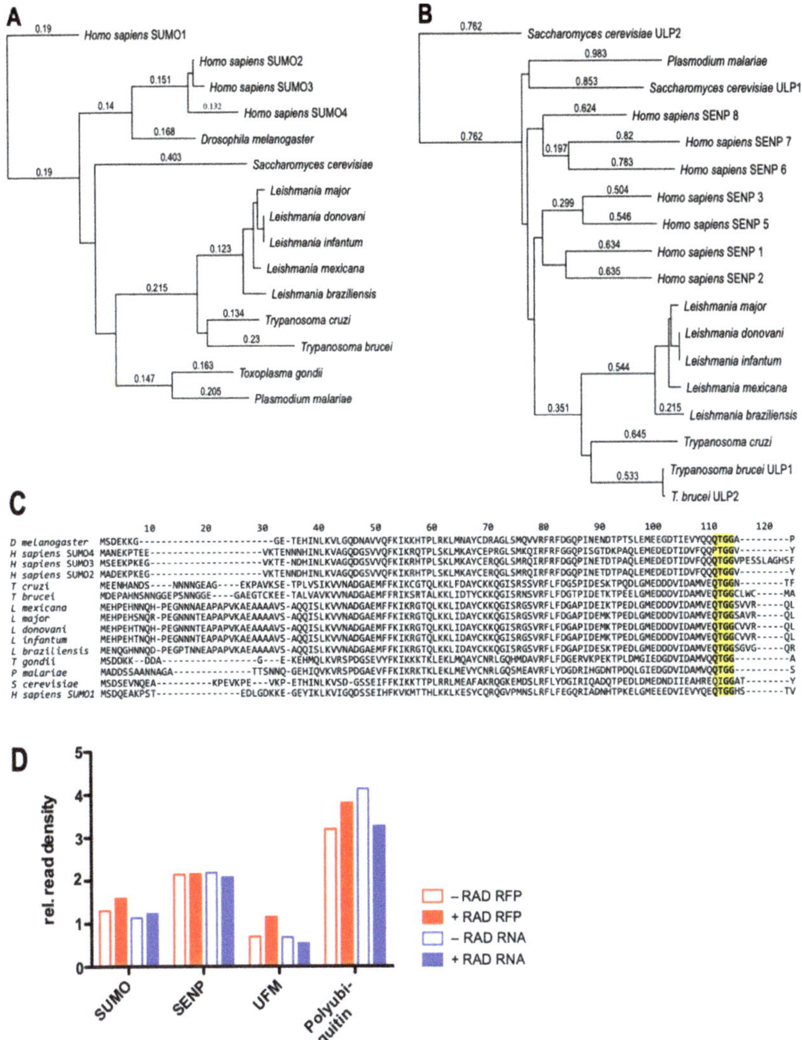

Figure 1. Conservation and expression of *SUMO* and *SENP* in *Leishmania*. (**A**) Phylogenetic analysis of SUMO proteins. Sequence alignment and tree building were done using the neighbor joining algorithm and best-fit analysis with Poisson correction. Numbers indicate amino acid sequence deviation. (**B**) Phylogenetic analysis of SENP proteins, performed as in (**A**). (**C**) Alignment of SUMO amino acid sequences, with the C-terminal di-glycine highlighted. (**D**) Gene expression analysis by ribosome profiling and RNA-seq analysis for *L. donovani* before (−RAD) and after (+RAD) radicicol-induced differentiation. Shown are relative read densities, normalized to the median read densities, for protein synthesis (RFP) and RNA abundance (RNA). Data collected from [5].

3.2. Replacement of L. donovani SUMO

To test the importance of *SUMO* for *L. donovani* viability and/or proliferation, we targeted *SUMO* for CRISPR/Cas9-mediated replacement, following an established protocol [34,39]. The 5′sgRNA- and 3′sgRNA-coding sequences, along with the upstream and downstream flanking primers were designed as shown in Figure 2A, together with two primer pairs to test for the presence of *SUMO*. The selection marker gene cassettes from plasmids pTPURO and pTBLAST were amplified using the upstream and downstream flanking primers to yield 1.9 kb PCR products (Figure 2B). Those, together with the 5′- and 3′-sgRNA-coding oligonucleotides (Figure 2C) were transfected into *L. donovani* expressing both the Cas9 recombinase and the T7 RNA polymerase (*L. donovani* (Cas9/T7-RNAP)). The transfectants were then selected under IC_{95} (95%-inhibiting concentration) for puromycin and blasticidin. Selected parasites were cloned by limiting dilution [42] and tested for the presence of *SUMO* by PCR with two independent primer pairs. Figure 2D shows that all tested clones remained positive for *SUMO*.

The success of CRISPR-mediated gene replacement is very dependent on a perfect match between gene sequences and the annealing sgRNA regions. We therefore tested whether the sgRNA pair was able to basepair with the *SUMO* coding sequence. For this, we repeated the transfection of sgRNA-coding oligonucleotides and selection marker cassettes in an *L. donovani* strain over expressing *SUMO* from episomal gene copies to create SUMO$^{-/-/+}$ parasites. In five out of six clones, we could verify the loss of the chromosomal *SUMO* alleles. This confirms the specificity of the sgRNAs and selection marker cassette amplificates. We conclude that replacement of *SUMO* is only possible in the presence of ectopic SUMO gene copies, giving strong evidence for an essential role of *SUMO* in viability and/or proliferative capacity of *L. donovani*.

As C-terminal processing by SENP/Ulp2 is thought to be critical for conjugation and polymerization of SUMO, but also for de-SUMOylation, we next targeted the putative SENP ortholog for replacement.

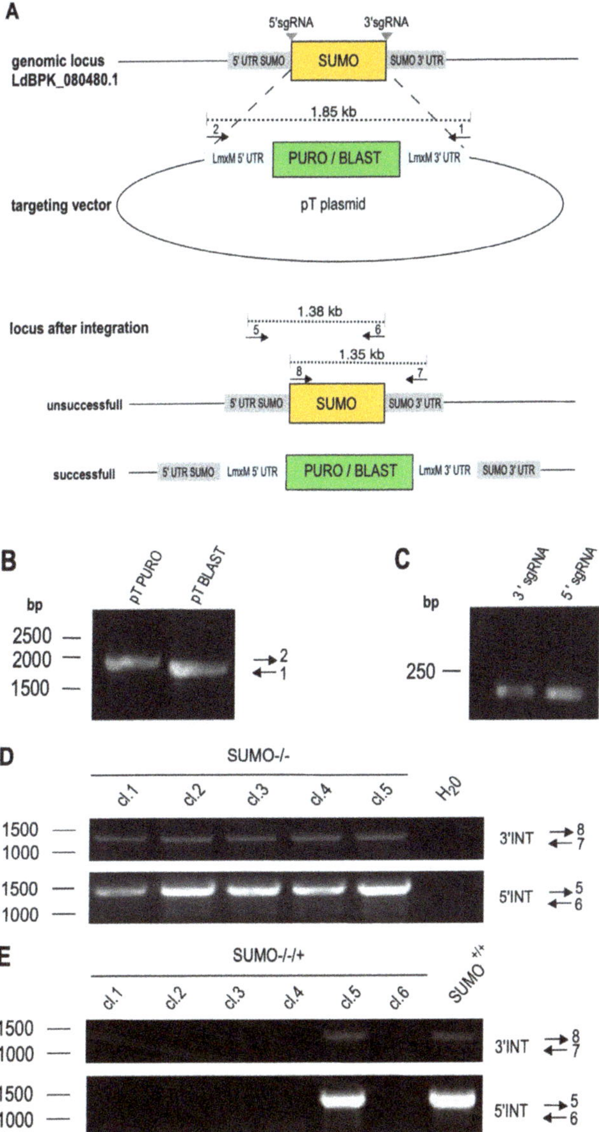

Figure 2. Replacement of *SUMO* in *Leishmania donovani*. (**A**) Schematic representation of LdBPK_080480.1 replacement using the CRISPR/Cas9 technology. *SUMO*-targeting sgRNAs (grey) and the replacement cassettes were PCR-amplified and transfected into a Cas9/T7-RNAP-expressing *L. donovani* strain. Two sets of genotyping primers were used to test for the presence of the gene of interest (GOI) (**B**) Gene-specific replacement cassettes amplified from pTPURO or pTBLAST vector were analyzed by agarose gel electrophoresis and ethidium bromide staining. The position of the DNA size marker is indicated on the left, the primers used are indicated on the right. (**C**) Amplified sgRNA-coding sequences were separated on a 1% agarose gel and stained with ethidium bromide. (**D**) Genotyping of putative gene replacement mutant clones with primer pairs 7+8 or 5+6. PCR products were analyzed by 1% agarose gel electrophoresis. Positions of DNA size markers are shown to the left; the primer pairs are indicated on the right. (**E**) Genotyping of gene replacement mutants in the SUMO over expression background (SUMO$^{-/-/+}$) indicated primer pairs.

3.3. Replacement of SENP

Again, we used the LeishGedit toolbox to design 5′- and 3′-sgRNAs. Selection marker cassettes were amplified from the pTPURO and pTBLAST plasmids with ends targeting the SENP UTR sequences (Figure 3A). A mix of amplified sgRNA coding DNA and amplified selection marker cassettes was then transfected into *L. donovani* (Cas9/T7-RNAP). The transfectants were then cultivated under puromycin/blasticidin double selection. Selected parasite populations were then subjected to limiting dilution to raise putative SENP$^{-/-}$ clones. RT-qPCR analysis of SENP RNA confirmed the lack of GOI-specific RNA for all selected clones, confirming them as null mutants (Figure 3B). Reintroduction of SENP as an episomal gene copy into clone#1 resulted in a massive over production of SENP RNA (SENP$^{-/-/+}$, Figure 3B). Given the confounding potential of Cas9 expression in the mutants, we analyzed them for Cas9 RNA as well (Figure 3C). Only *L. donovani* (Cas9/T7-RNAP) kept under the episome-specific antibiotic selection showed detectable levels of Cas9 RNA while the SENP$^{-/-}$ mutants had lost the expression plasmid during selection and cloning.

To confirm the loss of SENP on a genomic level, we also performed whole genome sequencing of genomic DNA (gDNA) from *L. donovani* wild type, *L. donovani* (Cas9/T7-RNAP), *L. donovani* SENP$^{-/-}$ cl.1 and *L. donovani* SENP$^{-/-}$ cl.2. Next generation sequencing reads were then aligned to *L. donovani* chromosome 26, using the Bowtie2 algorithm. As expected, both wild type and the Cas9/T7 strain showed uninterrupted read coverage over the SENP gene locus. Conversely, the SENP$^{-/-}$ cl.1 showed a complete lack of SENP-specific reads, while clone 2 showed minimal read coverage, possibly indicating a mosaic population (Figure 3D). However, RT-qPCR analysis (Figure 3B) did not show a low level SENP RNA presence. Still, we chose clone 1 for our further analyses.

3.4. SENP Processes the SUMO C Terminus

In the next step, we verified that SENP is indeed required for C-terminal processing of SUMO. We constructed plasmids for ectopic expression of SUMO either with an N-terminal 3×HA tag (Figure 4A) or with a C-terminal 3×HA tag (Figure 4B) and transfected them into *L. donovani* wild type and *L. donovani* SENP$^{-/-}$ cl.1. The cells were grown to mid-logarithmic density, collected by centrifugation and lysed in SDS sample buffer. Samples representing equal cell numbers of *L. donovani*, *L. donovani* SENP$^{-/-}$, *L. donovani* (3×HA-SUMO), *L. donovani* SENP$^{-/-}$ (3×HA-SUMO), *L. donovani* (SUMO-3×HA) and *L. donovani* SENP$^{-/-}$ (SUMO-3×HA) were separated by SDS-PAGE, blotted and stained with an anti-HA antibody (Figure 4C). No unspecific HA tag staining was observed for wild type and the SENP$^{-/-}$ mutant. Ectopic expression of 3×HA-SUMO in the wild type background resulted in a band corresponding to 25 kD, not the expected 16 kD of the triple-HA-tagged SUMO. The aberrant migration of SUMO in SDS-PAGE has been described before [43] and explains the observed band. We also observe numerous bands of higher molecular mass. Their spacing and varying intensities does not reflect the incremental size increases expected of SUMO homoconjugates, but rather suggests HA-tagged, SUMOylated substrate proteins. In the SENP$^{-/-}$ background, expression of the same transgene resulted in a slightly larger band, presumably representing the monomeric, non-processed 3×HA-SUMO. No larger HA-tagged bands were detectable, indicating that unprocessed SUMO is incapable of being conjugated to itself or to target proteins.

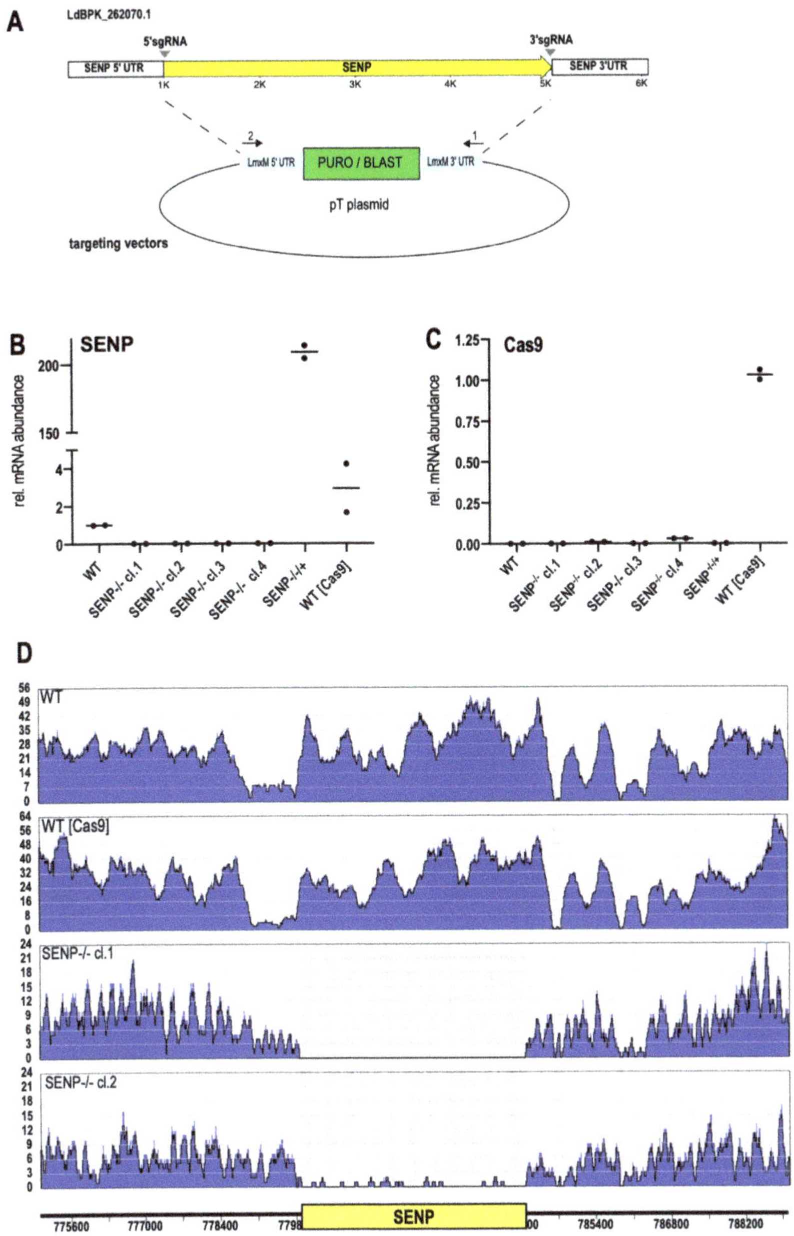

Figure 3. Replacement of SENP. (**A**) Schematic representation of LdBPK_262070 replacement using the CRISPR/Cas9 technology. *SUMO*-targeting sgRNAs (grey arrowheads) and replacement cassettes were PCR-amplified and transfected into *L. donovani* (Cas9/T7RNAP). (**B**,**C**) RT-qPCR of RNA from *L. donovani* wild type (WT), SENP$^{-/-}$ clones 1–4, SENP$^{-/-}$ cl.1[pCLN-SENP], and *L. donovani* (Cas9/T7RNAP). (**B**) SENP-specific RT-qPCR. (**C**) Cas9-specific RT-qPCR. $n = 2$. (**D**) Whole genome sequencing of *L. donovani* wild type (WT), *L. donovani* (Cas9/T7RNAP), SENP$^{-/-}$ cl.1 and SENP$^{-/-}$ cl.2. Sequence reads were aligned to *L. donovani* chromosome 26. The ruler shows the position of the SENP CDS; the numbers refer to the position within chromosome 26. Read coverage is shown in blue.

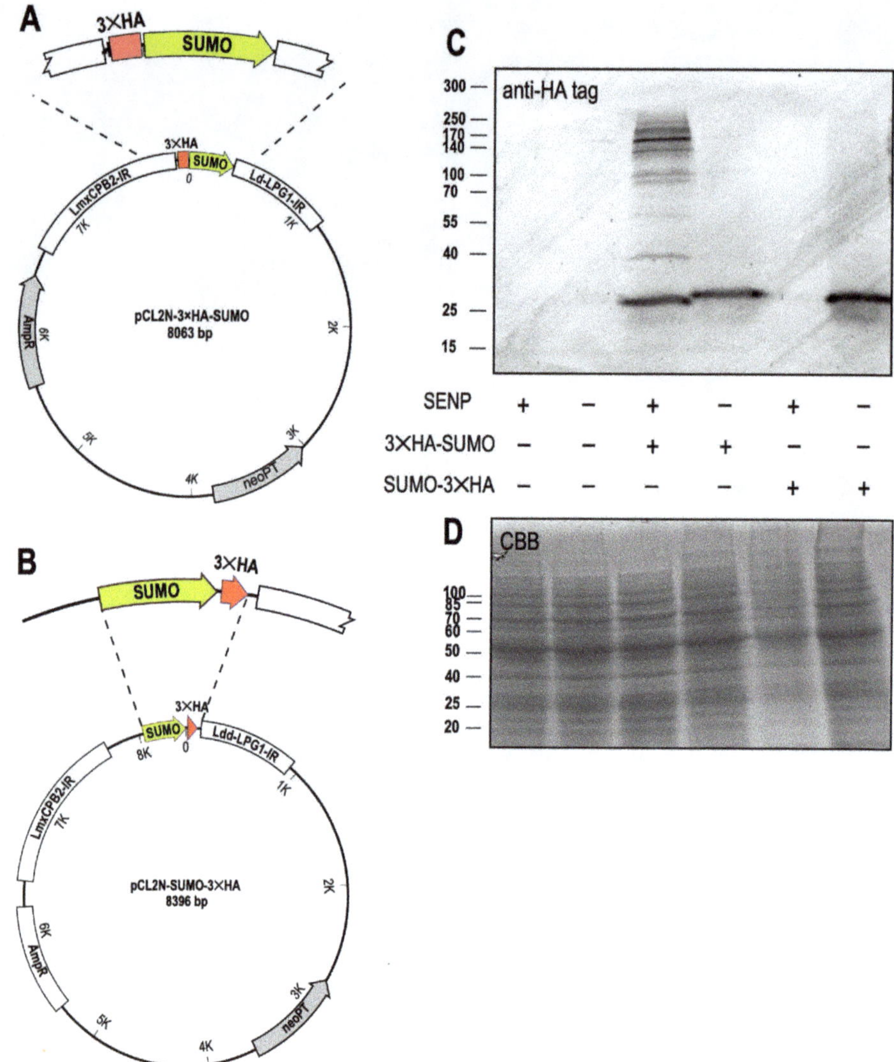

Figure 4. SUMO processing by SENP. (**A**) Schematic drawing of pCL2N-3×HA-SUMO, a plasmid for ectopic expression of SUMO with an N-terminal triple HA tag. (**B**) Schematic drawing of pCL2N-SUMO-3×HA, a plasmid for ectopic expression of SUMO with a C-terminal triple HA tag. (**C**) Western blot of *L. donovani* wild type or SENP$^{-/-}$ null mutants, expressing 3×HA-SUMO or SUMO-3×HA, probed with anti-HA tag antibodies. $n = 2$. (**D**) Coomassie brilliant blue (CBB) staining of replicate SDS-PAGE gel, serving as a loading control. The positions and masses of protein size markers are indicated on the left. Original Western blot and gel images can be seen in Figure S1.

No HA-tagged proteins are visible when the C-terminally tagged SUMO-3×HA is expressed in the wild type background. Expression of the same chimera in SENP$^{-/-}$ cells, by contrast, yields HA-tagged SUMO. This demonstrates that C-terminal processing of SUMO depends on SENP. We conclude that SENP is required for processing and conjugation of SUMO to itself and/or to other proteins, and establishes C-terminal cleavage of SUMO as a critical step for SUMOylation in *Leishmania*.

3.5. SENP-Dependent Processing Determines SUMO Localization

We next investigated the impact of SENP-dependent processing on the subcellular localization of SUMO. For this, promastigotes of six strains, *L. donovani* wild type, *L. donovani* SENP$^{-/-}$, *L. donovani* (3×HA-SUMO), *L. donovani* SENP$^{-/-}$ (3×HA-SUMO), *L. donovani* (SUMO-3×HA) and *L. donovani* SENP$^{-/-}$ (SUMO-3×HA), from logarithmic culture, were spread on glass slides, fixed and stained with DAPI and with anti-HA tag antibody/anti-mouse AlexaFluor 594, followed by immune fluorescence microscopy. As expected, *L. donovani* wild type and *L. donovani* SENP$^{-/-}$ showed no 3×HA-specific staining (Figure 5A,D). We also did not observe a 3×HA-specific signal in *L. donovani* (SUMO-3×HA; Figure 5B), likely due to the cleavage of the C-terminal 3×HA tag in wild type cells. *L. donovani* (3×HA-SUMO) cells showed overlapping staining by DAPI and anti-HA tag antibody, indicating a nuclear localization of 3×HA-SUMO in the wild type.

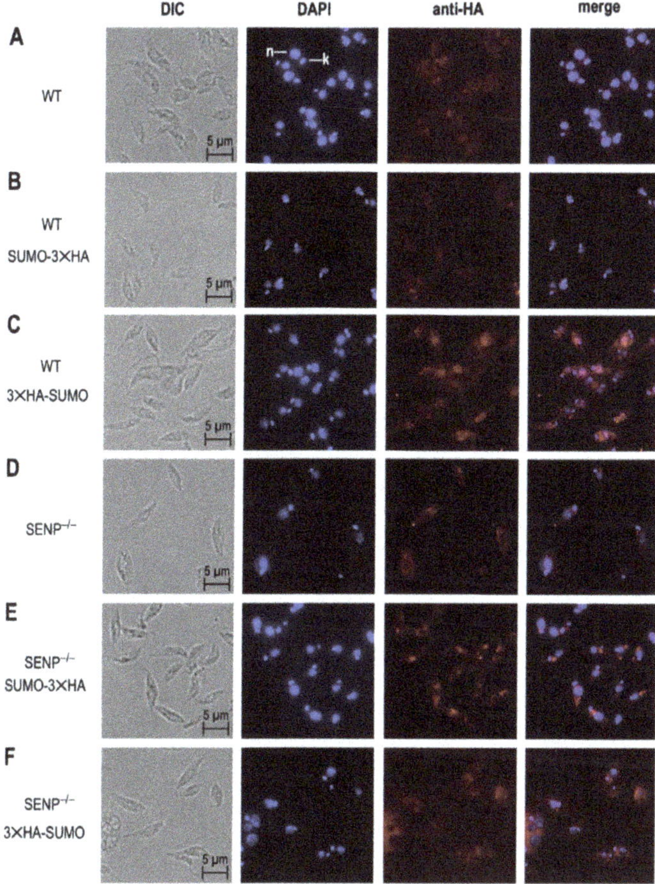

Figure 5. Subcellular localisation of HA-tagged SUMO. Wild type (WT) (**A**) or *L. donovani* SENP$^{-/-}$ (**D**) cells expressing SUMO-3×HA (**B,D**) or 3×HA-SUMO (**C,F**). Cells were visualized by differential interference contrast (DIC), DAPI staining of nucleus and kinetoplast, and mouse anti-HA antibody/anti-mouse AlexaFluor 594. DAPI and anti-HA images were merged with 50% transparence. Size markers (5 µm) are shown in the DIC images; nucleus (n) and kinetoplast (k) are pointed out in the top DAPI image.

L. donovani SENP$^{-/-}$ (SUMO-3×HA) and *L. donovani* SENP$^{-/-}$ (3×HA-SUMO) both showed cytoplasmic staining. We conclude from this that (i) SENP$^{-/-}$ mutants cannot cleave off the C-terminal 3×HA tag (Figure 5E) and (ii) SENP-mediated cleavage of the C terminus is essential for nuclear localization of 3×HA-SUMO. The lack of SENP therefore prevents C-terminal processing of SUMO, preventing SUMO from attaining or maintaining a nuclear localization.

3.6. Growth Phenotypes of SENP Null Mutants

Given its critical function in SUMOylation, we tested the impact of SENP on the growth of *L. donovani* at different temperatures. *L. donovani*, *L. donovani* (Cas9/T7-RNAP), SENP$^{-/-}$ clones 1 and 2 and the SENP$^{-/-/+}$ add-back strain were seeded at low density, and growth was then monitored over 72 h. Cell densities at 72 h were normalized, with wild type *L. donovani* set at 100% growth. At optimal growth conditions, 25 °C and pH 7.0, both SENP$^{-/-}$ null mutants showed a 50% reduced proliferation compared with wild type and the Cas9-expressing strain. This growth phenotype was reversed by ectopic SENP expression (Figure 6A). At 37 °C, we recorded less, but still significant growth reduction due to the loss of SENP (Figure 6B). This may indicate that SENP function and/or SUMO conjugation is more important at the lower temperature associated with the insect stage.

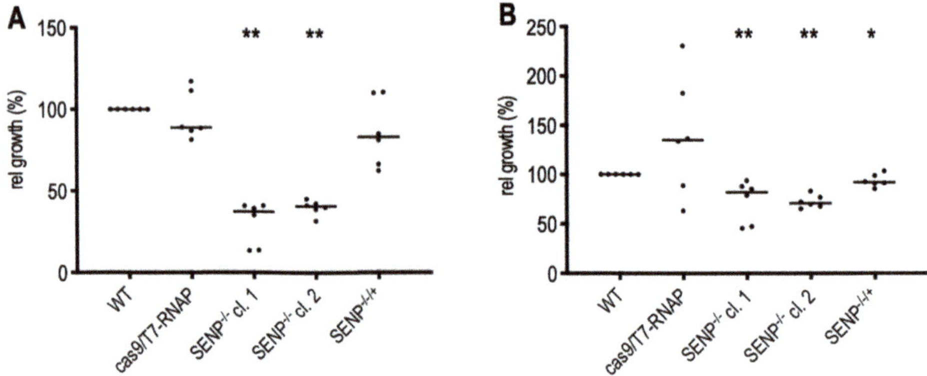

Figure 6. In vitro growth of wild type and mutant *L. donovani*. Cells were seeded at 5×10^5/mL and grown either at 25 °C/pH 7.0 (**A**) or at 37 °C/pH 7.0 (**B**) for 72 h. Final cell densities were normalized against wild type growth (100%). Bars show the median cell growth. $n = 6$ (3 biol. repeats, 2 techn. repeats each). ** = $p < 0.01$; * = $p < 0.05$ (U-test, two-sided).

We also tested the intracellular survival of SENP null mutants in mouse bone marrow-derived macrophages and found no differences in parasite loads compared with wild type parasites (A.B. and C.B., unpublished observations), consistent with a primary role for SENP in the promastigote stage.

4. Discussion

As a vector-transmitted parasite, *Leishmania* must adapt to vastly different environments, carbon sources, and antimicrobial defense mechanisms. This must be achieved without differentially regulated RNA synthesis [3,44,45]. Instead, *Leishmania* relies on modulated RNA stability [46], RNA processing [47] and inducible translation [2,5,48] as means of short-term gene expression control. Long-term adaption to changing environments, by contrast appears to be mediated by gene copy number variations, either by chromosomal aneuploidy [6,7] or by amplification of genes and gene clusters [36,49,50]. A third level of gene expression control are PTMs of proteins that may activate or inhibit activities or influence localization. Examples of PTMs are protein kinase mediated phosphorylation of threonine and serine side chains [8,11,13]. Side chain-specific modifications can impact on protein folding or protein–protein interactions. The covalent attachment of modifying polypeptides is another, as yet little understood mode of expression control in *Leishmania*. So far, only the impact of a ubiquitin fold modifier (UFM1) protein was demonstrated [17,51] in *L. donovani*. A similar modifier, small ubiquitin-like modifier (SUMO) was identified and characterized in *Trypanosoma* spp. where it is involved in surface antigen expression and nuclear organization [18,19,52]. Here we describe the *Leishmania* SUMO and SENP orthologs and characterize them by reverse genetic, biochemical and cell biological means.

To the best of our knowledge, SUMO is an essential gene in *L. donovani* promastigotes. Attempts to produce SUMO$^{-/-}$ null mutants by CRISPR-mediated gene editing failed while the same gene replacement tools were successfully employed in a strain carrying ectopic SUMO copies (Figure 2D,E), indicating that null mutants are either non-viable or non-proliferative as promastigotes *in vitro*. It was shown for higher eukaryotes that the SUMO pathways are essential during differentiation processes [53,54], but our literature search did not turn up reports of an outright SUMO gene replacement. This is probably also due to the presence of multiple SUMO genes in mammalian cells [55], which may confound reverse genetics approaches.

Unlike SUMO, SENP appears to be non-essential, albeit with a significant impact on promastigote proliferation at optimal growth temperature, with a smaller effect at mammalian tissue temperatures. Fittingly, the survival of amastigotes within mouse macrophages is unaffected by the loss of SENP. This may indicate an important role of SENP and its clients during logarithmic growth of *Leishmania* promastigotes, but less impact during the slow growth of intracellular amastigotes. Yet, with SUMO C-terminal processing abrogated by the loss of SENP (Figure 4C) and its nuclear localization severely reduced (Figure 5), it surprises that the effect of SENP loss is not equally deleterious as the loss of SUMO. Strong signals for C-terminally tagged SUMO in SENP null mutants (Figures 4C and 5) argue against a SUMO processing pathway using alternative proteases. One must therefore assume, that apart from its role as a conjugated protein modifier, SUMO must have additional, essential functions in *Leishmania*.

SUMO and its processing protease, SENP, are proteins with constitutive, above-average synthesis rates in *Leishmania*, indicating a need for abundance or a high turnover rate. Indeed, SUMO (LinJ.08.0480) showed little changes of abundance during promastigote-to-amastigote differentiation *in vitro* [21], and SENP (LinJ.26.2070) has a constitutive abundance too [24].

Immune fluorescence microscopy of tagged SUMO protein shows a nuclear, but not kinetoplast localization. This localization fully depends on SENP-mediated C-terminal processing (Figure 5). This result is in keeping with reports that show involvement of SUMO in nuclear organization and chromosome segregation [53]. Preliminary data (A.B.), however, show no impact of a SENP loss on the accessibility of *L. donovani* chromatin to micrococcal nuclease digest. This must be seen, however, in the context of the Trypanosomatida having a divergent chromatin structure and nuclear architecture. While the genomic DNA is assembled into 10 nm fibers of nucleosomes, these protozoa lack further condensation of chromosomes into 30 nm solenoid fibers [56]. The function of SUMO in the nucleus may therefore be diverged.

The affinity of HA-tagged SUMO for the nucleus is also a promising possibility to identify SUMOylated proteins from the cytoplasm and the nucleus via immune precipitation of SUMO-target conjugates and subsequent mass spectrometric analysis.

5. Conclusions

Leishmania parasites express proteins belonging to the SUMO protein modification pathway. The gene coding for SUMO is essential for growth and/or viability of *L. donovani* promastigotes, while the SENP processing enzyme is required for the C-terminal processing of SUMO and its nuclear localization, but dispensable for *L. donovani* viability. The SENP$^{-/-}$ null mutants show a 60% reduced growth at ambient temperature, but less impact at mammalian tissue temperature. No decrease of viability during *in vitro* infection can be observed, indicating a primary role for SENP-dependent SUMOylation in the fast growing promastigote stage. Additionally, the viability of SENP$^{-/-}$ null mutants hints at a vital importance of as yet unknown, SENP-independent functions of SUMO.

Supplementary Materials: The following are available online at http://www.mdpi.com/2073-4425/11/10/1198/s1, Table S1: Oligonucleotides used for targeting constructs; Table S2. Primers used for RT-qPCR and PCR; and Figure S1: Images of original Western blot and Coomassie Brilliant Blue-stained PA gel.

Author Contributions: A.B.: gene replacements, transgene expression and phenotype analyses, imaging, draft manuscript. C.K.-B.: gene , experimental design, imaging, draft manuscript, supervision. C.B.: NGS analysis, growth kinetics. M.S.: gene replacements. J.C.: study design, supervision, artwork, manuscript conception and finalization. All authors have read and agreed to the published version of the manuscript.

Funding: A.B. is funded by the Joachim Herz Graduate School of Infection Biology at the Bernhard Nocht Institute for Tropical Medicine, Hamburg. No further external funding was received.

Acknowledgments: We thank D. Çadar for the use of the Illumina MiSeq system.

Conflicts of Interest: The authors declare no conflict of interest.

References

1. Argaman, M.; Aly, R.; Shapira, M. Expression of heat shock protein 83 in Leishmania is regulated post- transcriptionally. *Mol. Biochem. Parasitol.* **1994**, *64*, 95–110. [CrossRef]

2. Brandau, S.; Dresel, A.; Clos, J. High constitutive levels of heat-shock proteins in human-pathogenic parasites of the genus Leishmania. *Biochem. J.* **1995**, *310*, 225–232. [CrossRef] [PubMed]

3. Clayton, C. Regulation of gene expression in trypanosomatids: Living with polycistronic transcription. *Open Biol.* **2019**, *9*, 190072. [CrossRef] [PubMed]

4. Bringaud, F.; Muller, M.; Cerqueira, G.C.; Smith, M.; Rochette, A.; El-Sayed, N.M.; Papadopoulou, B.; Ghedin, E. Members of a large retroposon family are determinants of post-transcriptional gene expression in Leishmania. *PLoS Pathog.* **2007**, *3*, 1291–1307. [CrossRef]

5. Bifeld, E.; Lorenzen, S.; Bartsch, K.; Vasquez, J.J.; Siegel, T.N.; Clos, J. Ribosome Profiling Reveals HSP90 Inhibitor Effects on Stage-Specific Protein Synthesis in Leishmania donovani. *mSystems* **2018**, *3*. [CrossRef]

6. Mannaert, A.; Downing, T.; Imamura, H.; Dujardin, J.C. Adaptive mechanisms in pathogens: Universal aneuploidy in Leishmania. *Trends Parasitol.* **2012**, *28*, 370–376. [CrossRef]

7. Laffitte, M.N.; Leprohon, P.; Papadopoulou, B.; Ouellette, M. Plasticity of the Leishmania genome leading to gene copy number variations and drug resistance. *F1000 Res.* **2016**, *5*, 2350. [CrossRef]

8. Wiese, M. *Leishmania* MAP kinases–familiar proteins in an unusual context. *Int. J. Parasitol.* **2007**, *37*, 1053–1062. [CrossRef]

9. Soulat, D.; Bogdan, C. Function of Macrophage and Parasite Phosphatases in Leishmaniasis. *Front. Immunol.* **2017**, *8*, 1838. [CrossRef]

10. Morales, M.; Watanabe, R.; Dacher, M.; Chafey, P.; Osorio y Fortéa, J.; Beverley, S.; Ommen, G.; Clos, J.; Hem, S.; Lenormand, P.; et al. Phosphoproteome dynamics reveals heat shock protein complexes specific to the *Leishmania* infectious stage. *Proc. Natl. Acad. Sci. USA* **2010**, *107*, 8381–8386. [CrossRef]

11. Mottram, J.C. cdc2-related protein kinases and cell cycle control in trypanosomatids. *Parasitol. Today* **1994**, *10*, 253–257. [CrossRef]

12. Hassan, P.; Fergusson, D.; Grant, K.M.; Mottram, J.C. The CRK3 protein kinase is essential for cell cycle progression of Leishmania mexicana. *Mol. Biochem. Parasitol.* **2001**, *113*, 189–198. [CrossRef]

13. Spath, G.F.; Drini, S.; Rachidi, N. A touch of Zen: Post-translational regulation of the Leishmania stress response. *Cell. Microbiol.* **2015**, *17*, 632–638. [CrossRef]

14. Paolantonacci, P.; Lawrence, F.; Lederer, F.; Robert-Gero, M. Protein methylation and protein methylases in Leishmania donovani and Leishmania tropica promastigotes. *Mol. Biochem. Parasitol.* **1986**, *21*, 47–54. [CrossRef]

15. Rosenzweig, D.; Smith, D.; Myler, P.J.; Olafson, R.W.; Zilberstein, D. Post-translational modification of cellular proteins during Leishmania donovani differentiation. *Proteomics* **2008**, *8*, 1843–1850. [CrossRef]

16. Gannavaram, S.; Sharma, P.; Duncan, R.C.; Salotra, P.; Nakhasi, H.L. Mitochondrial associated ubiquitin fold modifier-1 mediated protein conjugation in Leishmania donovani. *PLoS ONE* **2011**, *6*, e16156. [CrossRef]

17. Gannavaram, S.; Connelly, P.S.; Daniels, M.P.; Duncan, R.; Salotra, P.; Nakhasi, H.L. Deletion of mitochondrial associated ubiquitin fold modifier protein Ufm1 in Leishmania donovani results in loss of beta-oxidation of fatty acids and blocks cell division in the amastigote stage. *Mol. Microbiol.* **2012**, *86*, 187–198. [CrossRef]

18. Annoura, T.; Makiuchi, T.; Sariego, I.; Aoki, T.; Nara, T. SUMOylation of paraflagellar rod protein, PFR1, and its stage-specific localization in Trypanosoma cruzi. *PLoS ONE* **2012**, *7*, e37183. [CrossRef]

19. Klein, C.A.; Droll, D.; Clayton, C. SUMOylation in Trypanosoma brucei. *PeerJ* **2013**, *1*, e180. [CrossRef]

20. Saura, A.; Iribarren, P.A.; Rojas-Barros, D.; Bart, J.M.; Lopez-Farfan, D.; Andres-Leon, E.; Vidal-Cobo, I.; Boehm, C.; Alvarez, V.E.; Field, M.C.; et al. SUMOylated SNF2PH promotes variant surface glycoprotein expression in bloodstream trypanosomes. *EMBO Rep.* **2019**, *20*, e48029. [CrossRef]

21. Rosenzweig, D.; Smith, D.; Opperdoes, F.; Stern, S.; Olafson, R.W.; Zilberstein, D. Retooling Leishmania metabolism: From sand fly gut to human macrophage. *FASEB J.* **2008**, *22*, 590–602. [CrossRef] [PubMed]

22. Xu, Z.; Au, S.W. Mapping residues of SUMO precursors essential in differential maturation by SUMO-specific protease, SENP1. *Biochem. J.* **2005**, *386*, 325–330. [CrossRef] [PubMed]

23. Geiss-Friedlander, R.; Melchior, F. Concepts in sumoylation: A decade on. *Nat. Rev. Mol. Cell Biol.* **2007**, *8*, 947–956. [CrossRef] [PubMed]

24. Lahav, T.; Sivam, D.; Volpin, H.; Ronen, M.; Tsigankov, P.; Green, A.; Holland, N.; Kuzyk, M.; Borchers, C.; Zilberstein, D.; et al. Multiple levels of gene regulation mediate differentiation of the intracellular pathogen Leishmania. *FASEB J. Off. Publ. Fed. Am. Soc. Exp. Biol.* **2011**, *25*, 515–525. [CrossRef] [PubMed]

25. Hay, R.T. SUMO: A history of modification. *Mol. Cell* **2005**, *18*, 1–12. [CrossRef] [PubMed]

26. Hietakangas, V.; Anckar, J.; Blomster, H.A.; Fujimoto, M.; Palvimo, J.J.; Nakai, A.; Sistonen, L. PDSM, a motif for phosphorylation-dependent SUMO modification. *Proc. Natl. Acad. Sci. USA* **2006**, *103*, 45–50. [CrossRef]

27. Du, L.; Li, Y.J.; Fakih, M.; Wiatrek, R.L.; Duldulao, M.; Chen, Z.; Chu, P.; Garcia-Aguilar, J.; Chen, Y. Role of SUMO activating enzyme in cancer stem cell maintenance and self-renewal. *Nat. Commun.* **2016**, *7*, 12326. [CrossRef]

28. Yang, Y.; He, Y.; Wang, X.; Liang, Z.; He, G.; Zhang, P.; Zhu, H.; Xu, N.; Liang, S. Protein SUMOylation modification and its associations with disease. *Open Biol.* **2017**, *7*. [CrossRef]

29. Barak, E.; Amin-Spector, S.; Gerliak, E.; Goyard, S.; Holland, N.; Zilberstein, D. Differentiation of Leishmania donovani in host-free system: Analysis of signal perception and response. *Mol. Biochem. Parasitol.* **2005**, *141*, 99–108. [CrossRef]

30. Hombach, A.; Ommen, G.; Chrobak, M.; Clos, J. The Hsp90-Sti1 Interaction is Critical for *Leishmania donovani* Proliferation in Both Life Cycle Stages. *Cell. Microbiol.* **2013**, *15*, 585–600. [CrossRef]

31. Zirpel, H.; Clos, J. Gene Replacement by Homologous Recombination. *Methods Mol. Biol.* **2019**, *1971*, 169–188. [CrossRef] [PubMed]

32. Bartsch, K.; Hombach-Barrigah, A.; Clos, J. Hsp90 inhibitors radicicol and geldanamycin have opposing effects on Leishmania Aha1-dependent proliferation. *Cell Stress Chaperones* **2017**, *22*, 729–742. [CrossRef] [PubMed]

33. Green, M.R.; Sambrook, J.; Sambrook, J. *Molecular Cloning: A Laboratory Manual*, 4th ed.; Cold Spring Harbor Laboratory Press: Cold Spring Harbor, NY, USA, 2012.

34. Beneke, T.; Madden, R.; Makin, L.; Valli, J.; Sunter, J.; Gluenz, E. A CRISPR Cas9 high-throughput genome editing toolkit for kinetoplastids. *R. Soc. Open Sci.* **2017**, *4*, 170095. [CrossRef] [PubMed]

35. Choudhury, K.; Zander, D.; Kube, M.; Reinhardt, R.; Clos, J. Identification of a Leishmania infantum gene mediating resistance to miltefosine and SbIII. *Int. J. Parasitol.* **2008**, *38*, 1411–1423. [CrossRef]

36. Kröber-Boncardo, C.; Lorenzen, S.; Brinker, C.; Clos, J. Casein kinase 1.2 over expression restores stress resistance to Leishmania donovani HSP23 null mutants. *Sci. Rep.* **2020**, *10*, 15969. [CrossRef]

37. Krobitsch, S.; Brandau, S.; Hoyer, C.; Schmetz, C.; Hübel, A.; Clos, J. *Leishmania donovani* heat shock protein 100: Characterization and function in amastigote stage differentiation. *J. Biol. Chem.* **1998**, *273*, 6488–6494. [CrossRef]

38. Hombach, A.; Ommen, G.; MacDonald, A.; Clos, J. A small heat shock protein is essential for thermotolerance and intracellular survival of Leishmania donovani. *J. Cell Sci.* **2014**, *127*, 4762–4773. [CrossRef]

39. Beneke, T.; Gluenz, E. LeishGEdit: A Method for Rapid Gene Knockout and Tagging Using CRISPR-Cas9. *Methods Mol. Biol.* **2019**, *1971*, 189–210. [CrossRef]

40. Langmead, B.; Trapnell, C.; Pop, M.; Salzberg, S.L. Ultrafast and memory-efficient alignment of short DNA sequences to the human genome. *Genome Biol.* **2009**, *10*, R25. [CrossRef]

41. Mann, H.B.; Whitney, D.R. On a test of whether one of two random variables is stochastically larger than the other. *Ann. Math. Stat.* **1947**, *18*, 50–60. [CrossRef]

42. Ommen, G.; Lorenz, S.; Clos, J. One-step generation of double-allele gene replacement mutants in Leishmania donovani. *Int. J. Parasitol.* **2009**, *39*, 541–546. [CrossRef] [PubMed]

43. Johnson, E.S. Protein modification by SUMO. *Annu. Rev. Biochem.* **2004**, *73*, 355–382. [CrossRef] [PubMed]

44. Clayton, C.E. Life without transcriptional control? From fly to man and back again. *EMBO J.* **2002**, *21*, 1881–1888. [CrossRef] [PubMed]

45. Clayton, C.; Shapira, M. Post-transcriptional regulation of gene expression in trypanosomes and leishmanias. *Mol. Biochem. Parasitol.* **2007**, *156*, 93–101. [CrossRef]

46. Aly, R.; Argaman, M.; Halman, S.; Shapira, M. A regulatory role for the 5′ and 3′ untranslated regions in differential expression of hsp83 in Leishmania. *Nucleic Acids Res.* **1994**, *22*, 2922–2929. [CrossRef]

47. Zilka, A.; Garlapati, S.; Dahan, E.; Yaolsky, V.; Shapira, M. Developmental regulation of HSP83 in Leishmania: Transcript levels are controlled by the efficiency of 3? RNA processing and preferential translation is directed by a determinant in the 3′ UTR. *J. Biol. Chem.* **2001**, *11*, 11.

48. Clos, J.; Brandau, S.; Hoyer, C. Chemical stress does not induce heat shock protein synthesis in *Leishmania donovani*. *Protist* **1998**, *149*, 167–172. [CrossRef]

49. Ubeda, J.M.; Raymond, F.; Mukherjee, A.; Plourde, M.; Gingras, H.; Roy, G.; Lapointe, A.; Leprohon, P.; Papadopoulou, B.; Corbeil, J.; et al. Genome-wide stochastic adaptive DNA amplification at direct and inverted DNA repeats in the parasite *Leishmania*. *PLoS Biol.* **2014**, *12*, e1001868. [CrossRef] [PubMed]

50. Dumetz, F.; Cuypers, B.; Imamura, H.; Zander, D.; D'Haenens, E.; Maes, I.; Domagalska, M.A.; Clos, J.; Dujardin, J.C.; De Muylder, G. Molecular Preadaptation to Antimony Resistance in Leishmania donovani on the Indian Subcontinent. *mSphere* **2018**, *3*. [CrossRef]

51. Gannavaram, S.; Davey, S.; Lakhal-Naouar, I.; Duncan, R.; Nakhasi, H.L. Deletion of ubiquitin fold modifier protein Ufm1 processing peptidase Ufsp in L. donovani abolishes Ufm1 processing and alters pathogenesis. *PLoS Negl. Trop. Dis.* **2014**, *8*, e2707. [CrossRef]

52. Klein, U.R.; Nigg, E.A. SUMO-dependent regulation of centrin-2. *J. Cell Sci.* **2009**, *122*, 3312–3321. [CrossRef] [PubMed]

53. Nacerddine, K.; Lehembre, F.; Bhaumik, M.; Artus, J.; Cohen-Tannoudji, M.; Babinet, C.; Pandolfi, P.P.; Dejean, A. The SUMO pathway is essential for nuclear integrity and chromosome segregation in mice. *Dev. Cell* **2005**, *9*, 769–779. [CrossRef] [PubMed]

54. Wang, L.; Wansleeben, C.; Zhao, S.; Miao, P.; Paschen, W.; Yang, W. SUMO2 is essential while SUMO3 is dispensable for mouse embryonic development. *EMBO Rep.* **2014**, *15*, 878–885. [CrossRef] [PubMed]

55. Dohmen, R.J. SUMO protein modification. *Biochim. Biophys. Acta* **2004**, *1695*, 113–131. [CrossRef]

56. Hecker, H.; Betschart, B.; Bender, K.; Burri, M.; Schlimme, W. The chromatin of trypanosomes. *Int. J. Parasitol.* **1994**, *24*, 809–819. [CrossRef]

Article

Application of CRISPR/Cas9-Based Reverse Genetics in *Leishmania braziliensis*: Conserved Roles for HSP100 and HSP23

Vanessa Adaui [1,2,3,†], **Constanze Kröber-Boncardo** [1,†], **Christine Brinker** [1], **Henner Zirpel** [1,4], **Julie Sellau** [1], **Jorge Arévalo** [2], **Jean-Claude Dujardin** [3,4,5,6] and **Joachim Clos** [1,*]

[1] Bernhard Nocht Institute for Tropical Medicine, D-20359 Hamburg, Germany; vanessa.adaui@upc.edu.pe (V.A.); kroeber@bnitm.de (C.K.-B.); brinker@bnitm.de (C.B.); henner-444@gmx.net (H.Z.); sellau@bnitm.de (J.S.)
[2] Instituto de Medicina Tropical Alexander von Humboldt, Universidad Peruana Cayetano Heredia, Lima 15102, Peru; jorge.arevalo@upch.pe
[3] Centre for Research and Innovation, Faculty of Health Sciences, Universidad Peruana de Ciencias Aplicadas, Lima 15067, Peru; JCDujardin@itg.be
[4] City of Hope National Medical Center, Duarte, CA 91010, USA
[5] Institute of Tropical Medicine, 2000 Antwerp, Belgium
[6] Department of Biomedical Sciences, University of Antwerp, 2000 Antwerp, Belgium
[*] Correspondence: clos@bnitm.de; Tel.: +49-40-42818481
[†] These authors contributed equally to this work.

Received: 4 September 2020; Accepted: 25 September 2020; Published: 30 September 2020

Abstract: The protozoan parasite *Leishmania (Viannia) braziliensis (L. braziliensis)* is the main cause of human tegumentary leishmaniasis in the New World, a disease affecting the skin and/or mucosal tissues. Despite its importance, the study of the unique biology of *L. braziliensis* through reverse genetics analyses has so far lagged behind in comparison with Old World *Leishmania* spp. In this study, we successfully applied a cloning-free, PCR-based CRISPR–Cas9 technology in *L. braziliensis* that was previously developed for Old World *Leishmania major* and New World *L. mexicana* species. As proof of principle, we demonstrate the targeted replacement of a transgene (*eGFP*) and two *L. braziliensis* single-copy genes (*HSP23* and *HSP100*). We obtained homozygous Cas9-free *HSP23*- and *HSP100*-null mutants in *L. braziliensis* that matched the phenotypes reported previously for the respective *L. donovani* null mutants. The function of *HSP23* is indeed conserved throughout the Trypanosomatida as *L. major HSP23* null mutants could be complemented phenotypically with transgenes from a range of trypanosomatids. In summary, the feasibility of genetic manipulation of *L. braziliensis* by CRISPR–Cas9-mediated gene editing sets the stage for testing the role of specific genes in that parasite's biology, including functional studies of virulence factors in relevant animal models to reveal novel therapeutic targets to combat American tegumentary leishmaniasis.

Keywords: *Leishmania braziliensis*; reverse genetics; CRISPR–Cas9; gene targeting; phenotyping; heat shock proteins

---

1. Introduction

The protozoan parasite *Leishmania (Viannia) braziliensis* (henceforth: *L. braziliensis)* is the main causative agent of human tegumentary leishmaniasis in Latin America. Infection with *L. braziliensis* generally causes cutaneous lesions, with possible, severe, metastatic mucosal involvement, and it is difficult to cure with the first-line pentavalent antimonial drugs [1–4]. In spite of its importance, the biology of *L. braziliensis* has not been analysed extensively, in part due to the limited set of genetic manipulation tools developed or adapted to this species.

While Gene replacement using homologous recombination has proven a useful tool for testing gene function in Old World *Leishmania* spp. [5–7], yet—to our knowledge—no gene replacement analyses have been reported for *L. braziliensis*. However, a functional RNA interference (RNAi) machinery, predicted from the *L. braziliensis* genome sequence [8], was corroborated experimentally [9], allowing gene function analysis in this species [9,10]. The RNAi pathway and associated genes are absent in species of the *L.* (*Leishmania*) subgenus such as *L. major* and *L. donovani* [9]. However, RNAi-based gene knock-down is prone to off-target effects [11], which can confound phenotypic analyses.

Recently, the CRISPR (clustered regularly interspaced short palindromic repeats)–Cas9 (CRISPR-associated protein 9) technology is revolutionizing gene function studies in a wide range of organisms, due to its high efficiency, precision, relative simplicity, and versatility [12]. Using this tool, the Cas9 endonuclease can be directed to a specific genomic locus by a single guide RNA (sgRNA) to introduce a double-stranded break (DSB) in the target DNA [13]. DSBs compromise genomic integrity and are identified and repaired by the nuclear machinery by regulated and error-prone DNA repair pathways [14], and homologous donor DNA templates may be inserted introducing defined changes into the DNA near the DSB as part of the repair process [15].

CRISPR–Cas9-mediated gene targeting and gene editing (e.g., to generate point mutations, or add tags to endogenous genes) have been successfully developed and applied in kinetoplastids, including *Trypanosoma cruzi* [16], *T. brucei* [17,18], and several species of *Leishmania* [17,19–24], with the notable exception of New World *L.* (*Viannia*) species. This new technology has greatly improved the efficiency of gene targeting in *Leishmania* spp. over traditional homologous recombination-based gene replacement.

First, CRISPR–Cas9 allows the rapid generation of gene deletion or gene disruption mutants in the promastigote stage (within 1–2 weeks depending on the species); thus minimising the occurrence of compensatory adaptations in the parasites [21,25]. This is particularly the case when a gene required for optimal *in vitro* survival and/or growth is targeted [26], since *Leishmania* have the remarkable ability to adapt to environmental changes by chromosome copy number variations [27,28]. Second, the generation of CRISPR-derived null mutants is facilitated by the use of donor DNA repair cassettes (containing antibiotic selection markers) flanked by short homology arms targeting the gene of interest (GOI), in a single transfection [17,29]. Third, both single and multigene families can be targeted with this system [16,30], and it even allows simultaneous editing of multiple loci [24,30], as well as the identification of essential genes [20,30,31]. CRISPR gene editing also allows for *in situ* addition of flanking loxP sites to a gene of interest and the subsequent rapamycin-inducible gene deletion by dimerisable Cre (DiCre) recombinase [32,33]. This facilitates deletion of essential genes and observation of the cell biological and morphological effects on living cells in a time-dependent manner.

In the absence of a donor DNA repair template, *Leishmania* use microhomology-mediated end-joining (MMEJ) or single-strand annealing (SSA) to repair DSBs, both of which lead to deletions of various sizes that disrupt the targeted gene [20,30,31]. These DSB repair pathways (MMEJ and SSA) have a generally low efficiency in *Leishmania*, and SSA may result in unwanted deletions of adjacent genes [31]. Transfections of a donor DNA template to facilitate homology-directed repair significantly improves CRISPR–Cas9 gene targeting efficiency and specificity, and eases the identification of CRISPR-edited mutants in *Leishmania* [17,19,20,30,31].

In this study, we establish the CRISPR–Cas9 technology as an experimental tool for reverse genetics in *L. braziliensis* facilitating the generation of null mutants and the analysis of gene function in this important human pathogen. We applied a cloning-free, PCR-based CRISPR–Cas9 method that was used successfully in *Leishmania mexicana*, *L. major*, *L. donovani*, and *Trypanosoma brucei* for rapid and precise gene editing [17,21]. As a proof of principle, we first targeted an integrated transgene coding for enhanced green fluorescent protein (eGFP) and then replaced two single-copy genes of *L. braziliensis* encoding heat shock proteins HSP23 and HSP100. In addition, we show that functions of these genes are conserved in the *Viannia* subgenus of *Leishmania*.

2. Materials and Methods

2.1. Leishmania Strains and Culture

Promastigotes of the Peruvian *L. braziliensis* strain PER005 (MHOM/PE/01/LH2182(PER005)) [34] clone 2 (clone originally derived from a clinical isolate), *L. donovani* 1S (MHOM/SD/62/1S) [35], *L. major* 5-ASKH (MHOM/SU/73/5-ASKH) [36], and their genetically modified derived lines reported in this study were routinely grown at 25 °C in monophasic M199 medium (Sigma-Aldrich, München, Germany) supplemented with 20% heat-inactivated fetal calf serum (Sigma-Aldrich), 10 mg/L hemin, 100 µM adenine, 5 µM 6-biopterin, 40 mM HEPES (pH 7.4), 2 mM L-glutamine, 100 units/ml penicillin and 100 µg/mL streptomycin (hereafter referred as complete M199 medium) [37,38]. Cultures were subcultured to fresh medium every 3–4 days. Appropriate selection drugs were added to the medium when necessary as indicated below. The isolation and use of ex vivo macrophage progenitor cells from mice was duly registered with the Animal Protection Authority of the State of Hamburg and in accordance with the German Animal Protection Law.

2.2. Promastigote Cultivation

Promastigotes were grown in complete M199 medium in 25 cm^2 cell culture flasks. Cell density was monitored using a CASY® Cell Counter and Analyzer (Roche, Mannheim, Germany).

2.3. Transfections, Selection, and Cell Cloning

Electrotransfection of circular DNA was performed using a Bio-Rad Gene Pulser apparatus and electroporation conditions as described [39]. Briefly, promastigotes grown to mid-log phase were harvested by centrifugation (1251 *g*, 10 min, 4 °C), washed twice with ice-cold phosphate-buffered saline (PBS), once in pre-chilled electroporation buffer, and suspended in electroporation buffer at a density of 1×10^8 parasites/mL.

For the generation of double allele replacements and for the integration of linearised DNA constructs, cells were transfected following the Amaxa protocol as described previously [17,40]. Briefly, 1×10^7 promastigotes grown to mid- to late-log phase were harvested by centrifugation at 1251 *g* for 10 min (at RT), washed once with 1 × Tb-BSF electroporation buffer (90 mM NaHPO$_3$, 5 mM KCl, 0.15 mM CaCl$_2$, 50 mM HEPES, pH 7.3) [41] at RT, and suspended in 150 µL electroporation buffer per transfection. For gene editing, the cell suspension was mixed with the pooled unpurified PCR amplicons for the two single-guide RNA (sgRNA) templates and two donor DNAs (combined volume approximately 100 µL, heat-sterilised at 94 °C for 5 min before transfection) in a total volume of 250 µL. For integration of transgenes into the 18S SSU rRNA locus, cells were mixed with 2 µg of the *Swa*I-linearised DNA construct. Electroporation was performed in a 0.2 cm gap Gene Pulser electroporation cuvette (Bio-Rad, München, Germany) using one pulse with program X-001 in the Amaxa Nucleofector IIb device (Lonza, Basel, Switzerland). A mock transfection control without DNA was included to check the real transfection efficiency.

Following electroporation, cells were immediately transferred into 5 mL drug-free pre-warmed complete M199 medium in 25 cm^2 cell culture flasks. After parasite recovery at 25 °C for 16–20 h, the selection antibiotics were added at the indicated strain-specific concentrations. Nourseothricin (ClonNat, at 150 µg/mL for all parasite species; Werner BioAgents, Jena, Germany), hygromycin B (at 50 µg/mL for all parasite species; Roth, Karlsruhe, Germany), bleocin (at 5 µg/mL; Calbiochem, San Diego, CA, USA). Additionally, blasticidin (at 10 µg/mL for *L. donovani* and *L. major*; at 2.5 µg/mL for *L. braziliensis*; Roth, Karlsruhe, Germany) and puromycin (at 25 µg/mL for *L. donovani* and *L. major*; at 10 µg/mL for *L. braziliensis*; Sigma-Aldrich, München, Germany) were used to select for integration of the donor gene fragments. For *L. braziliensis*, double drug-resistant cell populations with the intended gene replacements were first selected at a lower selection pressure as indicated until they emerged in culture (about 2–3 weeks), followed by an increase in the selection pressure (at ~IC$_{99.7}$:

5 µg/mL blasticidin; ~IC_{96}: 20 µg/mL puromycin) to allow discrimination with the mock-transfected control cultures.

For cloning by limiting dilution, exponential log-phase cultures of the candidate *L. braziliensis* *HSP23-* and *HSP100*-null mutants were seeded in complete M199 medium at 0.5 cells per well in two 96-well microtitre plates, as described previously [39]. After 14 days, monitoring of wells for promastigote growth by light microscopy was started and continued until growth-positive wells were observed. The contents of positive wells were seeded into 2 ml complete M199 medium maintaining the drug pressure (blasticidin and puromycin at ~IC_{96}–$IC_{99.7}$) in 25 cm^2 cell culture flasks to expand the culture. Each population that emerged from an individual well was considered an individual clone.

2.4. Construction and Preparation of Recombinant DNA

HSP23-encoding genes of different kinetoplastid species including *L. donovani* (LdBPK_340230), *L. major* (LmjF.34.0210), *L. infantum* (LinJ.34.0230), *L. braziliensis* (LbrM.20.0220), *Trypanosoma brucei* (Tb927.10.2620), were amplified from species-specific genomic DNA using primer pairs that introduce a *Kpn*I and a *Bcl*I or *Bam*HI (for *L. infantum* only) restriction sites (Table S1). Fragments were subsequently ligated into the *Leishmania* expression plasmid pCL1S [42] previously digested with *Kpn*I and *Bgl*II.

2.5. PCR-Amplification of Targeting Constructs

For gene disruption in *L. braziliensis*, PCR amplification of sgRNA templates (using a common sgRNA scaffold primer) and of donor DNAs, the latter from pTBlast and pTPuro plasmids [17], was done using the Expand™ High Fidelity PCR System (Roche, Mannheim, Germany) and PCR conditions as described [40].

For gene disruption in *L. major*, sgRNA templates were amplified in a total volume of 20 µL using 1 × iProof high-fidelity PCR master mix (Bio-Rad, München, Germany), 2 µM G00 primer (sgRNA scaffold) and 2 µM LmHSP23-specific 3′sgRNA or 5′sgRNA primer (Table S1). Cycling conditions were 30 s at 98 °C followed by 35 cycles of 10 s at 98 °C, 30 s at 55 °C, 15 s at 72 °C, and a final elongation step of 10 min at 72 °C. The targeting fragments were amplified from 10 ng pTPuro or pTBlast plasmid in 1 × iProof mix (Bio-Rad) using 2 µM forward and reverse primers, 3% DMSO in a total volume of 25 µL. PCR steps were 3 min at 98 °C followed by 35 cycles of 30 s at 98 °C, 30 s at 65 °C, 30 s at 72 °C, and a final elongation step of 5 min at 72 °C.

2.6. Analytical PCR

To screen for target-gene disruption in drug-resistant transfectant cell lines, genomic DNA was isolated from non-clonal populations of *eGFP*-deletion mutants and analysed by PCR. Genomic DNA was isolated using ISOLATE II Genomic DNA Kit (Bioline, Luckenwalde, Germany).

To test for the presence of the *eGFP* ORF and integration of the drug-resistance genes (*BSD*, blasticidin-S deaminase; and *PAC*, puromycin N-acetyltransferase) in the *eGFP* mutants, 1 µL of isolated DNA was mixed with 1 × iProof high-fidelity PCR master mix (Bio-Rad), 0.4 µM each forward and reverse primers, and 12% DMSO in a 25.5 µL total volume. In parallel, a technical control PCR (to demonstrate the presence of DNA in the analysed samples) was performed by amplifying a fragment from the *L. donovani HSP23* or *L. braziliensis actin* ORFs. PCR steps were 3 min at 98 °C followed by 30 cycles of 30 s at 98 °C, 30 s at 60 °C, 30 s at 72 °C followed by a final elongation step for 5 min at 72 °C.

The *Leishmania* wild-type and parental cell lines were included as controls. 10 µL of each PCR reaction was run on a 1% agarose gel to check for the presence of the expected product. The list of primer pairs used is given in Table S1.

2.7. RNA Extraction, cDNA Synthesis, and Quantitative Real-Time PCR (qRT-PCR)

qRT-PCR was performed essentially as described [43]. Total RNA was isolated from 5×10^7 parasites using the InviTrap spin cell RNA mini kit (STRATEC Molecular GmbH, Berlin, Germany) according to manufacturer's instructions. First strand cDNA synthesis was performed using a mix of oligo-dT and random primers (QuantiTect Reverse Transcription kit, Qiagen, Hilden, Germany) following the manufacturer's protocol. Real-time qPCR reactions were performed in a 20 µL-reaction mixture consisting of 1 µL of cDNA sample, 0.5 µM each gene-specific forward and reverse primers, and 1 × DyNAmo Color Flash SYBR Green Master Mix (Thermo Fisher Scientific, Waltham, MA, USA). The primers used for amplification of the target and reference genes are listed in Table S1. Reactions were run on a Rotor-Gene™ RG 3000 Instrument (Corbett, Sydney, Australia) using the following thermal cycling conditions: an initial denaturation step at 95 °C for 7 min, followed by 35 cycles at 95 °C for 15 s, 69 °C for 20 s, and 71 °C for 30 s. After PCR amplification, a step at 95 °C for 1 min was included, followed by a melting curve analysis (67–95 °C, hold 60 s on the first step, hold 8 s on next steps). Data collection and analysis were performed with the Rotor-Gene real-time analysis software 6.1.81 (Corbett, Sydney, Australia). The normalised expression ratio was calculated using the $2^{-\Delta\Delta Cq}$ method [44].

2.8. Next Generation Sequencing

DNA library construction, next generation sequencing and data analyses were performed as described [45]. Paired sequence data were aligned against a novel long-read assembly of the *L. braziliensis* M2904 reference genome [46].

2.9. Western Blotting

Western blots were performed following established protocols [38,47].

2.10. Immunofluorescence Assays

Indirect immunofluorescence microscopy was performed as described [48].

2.11. Flow Cytometry Cell Analysis

For GFP quantification, 2×10^6 parasites were harvested (1251 *g*, 10 min, 4 °C), washed once in PBS, fixed in 4% paraformaldehyde in PBS for 20 min at RT, washed twice in PBS, resuspended in 150 µL PBS, and immediately analysed by flow cytometry. The Cas9–GFP-expressing parental cell lines served as positive controls. The Cas9-expressing lines, which were negative for GFP, were included as negative controls to assess background fluorescence. Flow cytometric measurements were performed with the Accuri™ C6 flow cytometer (BD Biosciences, Heidelberg, Germany). A total of 30,000 events were recorded and analysed with FlowJo™ software V 10 (Becton, Dickinson and Company, Ashland, OR, USA).

2.12. In Vitro Infection of Murine Bone Marrow-Derived Macrophages

In vitro infections and parasite load quantification were performed as described [49–51].

2.13. In Silico Procedures

In silico cloning, DNA and protein sequence analysis were performed using the MacVector software version 17.x (Mac Vector, Cambridge, United Kingdom). Post-acquisition processing of images was performed using the ImageJ Fiji Software (Version 2.0.0, https://fiji.sc). Composite figures for publication were prepared using the Intaglio software (Purgatory Design, Durango, CO, USA). Numerical data and statistical differences were analysed using Prism (version 8, GraphPad Software, San Diego, CA, USA). Statistical comparisons between groups in the promastigote growth experiments were conducted using one-way analysis of variance (ANOVA)/Kruskal–Wallis test with Dunn's post test. For comparison of intracellular parasite survival within macrophages, a ratio-paired, one-sided Student's *t*-test was applied to offset the variability between primary cell populations. Differences were considered significant at $p < 0.05$.

In silico design of primers to generate sgRNA templates and donor DNA was performed essentially as described [40]. Guide RNA sequences were designed using the Eukaryotic Pathogen CRISPR gRNA Design Tool (EuPaGDT, available at http://grna.ctegd.uga.edu) [52], using the default parameters (SpCas9: 20 nt gRNA length; PAM: NGG on 3′ end; off-target PAM: NAG, NGA). In addition, two guide RNA sequences targeting *eGFP* (*eGFP*-52-5′sgRNA and *eGFP*-553-5′sgRNA) were retrieved from the Addgene repository (deposited as gRNA1 and gRNA2 by Guigo, Johnson; available at https://www.addgene.org/search/all/) as they had been experimentally validated for use in CRISPR experiments. Target-specific sgRNA primers were then designed manually and contained the T7 promoter (for T7 RNA polymerase-driven in vivo transcription of the sgRNA), the 20 nt sgRNA target sequence, and a sequence complementary to the sgRNA scaffold [17].

To generate gene replacement mutants, target-specific sgRNA primers were produced at http://www.leishgedit.net [17] (for whole GOI disruption) or designed manually (for partial GOI disruption). Donor DNA primer sequences contained target-specific 30 nt homology flanks corresponding to sequences immediately adjacent to the sgRNA target sequence for DSB-mediated repair by homologous recombination and recognition sequences for the pT template plasmids and were generated at http://www.leishgedit.net (for whole GOI disruption) or designed manually (for partial GOI disruption).

Since the sgRNA and donor DNA sequences identified using the EuPaGDT and LeishGEdit online tools used the *L. braziliensis* reference genomes (M2904 and M2903) available in TriTrypDB (https://tritrypdb.org/tritrypdb/), we verified the specificity of each sgRNA and homology flanks (donor DNA) by alignment against the *L. braziliensis* PER005cl2 genome [46] (focussing on chromosomes 20 and 29 which harbour the genes of interest) using the MacVector™ software (Mac Vector, Cambridge, United Kingdom).

Oligonucleotides were ordered from Sigma-Aldrich (München, Germany). See Table S1 for a list of all primers.

3. Results

3.1. Optimisation and Validation of the CRISPR-Cas9 System in L. braziliensis

To test the feasibility and efficiency of sgRNA-guided, Cas9-mediated gene editing in *L. braziliensis*, we first targeted an integrated transgene coding for green fluorescent protein (eGFP). To this end, we generated a stable cell line of *L. braziliensis* expressing Cas9 and T7 RNAP from an episome (pTB007). The eGFP coding sequence was fused into the pIR-mcs3+ plasmid [53], and the linearised plasmid was transfected into *L. braziliensis*, leading to integration into the small subunit rRNA (18S) coding sequence (Figure 1A).

We confirmed the expression of Cas9 protein by Western blot analysis (Figure S1A) and the detection of *T7RNAP* mRNA by qRT-PCR (Figure S1B). To better assess the efficiency of CRISPR–Cas9-mediated gene editing in *L. braziliensis*, we included Old World *L. donovani* strain 1S for comparative purposes, since the latter has long been used as a model for homologous recombination and genetic complementation in our laboratory.

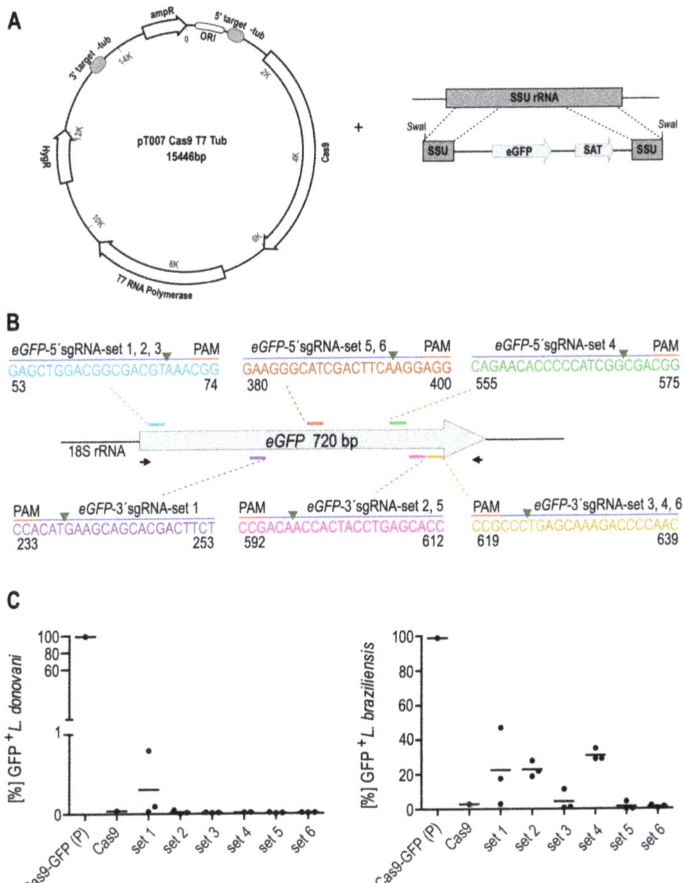

Figure 1. CRISPR–Cas9-mediated disruption of *eGFP* gene as proof-of-principle test in *L. braziliensis*. (**A**) Generation of Cas9–eGFP-expressing parasites. *Left panel*: plasmid pTB007 [17] bearing *hSpCas9* and *T7 RNAP* transgenes was transfected as circular episome into *L. braziliensis* PER005cl2 wild-type parasites. Transfectants were selected under Hygromycin B pressure. *Right panel*: schematic depiction of the double cross-over homologous recombination strategy to integrate the linearised pIR–*eGFP* construct into the SSU rRNA locus of *L. braziliensis* Cas9-expressing parasites. Regions shown are the SSU rRNA sequences on either ends resulting from *Swa*I restriction digest, the *eGFP* ORF, and the nourseothricine resistance gene ORF (*SAT*, encoding streptothricin-acetyltransferase). (**B**) Schematic representation of the *eGFP* locus and locations of the six 20-nt guide RNA sequences used for gene disruption; the guide sequence pairs with the DNA target (blue bar), directly upstream of a requisite 5'-NGG-3' adjacent motif (PAM). The green arrowhead indicates the predicted Cas9 cleavage sites. Only the coding strand is shown. Binding sites of primers used for genotyping of genetically engineered parasites are denoted by arrows. The PCR fragment size depended on the pair of single guide RNAs (sgRNAs) tested. Sets of sgRNAs tested: set 1 = *eGFP*-52-5'sgRNA and *eGFP*-253-3'sgRNA; set 2 = *eGFP*-52-5'sgRNA and *eGFP*-612-3'sgRNA; set 3 = *eGFP*-52-5'sgRNA and *eGFP*-639-3'sgRNA; set 4 = *eGFP*-553-5'sgRNA and *eGFP*-639-3'sgRNA; set 5 = *eGFP*-378-5'sgRNA and *eGFP*-612-3'sgRNA; set 6 = *eGFP*-378-5'sgRNA and *eGFP*-639-3'sgRNA. (**C**) Flow cytometry analysis of eGFP–Cas9-expressing parasites before and after transfection of *eGFP*-targeting sgRNAs. Efficiency of *eGFP* disruption using 6 different sets of sgRNAs in *L. donovani* (left panel) and *L. braziliensis* (right panel) as quantified by GFP expression. Each set of two sgRNAs was co-transfected with two donor DNAs; transfections were done in triplicate. Sets of sgRNAs tested (labelled as set 1 to 6 in the graphs) consisted of pairs as described in Figure 1B. P, parental cell line Cas9/T7/eGFP. The gating scheme, a representative histogram, and all FACS plots showing the percentage of GFP-positive cells are shown in Supplemental Figures S2 and S3.

The *L. braziliensis* and *L. donovani* parental cell lines (Cas9/T7/GFP) were co-transfected with a pair of *eGFP*-targeted sgRNAs and corresponding donor DNA cassettes (i.e., homologous repair templates) to facilitate homology-directed repair [54,55]. Six different sets of dual sgRNAs and donor DNAs (Figure 1B; Table S1) were tested in triplicate. Transfectants were subjected to blasticidin and puromycin drug selection. At this point, drug selection (hygromycin B) for maintenance of the pTB007 episome encoding Cas9 and T7 RNAP and nourseothricine selection for the integrated pIR-mcs-*eGFP* were stopped.

In *L. donovani* 1S, the antibiotic selection pressure with the drug-selectable markers was kept constant throughout the selection period (10 µg/mL blasticidin, 25 µg/mL puromycin), following the optimised conditions established previously for this parasite strain in our group (data not shown). Survival of *L. donovani* double drug-resistant transfectants became apparent 6–10 days after transfection. Transfectants with *eGFP*-targeted sgRNAs set 5 and set 6 were the first to emerge in culture (6 and 9 days after transfection, respectively). Candidate *eGFP* replacement populations were passaged at least twice before analysing the gene disruption outcome by flow cytometry. Each of the 6 pairs of sgRNAs resulted in highly efficient reduction of GFP expression (Figure 1C, left panel; Figure S2). PCR analysis of genomic DNA with primers amplifying the entire *eGFP* ORF showed no detectable band corresponding to the *eGFP* transgene in all selected *L. donovani* lines, but bands of higher size appeared, indicating the integration of the donor repair cassettes (Figure S4B, left panel), as expected (Figure S4A). This was verified with *BSD* and *PAC* gene-specific primers (Figure S4B, left panel) and confirmed the high efficiency of CRISPR–Cas9-mediated *eGFP* disruption in *L. donovani*.

In *L. braziliensis* PER005cl2 we first established the suitable concentrations of antibiotic selection through titration curves for 7 days (Figure S5). On this basis we decided to subject the parasites at first to the lowest concentrations of antibiotics that had a growth inhibitory effect, i.e., blasticidin at 2.5 µg/mL (~IC_{85}) and puromycin at 10 µg/mL (~IC_{65}). The first *L. braziliensis* drug-resistant transfectants to emerge in culture, as in *L. donovani*, were those transfected with *eGFP*-targeted sgRNAs set 5 (12–14 days after transfection) and set 6 (14 days after transfection). Transfectants with the other *eGFP* sgRNA sets (1, 2, 3 and 4) emerged 18–22 days after transfection. Candidate *eGFP* replacement populations were passaged at least twice and then analysed by flow cytometry as non-clonal populations. By flow cytometric analysis, sgRNAs sets 5 and 6 were the most efficient to abrogate the eGFP expression (0.02–4.30% GFP-positive cells), whereas sgRNA set 3 was slightly less efficient (0.69–11.4% GFP-positive cells). The sgRNAs sets 1, 2 and 4 were the least efficient (2.91–47.00% GFP-positive cells) (Figure 1C, right panel; Figure S3). Genomic DNAs from these parasite populations were examined by PCR confirming a complete loss of the *eGFP* transgene only in three selected *L. braziliensis* lines (*eGFP*-null mutants 5.1, 5.3 and 6.3) (not shown), which were transfected with the most potent sgRNAs, sets 5 and 6. For the other selected *L. braziliensis* lines, a band corresponding to the unmodified *eGFP* gene was still detected with varying intensities (Figure S4, right panel, for *eGFP* mutants 3.1, 3.2, and 3.3). PCR analysis with *eGFP* gene-specific primers also showed bands of higher size indicating the integration of the donor repair cassettes in the *L. braziliensis eGFP* mutants (Figure S4B, right panel), as expected (Figure S4A). While the blasticidin replacement cassette was confirmed to be integrated in all *L. braziliensis* selected lines by PCR analysis with *BSD*-specific primers (Figure S4B, right panel), the puromycin replacement cassette was detected in twelve out of 18 selected *L. braziliensis* lines, as assessed using *PAC*-specific primers (Figure S4B, right panel). This outcome reflected the moderate antibiotic selective pressure used to generate the *L. braziliensis eGFP* mutants.

At day 35 after transfection of the *L. braziliensis* Cas9/T7/eGFP parental cell line, inspection of the two *L. braziliensis* mock-transfected controls showed minimal growth. To impose a more stringent dual antibiotic selection, the mock cultures and selected *eGFP* mutants were passaged in complete M199 medium with blasticidin at 5 µg/mL (~$IC_{99.7}$) and puromycin at 20 µg/mL (~IC_{96}). The mock-transfected cultures succumbed to the antibiotic pressure within 4 days, while the *eGFP* mutant populations proliferated. This double antibiotic selection regimen was used in all subsequent experiments.

3.2. CRISPR–Cas9-Mediated Disruption of Endogenous HSP23 and HSP100 Genes in L. braziliensis

Next, we tested the applicability of the PCR-based CRISPR–Cas9 method on two endogenous, single-copy genes of *L. braziliensis* encoding the heat shock proteins HSP23 and HSP100. Both genes were successfully replaced in Old World *Leishmania* spp, using homologous recombination, giving rise to conditional phenotypes [47,56,57]. Previous work in *L. donovani* showed that *HSP23* null mutants are sensitive to temperature and chemical stresses. In *L. major* and *L. donovani*, Δ*clpB* (*HSP100*) null mutants showed loss of virulence in vitro and in vivo. We sought to replicate those findings in *L. braziliensis* to assess the practical application of CRISPR–Cas9-mediated genetic manipulation in this parasite species. First, we tested the fitness of *L. braziliensis* (Cas9/T7) cells by *in vitro* growth analysis (Figure S1C) and found slightly increased proliferation compared with wild type cells, thus excluding overt, detrimental effects of Cas9 expression.

For disruption of each targeted GOI, the *L. braziliensis* Cas9/T7 parental cell line was transfected in parallel with four different sets of sgRNAs and donor DNAs (see Table S1 for nucleotide sequences). Double drug-resistant cell populations for both targeted genes emerged in culture at day 18 post transfection, and were then subjected to a higher drug selection pressure, as established for *eGFP* deletion.

3.2.1. LbrHSP23 Gene Replacement

Three pairs of sgRNAs targeted different sites within the *LbrHSP23* ORF (Figure 2A), while a fourth pair of sgRNAs was designed to create DSBs upstream and downstream of the GOI coding region for whole-gene deletion (not shown). Putative *HSP23*-null mutants were obtained with sgRNAs sets 1 and 2 (Figure 2A), both of which disrupted the alpha-crystallin domain of HSP23, a conserved signature feature of the small heat shock protein family [58]. Transfection with sgRNAs set 3, which targeted the C terminal part of LbrHSP23, did not generate viable cells after double selection. No *LbrHSP23* whole-gene deletion mutants could be obtained with sgRNAs set 4, either. Later analysis revealed a one-base pair mismatch between primer P4-LbrHsp23–3′sgRNA (Table S1) and the *L. braziliensis* strain PER005 *HSP23* gene, explaining the lack of success for sgRNA set 4.

From the transfections with sgRNAs sets 1 and 2, three cell populations emerged: one with set 1 at day 18 post-transfection, and two with set 2, at day 18 and 25 post-transfection, respectively. From these three populations, clones were raised and expanded. Three clones were then subjected to whole genome sequencing: $HSP23^{-/-}$ cl.1 and cl.2, from transfection with sgRNAs set 2; and $HSP23^{-/-}$ cl.3, derived from the transfection with sgRNAs set 1. NGS analysis verified a lack of sequence reads for the targeted gene regions (Figure 2B), confirming site-specific disruption of the *LbrHSP23* ORF. Moreover, the precise integration of both drug-resistance cassettes in these $HSP23^{-/-}$ mutant clones was also verified (Figure S6). Western blot analysis using specific antibodies [47] failed to detect HSP23 protein in the $HSP23^{-/-}$ mutants (Figure 2C), confirming the null mutants on the genomic and proteomic levels.

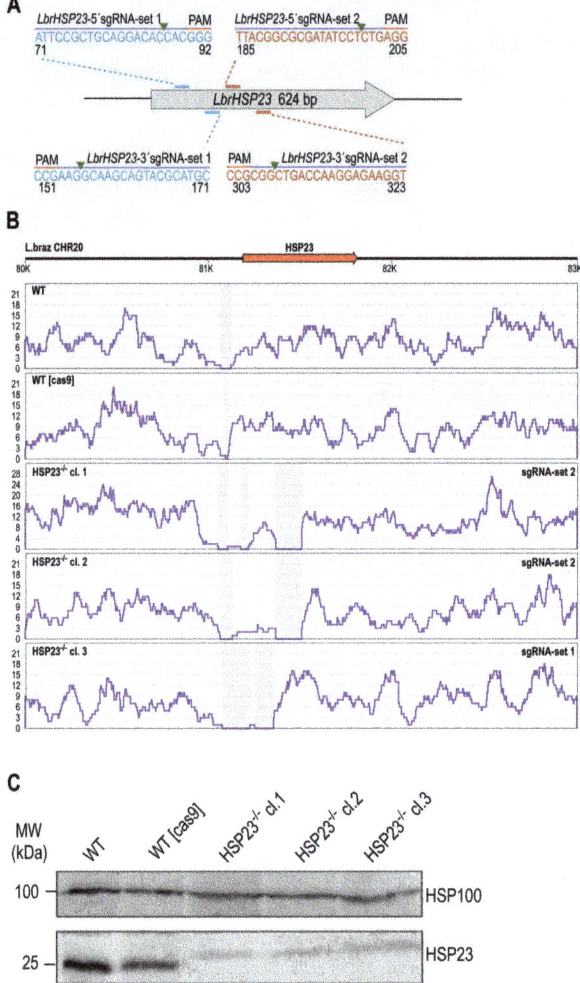

Figure 2. CRISPR–Cas9-mediated disruption of the endogenous *HSP23* gene in *L. braziliensis*. (**A**) Schematic representation of the *LbrHSP23* locus depicting the locations of 20-nt guide sequences that worked efficiently to disrupt the *LbrHSP23* ORF. Two sets of sgRNAs were tested (set 1 and set 2): set 1 = *LbrHSP23*-70-5'sgRNA and *LbrHSP23*-171-3'sgRNA; set 2 = *LbrHSP23*-183-5'sgRNA and *LbrHSP23*-323-3'sgRNA. Both pairs are designed to disrupt the conserved functional alpha-crystallin domain of HSP23 (amino acid positions 6–104). The guide sequence pairs with the DNA target (blue bar) directly upstream of a requisite 5'–NGG–3' adjacent motif (PAM). The green arrowhead indicates the predicted Cas9 cleavage sites. Only the coding strand sequence is shown. (**B**) NGS analysis of the *HSP23* locus after CRISPR–Cas9-mediated gene replacement. Genomic DNA of *L. braziliensis* PER005cl2 wild-type parasites (WT), the parental cell line WT [Cas9] and *HSP23*$^{-/-}$ mutant clones was isolated and subjected to NGS analysis. Resulting NGS reads were aligned to the *HSP23* gene locus (LbrM.20.0220) in the *L. braziliensis* M2904 reference genome using the Bowtie 2 algorithm. The read coverages (Y-axis) for the gene locus are shown in blue. The arrow represents the position and direction of the coding sequence. The X-axis numbering refers to the nucleotide position (bp) on chromosome 20. Grey-shaded areas denote lack of aligned reads. (**C**) Verification of *HSP23* gene replacement by Western blot analysis. 1×10^7 cells of WT, WT [Cas9], and of 3 *HSP23*$^{-/-}$ clones were lysed and the cell lysates were analysed by SDS-PAGE and Western blot using anti-HSP23 (1/500, lower panel). Anti-HSP100 (1/1000, upper panel) was used as loading control. MW = Molecular weight in kilodalton.

3.2.2. *LbrHSP100* Gene Replacement

sgRNA selection and replacement of the *LbrHSP100* gene were done following the same strategy. We obtained putative *LbrHSP100*-null mutants with sgRNAs set 3, targeting sequences in the N terminus of *LbrHSP100* ORF and set 4, targeting 5′ and 3′ non-coding sequences flanking the ORF for whole-gene deletion (Figure 3A). One cell population each emerged from the transfections and gave rise to multiple clones. Two *HSP100*$^{-/-}$ clones obtained with sgRNAs sets 3 and 4, respectively, were then selected for further genetic and phenotypic characterisation. NGS analysis indeed confirmed the target-specific disruption of the *LbrHSP100* ORF and the on-target integration of both drug resistance cassettes at the predicted genomic sites for both *HSP100*$^{-/-}$ mutants (Figure 3B; Figure S7). Western blot analysis using HSP100-specific antibodies [38] confirmed the lack of HSP100 in both mutants (Figure 3C).

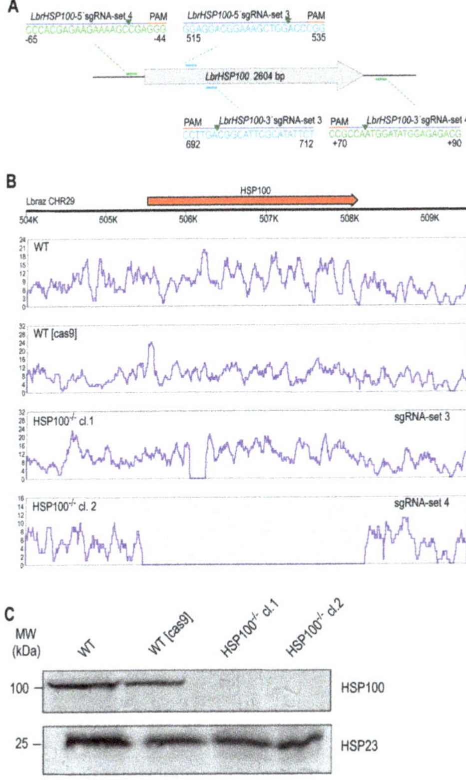

Figure 3. CRISPR–Cas9-mediated disruption of the endogenous *HSP100* gene in *L. braziliensis*. (**A**) For targeting *LbrHSP100* (LbrM.29.1350), two sets of sgRNAs tested (set 3 and set 4) worked efficiently. sgRNAs set 3 (*LbrHSP100*-513-5′sgRNA and *LbrHSP100*-712-3′sgRNA) targeted disruption of the *LbrHSP100* ORF in the N terminus. sgRNAs set 4 targeted 5′ and 3′ non-coding flanking sequences for *LbrHSP100* whole-gene deletion. Two cloned *L. braziliensis HSP100*$^{-/-}$ lines were studied, *HSP100*$^{-/-}$ cl.1 and *HSP100*$^{-/-}$ cl.2, derived from transfection of set 3 or set 4 of *LbrHSP100*-targeting sgRNAs, respectively. (**B**) Whole genome sequencing of *HSP100*-null mutant lines. Sequence reads from each analysed strain were aligned to the reference DNA sequence consisting of chromosome 29 of *L. braziliensis* M2904 reference genome using Bowtie 2 software. The Y-axis represents the number of reads and the X-axis shows the nucleotide position (bp) on chromosome 29. Grey shaded areas denote complete lack of aligned reads. (**C**) Verification of *HSP100*-null mutants by Western blot analysis using anti-HSP100 (1/1000) antibody. Anti-HSP23 antibody (1/500) served as loading control. MW = Molecular weight in kilodalton.

To assess the fate of the Cas9/T7 construct (pTB007 episome) in the CRISPR-derived null mutants, we analysed Cas9 expression on the mRNA and protein levels by qRT-PCR and Western blot, respectively. Cas9 protein was undetectable in the three $HSP23^{-/-}$ and two $HSP100^{-/-}$ mutant clones (Figure S8A,B), alleviating concerns over phenotypic, off-target Cas9 effects.

3.3. L. braziliensis HSP23- and HSP100-Null Mutant Phenotypes Resemble Those Described for Old World Leishmania

For the phenotype analysis, we first attempted to create gene add-back parasites for both null mutants. In the $HSP23^{-/-}$ mutants, we introduced the *LbrHSP23* transgene for integration into the 18S SSU rRNA locus, using the pIRmcs3+ vector [53], or as episome, using the over expression plasmid pCL1S-*LbrHSP23*. To generate the *HSP100* add-back cell lines, the $HSP100^{-/-}$ mutants were transfected with the pIRmcs3+ vector harbouring *LbrHSP100* for genomic integration. Despite several attempts with different experimental conditions (data not shown), we could not generate any of the intended gene add-back cell lines. We suspect that the selection marker gene, coding for streptothricine N-acetyl transferase (SAT), was not stably expressed, possibly due to the known RNAi activity in *L. braziliensis* [9]. Ectopic gene expression from integrated and episomal transgenes is unpredictable in *L. braziliensis* (V.A., unpublished observations, and [59]).

We nevertheless proceeded to test the growth phenotypes of the *L. braziliensis* $HSP23^{-/-}$ and $HSP100^{-/-}$ null mutants under various *in vitro* growth conditions compared with the wild-type and with Cas9-expressing cells. Cell density on day 4 (stationary phase) was analysed and displayed as percentage of growth relative to the wild type (set at 100%). Under optimal *in vitro* growth conditions for promastigotes (25 °C, pH 7.4), the *L. braziliensis* PER005cl2 wild-type strain achieved a median 24.9-fold growth (2.49×10^7 cells/ml). Two $HSP23^{-/-}$ null mutants, $HSP23^{-/-}$ cl.2 and $HSP23^{-/-}$ cl.3, grew at rates similar to the wild type (median relative growth: 85.0% for $HSP23^{-/-}$ cl.2 and 93.4% for $HSP23^{-/-}$ cl.3; Fig. 4A). $HSP23^{-/-}$ cl.1 displayed a 20% elevated proliferation, similar to the Cas9-expressing cells. The *HSP100*-null mutants showed proliferation rates (median relative growth: 86.1% for $HSP100^{-/-}$ cl.1 and 81.8% for $HSP100^{-/-}$ cl.2) comparable to those of the wild type (Figure 4A). Therefore, we see no growth phenotype for $HSP23^{-/-}$ and $HSP100^{-/-}$ null mutants under optimal culture conditions. This is in keeping with earlier findings about the significance of HSP100 and HSP23 in the promastigote [47,56]. Stable Cas9 expression from the pTB007 episome increased the growth rate of *L. braziliensis* promastigotes at 25 °C (Figure S1C), leading to a higher cell density in late-log phase (day 3; $p = 0.004$, U test) and in stationary phase (day 4; $p = 0.015$, U test) compared to the wild-type parasites, likely reflecting a positive effect on cell proliferation, similar to previous observations [21].

Next, we repeated the analysis at 30 °C, the upper temperature limit for *L. braziliensis* growth *in vitro* [60]. Proliferation of the *L. braziliensis* PER005cl2 wild-type strain was slowed considerably at 30 °C, reaching a median of 4.9×10^6 cells/ml at day 4 (4.9-fold growth). The *L. braziliensis* $HSP23^{-/-}$ null mutants, particularly $HSP23^{-/-}$ cl.2 and $HSP23^{-/-}$ cl.3, were sensitive to the 30 °C cultivation temperature and did not proliferate (Figure 4B). This temperature-sensitive phenotype is in line with previous work with *L. donovani* $HSP23^{-/-}$ null mutants [47]. We also tested the cell integrity of the *L. braziliensis* $HSP23^{-/-}$ null mutants at 30 °C. As shown by immunofluorescence microscopy (Figure 4C), all three *L. braziliensis* HSP23-null mutants showed abnormally rounded, swollen and irregular shapes, and formed cell aggregates indicating cellular damage. These changes were not observed in the control cells, *L. braziliensis* wild type and Cas9-expressing cells, which presented as individual, well defined cells.

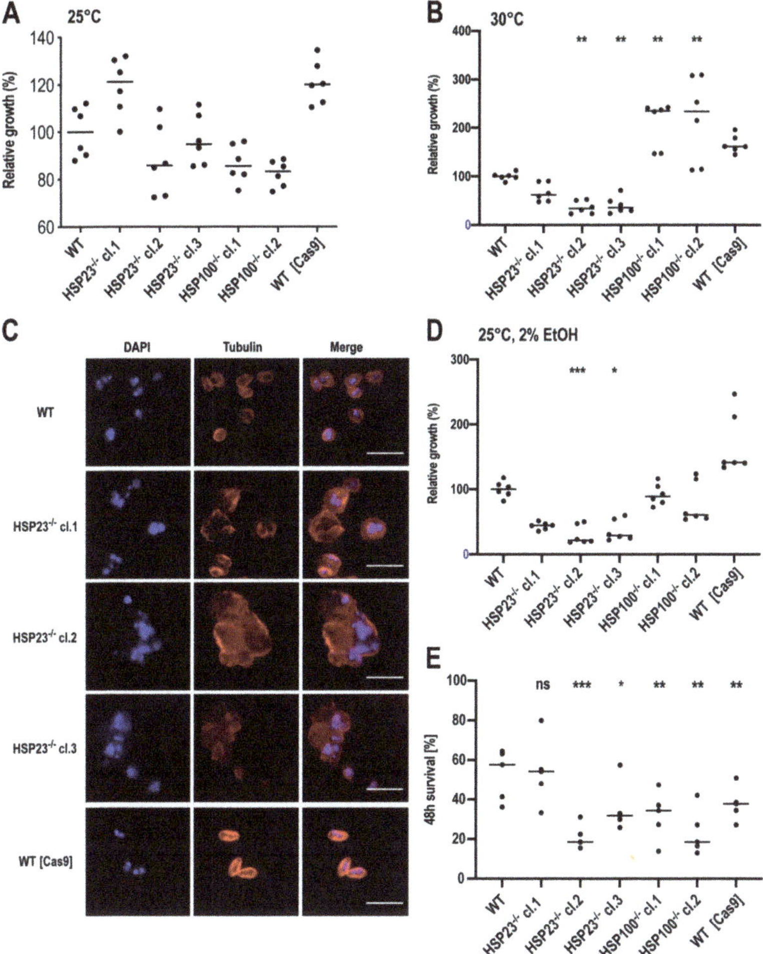

Figure 4. Phenotypic analyses of *L. braziliensis HSP23*$^{-/-}$ and *HSP100*$^{-/-}$ clones. For growth curves, promastigotes of WT, WT (Cas9), *HSP23*$^{-/-}$ clones, and *HSP100*$^{-/-}$ clones were seeded at a density of 1×10^6 parasites/mL into 5 ml of complete M199 medium and grown for 4 days. Cell density was measured on day 4 and is shown as a percentage of WT cell density (set at 100%). Parasites were grown at 25 °C (**A**) and 30 °C (**B**). The *HSP23*$^{-/-}$ clones incubated for 4 days at 30 °C were also stained with mouse anti-tubulin antibody (1/4000) and DAPI (1/50) (**C**). Images were taken on an EVOS FL Auto Cell Imaging System and processed using the ImageJ Software (https://fiji.sc). Scale bar: 10μm. Additional cultures were grown at 25 °C and pH 7.4 with the addition of 2% ethanol (**D**). The horizontal black lines in panels A, B, and D indicate the median of 6 biological samples from 3 separate experiments. Significance was tested using the Kruskal–Wallis test; * $p < 0.05$, ** $p < 0.01$, *** $p < 0.001$. (**E**) Primary mouse bone-marrow-derived macrophages were differentiated and infected with stationary-phase promastigotes of WT, WT [Cas9], *HSP23*$^{-/-}$ clones, and *HSP100*$^{-/-}$ clones at a MOI of 1:8 (macrophage-to-parasite ratio). After 4 h, free parasites were washed away and the infected macrophage cultures were further incubated at 34 °C under 5% CO_2 for 44 h. Genomic DNA from *Leishmania*-infected macrophages was isolated at 4.5 h and at 48 h post-infection, and parasite load was determined by TaqMan qPCR quantifying parasite *actin* gene DNA relative to host macrophage *actin* gene DNA. Shown is intracellular parasite survival [%] after 48 h, with the bar indicating the median of $n = 5$. Ratio-paired, one-sided Student's *t*-test: * $p < 0.05$, ** $p < 0.01$, *** $p < 0.001$ between data pairs. ns = not significant.

Conversely, the *L. braziliensis HSP100*[-/-] null mutants were fully viable and proliferating at 30 °C, even exhibiting a significant growth advantage over the wild type (Figure 4B). This temperature tolerance of the *L. braziliensis HSP100*[-/-] null mutants matches previous findings from phenotype analyses of *L. donovani HSP100*[-/-] null mutants [57], but contrasts with the phenotype of *L. major HSP100*[-/-] null mutants, which were hypersensitive at the upper limit of growth temperature [56]. Lastly, the Cas9-expressing cells grown at 30 °C also showed an elevated growth without reaching statistical significance (Figure 4B).

We next tested the *L. braziliensis HSP23*[-/-] and *HSP100*[-/-] null mutants for tolerance to sublethal ethanol concentrations, a trigger of the unfolded protein response, a stress signalling pathway of the endoplasmic reticulum (ER) that is related to the heat shock response [61,62]. Treatment with 2% ethanol caused growth reduction for all three *L. braziliensis HSP23*[-/-] null mutants (Figure 4D). This increased sensitivity of *L. braziliensis HSP23*[-/-] null mutants to a chemical stressor (i.e., ER stress-sensitive phenotype) is in agreement with previous work in *L. donovani HSP23*[-/-] mutants [47], further supporting the involvement of HSP23 in protecting *Leishmania* against protein misfolding stress. The *HSP100*[-/-] null mutants were not affected by exposure to 2% ethanol (Figure 4D). Again, the Cas9-expressing cells showed a slightly increased growth compared to the wild type (Figure 4D).

Lastly, we tested the ability of the wild type and mutant strains to survive inside macrophages. Primary mouse bone marrow-derived macrophages were differentiated and infected *in vitro* at a parasite to macrophage ratio of 8:1 using stationary-phase promastigotes. The parasite load was evaluated by qPCR [50] at 48 h post infection relative to the parasite load after 4.5 h of parasite internalisation.

The average percentage of surviving *L. braziliensis* PER005cl2 wild-type parasites within macrophages at 48 h post-infection was 52.6 ± 13.0% (Figure 4E). The loss of HSP100 had a significant impact on the intracellular survival of the two *L. braziliensis HSP100* null mutants. The effect was more pronounced for the whole-gene deletion mutant (*HSP100*[-/-] cl.2; mean survival ± SD: 23.4 ± 11.7%) than for the partial gene disruption (*HSP100*[-/-] cl.1; 32.0 ± 12.4%) (Figure 4E). The impaired ability of these *L. braziliensis HSP100*[-/-] null mutants for intracellular survival in *in vitro*-infected mouse macrophages was also documented for *L. major* and *L. donovani HSP100*-null mutants [56,57].

The ability to survive in macrophages was affected in only two *L. braziliensis HSP23*[-/-] mutants (*HSP23*[-/-] cl.2: mean survival ± SD: 21.2 ± 6.1%; *HSP23*[-/-] cl.3: 35.6 ± 12.5%)(Figure 4E), whereas the *HSP23*[-/-] cl.1 was able to survive intracellularly (54.1 ± 16.8%) at a rate similar to the wild-type parasites (Figure 4E). The reduced survival of *HSP23*[-/-] cl.2 and cl.3 matches the poor growth of these clones at elevated temperature and under ethanol stress (Figure 4B,D) and is in line with previous work performed with a *L. donovani HSP23*-null mutant [47].

In a first attempt to investigate possible genomic adaptations in the mutants as cause for varying phenotypes, we evaluated aneuploidy patterns. Using the NGS sequence reads from the WGS analysis and quantifying normalised sequence read densities for individual chromosomes in *L. braziliensis* WT cells, WT [Cas9] cells, three *HSP23*[-/-] mutant clones and two *HSP100*[-/-] mutant clones, we calculated chromosome ploidies (Figure S9A). Indeed, we found profound differences between *L. braziliensis HSP23*[-/-] mutants themselves and compared to the other parasite strains. *HSP23*[-/-] clone 1 is trisomic for chromosome 30 and shows intermediate somy (2.56) for chromosome 4. *HSP23*[-/-] clone 2 shows a marked increase of chromosome 2 ploidy (4.82). *HSP23*[-/-] clone 3 shows strong amplification (4.6) of chromosome 14, trisomies for chromosomes 18, 33 and 34, and a slight (2.39) increase for chromosome 4, which was also partly amplified in *HSP23*[-/-] clone 1. The strong increase of chromosome 2 sequence reads for *HSP23*[-/-] clone 2 is due to an apparent amplification of a ~20,000 bp region between positions 260,000 and 280,000 (Figure S9C). The amplified region contains mostly copies of a SLACS retrotransposon (LbrM.02.0550), and a possible context with the loss of HSP23 is not obvious.

All three *L. braziliensis HSP23$^{-/-}$* clones, but also the Cas9-expressing strain were trisomic for chromosome 26, possibly causing the minor fitness gain observed for the Cas9 strain.

3.4. Complementation Studies in L. major HSP23-Null Mutants Indicate a Conserved Function in Thermotolerance for Trypanosomatid HSP23

The failure to establish ectopic HSP23 expression in the *L. braziliensis HSP23$^{-/-}$* clones precluded a conclusive correlation between loss of HSP23 and the observed phenotypes. To complement this, we also produced CRISPR-derived *L. major HSP23$^{-/-}$* null mutants, following the same experimental strategies. Three selected *L. major HSP23$^{-/-}$* null mutant clones (*LmjHSP23$^{-/-}$* cl.1–cl.3) were analysed by whole genome sequencing, confirming the successful replacement of the *LmjHSP23* gene (Figure S10A) and the correct integration of both drug-resistance cassettes (Figure S11). Further verification by Western blot analysis using HSP23-specific antibodies showed a lack of the HSP23 protein in all *L. major HSP23$^{-/-}$* null mutants (Figure S10B). From these clones, we selected *LmjHSP23$^{-/-}$* cl.1 for genetic complementation and phenotypic analyses. We introduced the *LmjHSP23* transgene as episome to generate a *LmjHSP23* add-back cell line. In vitro, at optimal growth conditions for promastigotes (25 °C, pH 7.4), the null mutant showed a 50% reduced growth compared with wild-type cells (Figure 5A). This reduced growth of the null mutant could be restored to near-wild type levels by the *LmjHSP23* transgene, but not by the empty expression plasmid pCL1S (Figure 5A). At 34 °C, a temperature relevant for dermotropic *Leishmania* species, the *LmjHSP23$^{-/-}$* cl.1 mutant promastigotes were severely affected and did not proliferate (Figure 5B). This temperature-sensitive phenotype was rescued by the *LmjHSP23* transgene (Figure 5B), similar to what was reported for *L. donovani HSP23$^{-/-}$* mutants [47]. We also tested the *LmjHSP23$^{-/-}$* cl.1 mutant for tolerance to sublethal ethanol stress. A 2% ethanol exposure caused growth inhibition in the null mutant (Figure 5C), but not in the *LmjHSP23$^{-/+}$* (LmjHSP23) parasites (Figure 5C). Thus, we established *LmjHSP23$^{-/-}$* cl.1 as a suitable host strain for the functional complementation with various trypanosomatid *HSP23* genes.

A similar ploidy analysis was also performed for *L. major* WT, *L. major* WT [Cas9] and the two *L. major HSP23$^{-/+}$* clones, 1 and 2 (Figure S9B). Except for a very minor increase for chromosomes 5, 6, and 8, no karyotypic changes could be observed.

The *LmjHSP23$^{-/-}$* cl.1 mutant was transfected with pCL1S bearing the *L. donovani*, *L. infantum*, *L. major*, *L. braziliensis*, or *T. brucei HSP23* orthologs, respectively. Ectopic expression of these transgenes was verified at the RNA level using qRT-PCR analysis with *HSP23* species-specific primers, showing varying rates of over expression (Figure S12A). We also verified the HSP23 protein level by Western blot analysis using specific antibodies raised against *L. donovani* HSP23 [47]. Over expression was confirmed for all *Leishmania* HSP23 homologs, except for the putative *T. brucei* HSP23 (Figure S12B), the latter likely due to low amino acid sequence conservation (36%) between the *L. donovani* and *T. brucei* HSP23 homologs.

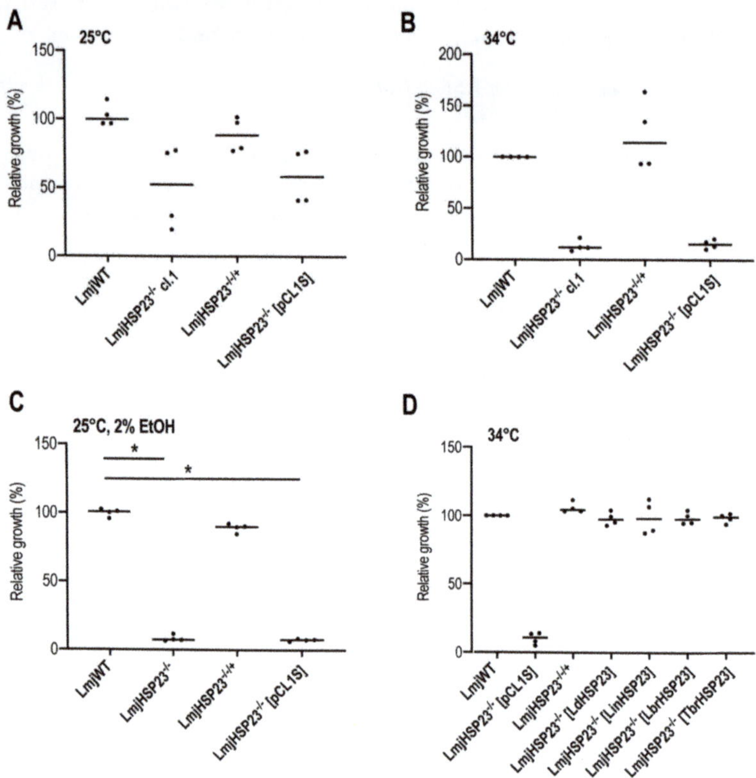

Figure 5. Phenotypic analysis of *L. major HSP23*$^{-/-}$ mutants and complementation strains. 1×10^6 or 5×10^6 parasites/ml were seeded in 10 ml complete M199 medium and parasite density was assessed at day 4. Parasites were grown at 25 °C (**A**), 34 °C (**B**), and 25 °C with 2% EtOH (**C**). Cell density is shown as percentage of WT (set at 100%). (**D**) Complementation studies in *LmjHSP23*$^{-/-}$ mutants. Null mutants were transfected with the pCL1S over expression vector harbouring the *HSP23* gene of *L. major, L. donovani, L. infantum, L. braziliensis,* and *Trypanosoma brucei* or with the empty vector only. Complementation populations were subjected to growth experiments at 34 °C. Cell density was assessed at day 4 and is shown normalised to *Lmj* WT growth (set at 100%). * = $p < 0.05$.

We then tested whether the temperature-sensitive phenotype of the *L. major HSP23*$^{-/-}$ mutant could be complemented by the HSP23-encoding, orthologous genes from other *Leishmania* species and the closely related *Trypanosoma brucei*. These supposed HSP23 homologs share between 36% and 99% amino acid sequence identity (Table S2). At 34 °C, all trypanosomatid *HSP23* transgenes restored growth of *L. major HSP23*-null mutants to wild-type levels, abrogating the mutant phenotype (Figure 5D). This shows that all trypanosomatid HSP23 homologs share the same functionality, conferring protection against heat stress, and likely maintaining protein folding homeostasis in trypanosomatid organisms. Furthermore, the functional conservation of HSP23 homologs among the Trypanosomatidae confirms the phenotypes we observed in the *L. braziliensis HSP23*-null mutants, since LbrHSP23 expression can restore thermotolerance to the *L. major HSP23*$^{-/-}$ mutant.

4. Discussion

The protozoan parasite *Leishmania braziliensis* is one of the most pathogenic dermotropic *Leishmania* species circulating in the Americas, where it is the main cause of cutaneous and mucocutaneous leishmaniasis [4,63]. Despite its prevalence and importance to public health, *L. braziliensis* has been less studied and is therefore less experimentally developed compared to Old World *Leishmania* species such as *L. major* and *L. donovani*, which have been traditionally used as models for studying the biology of these obligate intracellular parasites. Given that *L. braziliensis* is a member of the subgenus *Viannia*, with a considerable phylogenetic distance to the Old World species and even to the Central and South American *L. mexicana* complex, conservation of gene function between the subgenera may not be assumed automatically, and may require experimental confirmation by reverse genetics.

One of the main approaches for genetic modification of *Leishmania* parasites to probe gene function has been the generation of gene replacement mutants by homologous recombination-mediated replacement [5,64], which allows the creation of null mutants and their subsequent phenotypic analysis [6,65]. While this has proven a powerful genetic tool in Old World *Leishmania* spp., but also in Central American *L. mexicana* [66], our literature search did not turn up any work regarding homologous recombination-based gene replacement in *L. braziliensis*. Studies reporting on the use of homologous recombination in *L. braziliensis* demonstrate the generation of stable transgenic parasite lines from integration of DNA constructs into the SSU rDNA genomic locus. These include *L. braziliensis* lines expressing reporter genes, e.g., luciferase or eGFP, which hold potential for parasite tracking and monitoring effects of antileishmanial compounds *in vitro* and *in vivo* [67–69], and over expressing parasite lines for the analysis of gene products, e.g., to assess antimony susceptibility and resistance mechanisms [70–72]. Moreover, circular extrachromosomal cosmids can be stably introduced into *L. braziliensis* to over-express stretches of genomic DNA and connect the over expression phenotypes to biological processes such as virulence [73] and antimony resistance [59]. The experimental proof that *L. braziliensis* is a RNAi-competent species started the development of RNAi-based gene knockdown strategies for the loss-of-function phenotyping of genes in this species [9,10]. More recently, the CRISPR–Cas9 technology, with its advantages of being less time-consuming than traditional gene targeting and less susceptible to off-target effects than RNAi-based approaches [74], has added to the genetic toolbox that is available for the study of *Leishmania* spp. [19,20], allowing researchers to investigate gene functions with unprecedented ease, accuracy, efficiency, and scale in biological contexts [17,25,29,40].

In this study, we report the application of CRISPR–Cas9-mediated gene editing to the efficient and precise disruption of two endogenous, non-essential, single-copy genes and one integrated transgene in *L. braziliensis*. We opted for a CRISPR–Cas9, molecular cloning-free method developed for the use in *Leishmania* that relies on T7 RNAP-based expression of sgRNAs in vivo [17]. For this, we first generated a parental *L. braziliensis* cell line expressing Cas9 and T7 RNAP. Since plasmid pTB007 was designed for integration of both transgenes into the *L. major* beta-tubulin locus [17], we transfected pTB007 as stable, circular episome under hygromycin B selection. This episome was well tolerated by *L. braziliensis* strain PER005cl2 used in this study and was stably maintained for several months, with no apparent Cas9 toxicity during in vitro promastigote passage, indicating that this episomal transgene could be maintained without inducing deleterious RNAi effects in *L. braziliensis*.

For our study, we used a cloned *L. braziliensis* strain, derived from a clinical isolate, whose entire genome had been sequenced [46].This allowed us to select correct, highly specific sgRNA templates and donor DNAs for precise, targeted gene editing with no predicted off-target mutations. The original clinical isolate from which PER005cl2 strain is derived, was shown to be infective for primary mouse peritoneal macrophages [34], within which it is sensitive to pentavalent antimony. Furthermore, this isolate was confirmed not to harbour *Leishmaniavirus* LRV1 [75], a cytoplasmic double-stranded RNA virus frequently found as endosymbiont in *Leishmania* (*Viannia*) species [75–77], and which appears to enhance virulence and persistence of its *Leishmania* host [78,79].

We first targeted an eGFP coding sequence inserted into the SSU rRNA coding gene(s) of the *L. braziliensis* parental Cas9/T7 cell line. We applied double antibiotic selection after CRISPR targeting, using increasing antibiotic pressure at two time points, i.e., predetermined minimal effective concentrations of antibiotics at 24 h post-transfection and until transfectants emerged in culture, followed by higher antibiotic selection pressure to enrich for homozygously edited cells, and found this to be an effective strategy. The *eGFP* editing in *L. braziliensis* was assessed at the cell population level and compared to that achieved in *L. donovani*. Overall, we observed a different activity for the same pairs of sgRNAs in the two *Leishmania* species studied. While all 6 sgRNA sets that targeted sites within the *eGFP* gene were highly active in *L. donovani*, they had a wide range of efficiency in *L. braziliensis*. The most active sgRNAs (sets 5 and 6) were the same in *L. braziliensis* and *L. donovani*, indicating that the sgRNA sequence had an impact on the gene targeting efficiency. This is in line with a recent study that tested the efficiency of three gRNAs targeting identical sequences of the miltefosine transporter gene in *L. donovani*, *L. major*, and *L. mexicana*, and found the relative gRNA activity to be the same [31]. Studies in other systems revealed that sgRNA sequence features such as position-specific nucleotide composition, GC content, motifs located in the sgRNA "seed" region, and secondary structures of sgRNAs contribute to sgRNA efficacy [80–84].

The different gene targeting efficiencies of the same sgRNA sets observed for *L. braziliensis* and *L. donovani* may be due to different factors. First, the presence of an active RNAi machinery in *L. braziliensis* [9] may have an effect on ectopic Cas9 and T7 RNAP expression from episomal DNA constructs in this species, as was shown before [59]. Second, there may be differences in the T7-dependent expression level of different sgRNAs and Cas9 among *Leishmania* species [31]. We have used T7 RNAP-driven in vivo expression of sgRNA templates that were delivered to the *Leishmania* parental Cas9/T7 cell lines by transient transfection [17]. Variation of T7 RNAP-mediated transcription may lead to different intracellular levels of sgRNA that may limit the efficiency of Cas9-dependent DNA cleavage. A recent study suggested that a threshold level for both Cas9 and sgRNA expression is required for an efficient CRISPR-mediated gene knockout, which in turn is determined by the specific potency of a given sgRNA [85]. In keeping with this, increased sgRNA expression and maturation dramatically improved the efficiency of CRISPR–Cas9 mutagenesis in *Candida albicans* [86]. Thirdly, DSB repair efficiency may differ between *Leishmania* species [31]. Fourth, small variations in the intrinsic antibiotic sensitivity of different *Leishmania* species and strains may cause differences in transgene copy numbers, both for the integrated *GFP* gene and for the Cas9/T7-RNAP construct, leading to different efficiencies. Lastly, other factors playing a role in the biology of the *Leishmania* species studied may also play a role, such as variations of chromatin structure.

In our experiments, the copy numbers of *eGFP* within the SSU rRNA gene units of the *L. braziliensis* Cas9/T7/eGFP parental cell line were not determined. Assuming one copy of *eGFP* present per genome in the *L. braziliensis* Cas9/T7/eGFP, as shown in a recent study focused on the same species [87], our results suggest that the *eGFP*-specific sgRNA sets 1, 2, 3, and 4 generated mono-allelic edits, i.e., single-allele replacements, whereas the most efficient sgRNAs, sets 5 and 6, generated mostly double-allelic edits.

We were also able to efficiently disrupt two non-essential, endogenous, single-copy genes of *L. braziliensis* encoding the heat shock proteins HSP23 and HSP100. We obtained double-allelic, Cas9-free *HSP23^{-/-}* and *HSP100^{-/-}* null mutants. The in vitro phenotypes of the *L. braziliensis* HSP23- and HSP100-null mutants were assessed and compared to the wild-type strain, since gene add-back variants could not be obtained. Nevertheless, the analysis of independently cloned mutant cell lines revealed largely consistent phenotypes, strengthening the correlation between the disruption of the target gene and the loss-of-function phenotypes. This was further supported by the complementation studies carried out in the *L. major* HSP23-null mutant, which demonstrated functional homology between the HSP23 genes of the Trypanosomatidae. Furthermore, the rapid loss of the Cas9 episome in the absence of antibiotic selection is important when evaluating the phenotype, as the WT [Cas9] strain which was kept under selection showed a divergent phenotype from the wild type. We would therefore refrain from using genomic integration constructs for the expression of Cas9.

We do not know the reason behind the different capacity of intracellular amastigotes from the three studied *L. braziliensis HSP23^{-/-}* mutants to survive inside macrophages. All parasite strains/clones were subjected to the same *in vitro* culture, electroporation, cloning, antibiotic selection, and stress conditions. They had similar passage numbers before phenotype analyses, and their phenotypes were investigated in parallel in all assays. Moreover, the CRISPR–Cas9 components were no longer present when single-cell cloning was performed. We suspected that the mutant clones might have undergone some level of genetic adaptation, e.g., via spontaneous mosaic aneuploidy followed by selection for vitality. We observed a similar, spontaneous loss of phenotype for a *L. donovani HSP23*$^{-/-}$ clone, due to amplification of the gene coding for casein kinase 1.2 [45]. We indeed found ploidy changes that were specific to the *L. braziliensis HSP23*$^{-/-}$ mutants. One of those, a trisomy of chromosome 34, which harbours the casein kinase 1.2 gene in *L. braziliensis*, may have a similar effect as in *L. donovani*.

Lastly, an average of 37.7% (± 8.6%) *L. braziliensis* Cas9-expressing cells were able to survive inside macrophages (Figure 4E). Those cells show a trisomy for chromosome 26, similar to all three *L. braziliensis HSP23*$^{-/-}$ clones (Figure S9A). This trisomy is absent from the wild type and from the two *L. braziliensis HSP100*$^{-/-}$ clones.

5. Conclusions

Leishmania (V.) braziliensis is amenable to reverse genetics using a CRISPR–Cas9 protocol as shown in this work. Gene replacement occurs exclusively at the predicted sites. As is known, ectopic expression of the genes of interest presents a problem, due to the effects of RNAi in the *Viannia* subgenus. The functions of at least two amastigote-specific heat shock proteins, HSP100 and HSP23, are conserved between Old World and New World leishmaniae and likely in *T. brucei* as well. With a workable protocol for gene replacement now in place, urgent questions pertaining to the biology of the *Viannia* subgenus can now be addressed by means of reverse genetics.

Supplementary Materials: The following are available online at http://www.mdpi.com/2073-4425/11/10/1159/s1, Figure S1: Detection of Cas9 and T7 RNAP in the *L. braziliensis* parental cell line; Figure S2: Flow cytometric analysis results for *L. donovani* eGFP-Cas9-expressing promastigotes transfected with *eGFP*-targeting sgRNAs; Figure S3: Flow cytometric analysis results for *L. braziliensis* eGFP-Cas9-expressing promastigotes transfected with *eGFP*-targeting sgRNAs; Figure S4: PCR verification of *eGFP* gene replacement; Figure S5: Titration of selection antibiotics on *L. braziliensis* promastigotes; Figure S6: Verification of the *L. braziliensis HSP23* gene replacement by the respective resistance cassettes; Figure S7: Verification of the *L. braziliensis HSP100* gene replacement by the respective resistance cassettes; Figure S8: Cas9 expression in *L. braziliensis HSP23*- and *HSP100*-null mutants; Figure S9: Karyotype analysis of *L. braziliensis* and *L. major* strains; Figure S10: Verification of *L. major HSP23*-null mutants; Figure S11: Verification of replacement cassette integration into the *L.major HSP23* locus; Figure S12: Verification of *L.major HSP23*$^{-/-}$ complementation lines; Table S1: List of primers used in this study; Table S2: Sequence identity analysis of trypanosomatid HSP23 proteins.

Author Contributions: V.A.: Study conception: generation and phenotype analysis of *L. braziliensis* mutants, manuscript preparation; C.K.-B.: Study conception, generation and phenotype analysis of *L. major* mutants, manuscript preparation; C.B.: Gene replacements: phenotype analysis, in vitro infections, next generation sequencing; H.Z.: Experimental design, data analysis; J.S.: FACS analysis; J.A.: Study conception: manuscript preparation; J.-C.D.: Study conception, manuscript preparation; J.C.: Study conception: experimental design, supervision, manuscript preparation. All authors have read and agreed to the published version of the manuscript.

Funding: V.A. was supported by a Humboldt Research Fellowship for Postdoctoral Researchers from the Alexander von Humboldt Foundation, Germany, during the study period. The publication of this article was funded by the Open Access Fund of the Leibniz Association.

Acknowledgments: We thank Andrea MacDonald and Dorothea Zander–Dinse (Bernhard Nocht Institute for Tropical Medicine, Hamburg, Germany) for technical assistance; Hideo Imamura (Institute of Tropical Medicine, Antwerp, Belgium), for providing the genome sequence of the *L. braziliensis* strain PER005cl2 and the novel long–read assembly of the *L. braziliensis* M2904 reference genome; Eva Gluenz and Tom Beneke (University of Oxford, UK), for providing the CRISPR–Cas9 toolkit plasmids developed for *Leishmania* spp.

Conflicts of Interest: The authors are not aware of any conflict of interest.

References

1. Marsden, P.D. Mucosal leishmaniasis ("espundia" Escomel, 1911). *Trans R Soc. Trop. Med. Hyg.* **1986**, *80*, 859–876. [CrossRef]
2. Amato, V.S.; Tuon, F.F.; Siqueira, A.M.; Nicodemo, A.C.; Neto, V.A. Treatment of mucosal leishmaniasis in Latin America: systematic review. *Am. J. Trop. Med. Hyg.* **2007**, *77*, 266–274. [CrossRef] [PubMed]

3. Arevalo, J.; Ramirez, L.; Adaui, V.; Zimic, M.; Tulliano, G.; Miranda-Verastegui, C.; Lazo, M.; Loayza-Muro, R.; De Doncker, S.; Maurer, A.; et al. Influence of Leishmania (Viannia) species on the response to antimonial treatment in patients with American tegumentary leishmaniasis. *J. Infect. Dis.* **2007**, *195*, 1846–1851. [CrossRef] [PubMed]

4. Reithinger, R.; Dujardin, J.C.; Louzir, H.; Pirmez, C.; Alexander, B.; Brooker, S. Cutaneous leishmaniasis. *Lancet Infect. Dis.* **2007**, *7*, 581–596. [CrossRef]

5. Cruz, A.; Beverley, S.M. Gene replacement in parasitic protozoa. *Nature* **1990**, *348*, 171–173. [CrossRef]

6. Cruz, A.; Coburn, C.M.; Beverley, S.M. Double targeted gene replacement for creating null mutants. *Proc. Natl. Acad. Sci. USA* **1991**, *88*, 7170–7174. [CrossRef]

7. Zirpel, H.; Clos, J. Gene Replacement by Homologous Recombination. *Methods Mol. Biol.* **2019**, *1971*, 169–188.

8. Peacock, C.S.; Seeger, K.; Harris, D.; Murphy, L.; Ruiz, J.C.; Quail, M.A.; Peters, N.; Adlem, E.; Tivey, A.; Aslett, M.; et al. Comparative genomic analysis of three Leishmania species that cause diverse human disease. *Nat. Genet.* **2007**, *39*, 839–847. [CrossRef]

9. Lye, L.F.; Owens, K.; Shi, H.; Murta, S.M.; Vieira, A.C.; Turco, S.J.; Tschudi, C.; Ullu, E.; Beverley, S.M. Retention and loss of RNA interference pathways in trypanosomatid protozoans. *PLoS Pathog.* **2010**, *6*, e1001161. [CrossRef]

10. De Paiva, R.M.; Grazielle-Silva, V.; Cardoso, M.S.; Nakagaki, B.N.; Mendonca-Neto, R.P.; Canavaci, A.M.; Souza Melo, N.; Martinelli, P.M.; Fernandes, A.P.; daRocha, W.D.; et al. Amastin Knockdown in Leishmania braziliensis Affects Parasite–Macrophage Interaction and Results in Impaired Viability of Intracellular Amastigotes. *PLoS Pathog.* **2015**, *11*, e1005296. [CrossRef]

11. Jackson, A.L.; Bartz, S.R.; Schelter, J.; Kobayashi, S.V.; Burchard, J.; Mao, M.; Li, B.; Cavet, G.; Linsley, P.S. Expression profiling reveals off–target gene regulation by RNAi. *Nat Biotechnol.* **2003**, *21*, 635–637. [CrossRef] [PubMed]

12. Knott, G.J.; Doudna, J.A. CRISPR–Cas guides the future of genetic engineering. *Science* **2018**, *361*, 866–869. [CrossRef] [PubMed]

13. Jinek, M.; Chylinski, K.; Fonfara, I.; Hauer, M.; Doudna, J.A.; Charpentier, E. A programmable dual–RNA–guided DNA endonuclease in adaptive bacterial immunity. *Science* **2012**, *337*, 816–821. [CrossRef] [PubMed]

14. Ceccaldi, R.; Rondinelli, B.; D'Andrea, A.D. Repair Pathway Choices and Consequences at the Double–Strand Break. *Trends Cell Biol.* **2016**, *26*, 52–64. [CrossRef]

15. Bibikova, M.; Beumer, K.; Trautman, J.K.; Carroll, D. Enhancing gene targeting with designed zinc finger nucleases. *Science* **2003**, *300*, 764. [CrossRef]

16. Peng, D.; Kurup, S.P.; Yao, P.Y.; Minning, T.A.; Tarleton, R.L. CRISPR–Cas9–mediated single–gene and gene family disruption in Trypanosoma cruzi. *mBio* **2014**, *6*, e02097-14. [CrossRef]

17. Beneke, T.; Madden, R.; Makin, L.; Valli, J.; Sunter, J.; Gluenz, E. A CRISPR Cas9 high–throughput genome editing toolkit for kinetoplastids. *R Soc. Open Sci.* **2017**, *4*, 170095. [CrossRef]

18. Vasquez, J.J.; Wedel, C.; Cosentino, R.O.; Siegel, T.N. Exploiting CRISPR–Cas9 technology to investigate individual histone modifications. *Nucleic Acids Res.* **2018**, *46*, e106. [CrossRef]

19. Sollelis, L.; Ghorbal, M.; MacPherson, C.R.; Martins, R.M.; Kuk, N.; Crobu, L.; Bastien, P.; Scherf, A.; Lopez-Rubio, J.-J.; Sterkers, Y. First efficient CRISPR–Cas9–mediated genome editing in *Leishmania* parasites. *Cell. Microbiol.* **2015**, *17*, 1405–1412. [CrossRef]

20. Zhang, W.W.; Matlashewski, G. CRISPR–Cas9–Mediated Genome Editing in *Leishmania donovani*. *MBio* **2015**, *6*, e00861. [CrossRef]

21. Martel, D.; Beneke, T.; Gluenz, E.; Spath, G.F.; Rachidi, N. Characterisation of Casein Kinase 1.1 in Leishmania donovani Using the CRISPR Cas9 Toolkit. *Biomed. Res. Int.* **2017**, *2017*, 4635605. [CrossRef] [PubMed]

22. Soares Medeiros, L.C.; South, L.; Peng, D.; Bustamante, J.M.; Wang, W.; Bunkofske, M.; Perumal, N.; Sanchez-Valdez, F.; Tarleton, R.L. Rapid, Selection-Free, High–Efficiency Genome Editing in Protozoan Parasites Using CRISPR–Cas9 Ribonucleoproteins. *mBio* **2017**, *8*. [CrossRef]

23. Fernandez-Prada, C.; Sharma, M.; Plourde, M.; Bresson, E.; Roy, G.; Leprohon, P.; Ouellette, M. High–throughput Cos–Seq screen with intracellular Leishmania infantum for the discovery of novel drug–resistance mechanisms. *Int. J. Parasitol. Drugs Drug Resist.* **2018**, *8*, 165–173. [CrossRef] [PubMed]

24. Ishemgulova, A.; Hlavacova, J.; Majerova, K.; Butenko, A.; Lukes, J.; Votypka, J.; Volf, P.; Yurchenko, V. CRISPR/Cas9 in Leishmania mexicana: A case study of LmxBTN1. *PloS ONE* **2018**, *13*, e0192723. [CrossRef] [PubMed]

25. Bryant, J.M.; Baumgarten, S.; Glover, L.; Hutchinson, S.; Rachidi, N. CRISPR in Parasitology: Not Exactly Cut and Dried! *Trends Parasitol* **2019**, *35*, 409–422. [CrossRef]

26. Cruz, A.K.; Titus, R.; Beverley, S.M. Plasticity in chromosome number and testing of essential genes in Leishmania by targeting. *Proc. Natl. Acad. Sci. USA* **1993**, *90*, 1599–1603. [CrossRef]

27. Sterkers, Y.; Crobu, L.; Lachaud, L.; Pages, M.; Bastien, P. Parasexuality and mosaic aneuploidy in Leishmania: alternative genetics. *Trends Parasitol.* **2014**, *30*, 429–435. [CrossRef]

28. Dumetz, F.; Imamura, H.; Sanders, M.; Seblova, V.; Myskova, J.; Pescher, P.; Vanaerschot, M.; Meehan, C.J.; Cuypers, B.; De Muylder, G.; et al. Modulation of Aneuploidy in Leishmania donovani during Adaptation to Different In Vitro and In Vivo Environments and Its Impact on Gene Expression. *MBio* **2017**, *8*. [CrossRef]

29. Duncan, S.M.; Jones, N.G.; Mottram, J.C. Recent advances in Leishmania reverse genetics: Manipulating a manipulative parasite. *Mol. Biochem. Parasitol.* **2017**, *216*, 30–38. [CrossRef]

30. Zhang, W.W.; Lypaczewski, P.; Matlashewski, G. Optimized CRISPR–Cas9 Genome Editing for Leishmania and Its Use To Target a Multigene Family, Induce Chromosomal Translocation, and Study DNA Break Repair Mechanisms. *mSphere* **2017**, *2*. [CrossRef]

31. Zhang, W.W.; Matlashewski, G. Single–Strand Annealing Plays a Major Role in Double–Strand DNA Break Repair following CRISPR–Cas9 Cleavage in Leishmania. *mSphere* **2019**, *4*. [CrossRef] [PubMed]

32. Damasceno, J.D.; Reis-Cunha, J.; Crouch, K.; Beraldi, D.; Lapsley, C.; Tosi, L.R.O.; Bartholomeu, D.; McCulloch, R. Conditional knockout of RAD51–related genes in Leishmania major reveals a critical role for homologous recombination during genome replication. *PLoS Genet.* **2020**, *16*, e1008828. [CrossRef] [PubMed]

33. Yagoubat, A.; Crobu, L.; Berry, L.; Kuk, N.; Lefebvre, M.; Sarrazin, A.; Bastien, P.; Sterkers, Y. Universal highly efficient conditional knockout system in Leishmania, with a focus on untranscribed region preservation. *Cell. Microbiol.* **2020**, *22*, e13159. [CrossRef] [PubMed]

34. Yardley, V.; Ortuno, N.; Llanos-Cuentas, A.; Chappuis, F.; Doncker, S.D.; Ramirez, L.; Croft, S.; Arevalo, J.; Adaui, V.; Bermudez, H.; et al. American tegumentary leishmaniasis: Is antimonial treatment outcome related to parasite drug susceptibility? *J. Infect. Dis.* **2006**, *194*, 1168–1175. [CrossRef]

35. Rosenzweig, D.; Smith, D.; Opperdoes, F.; Stern, S.; Olafson, R.W.; Zilberstein, D. Retooling Leishmania metabolism: from sand fly gut to human macrophage. *FASEB J* **2008**, *22*, 590–602. [CrossRef] [PubMed]

36. Al-Jawabreh, A.; Diezmann, S.; Muller, M.; Wirth, T.; Schnur, L.F.; Strelkova, M.V.; Kovalenko, D.A.; Razakov, S.A.; Schwenkenbecher, J.; Kuhls, K.; et al. Identification of geographically distributed sub–populations of Leishmania (Leishmania) major by microsatellite analysis. *BMC Evol. Biol.* **2008**, *8*, 183. [CrossRef]

37. Kapler, G.M.; Coburn, C.M.; Beverley, S.M. Stable transfection of the human parasite *Leishmania major* delineates a 30–kilobase region sufficient for extrachromosomal replication and expression. *Mol. Cell. Biol.* **1990**, *10*, 1084–1094. [CrossRef] [PubMed]

38. Krobitsch, S.; Brandau, S.; Hoyer, C.; Schmetz, C.; Hübel, A.; Clos, J. *Leishmania donovani* heat shock protein 100: characterization and function in amastigote stage differentiation. *J. Biol. Chem.* **1998**, *273*, 6488–6494. [CrossRef]

39. Ommen, G.; Lorenz, S.; Clos, J. One–step generation of double–allele gene replacement mutants in Leishmania donovani. *Int. J. Parasitol.* **2009**, *39*, 541–546. [CrossRef]

40. Beneke, T.; Gluenz, E. LeishGEdit: A Method for Rapid Gene Knockout and Tagging Using CRISPR–Cas9. *Methods Mol. Biol.* **2019**, *1971*, 189–210.

41. Schumann Burkard, G.; Jutzi, P.; Roditi, I. Genome–wide RNAi screens in bloodstream form trypanosomes identify drug transporters. *Mol. Biochem. Parasitol.* **2011**, *175*, 91–94. [CrossRef] [PubMed]

42. Bartsch, K.; Hombach-Barrigah, A.; Clos, J. Hsp90 inhibitors radicicol and geldanamycin have opposing effects on Leishmania Aha1–dependent proliferation. *Cell Stress Chaperones* **2017**, *22*, 729–742. [CrossRef] [PubMed]

43. Choudhury, K.; Zander, D.; Kube, M.; Reinhardt, R.; Clos, J. Identification of a Leishmania infantum gene mediating resistance to miltefosine and SbIII. *Int. J. Parasitol.* **2008**, *38*, 1411–1423. [CrossRef] [PubMed]

44. Livak, K.J.; Schmittgen, T.D. Analysis of relative gene expression data using real–time quantitative PCR and the 2(–Delta Delta C(T)) Method. *Methods* **2001**, *25*, 402–408. [CrossRef]

45. Kröber-Boncardo, C.; Lorenzen, S.; Brinker, C.; Clos, J. Casein Kinase 1.2 Over Expression Restores Stress Resistance toLeishmania donovaniHSP23 Null Mutants. *Sci. Rep.* **2020**, *10*. in press.

46. Van den Broeck, F.; Savill, N.J.; Imamura, H.; Sanders, M.; Maes, I.; Cooper, S.; Mateus, D.; Jara, M.; Adaui, V.; Arevalo, J.; et al. Ecological divergence and hybridization of Neotropical Leishmania parasites. *Proc. Natl. Acad. Sci. USA* **2020**, *10*, 210. [CrossRef]

47. Hombach, A.; Ommen, G.; MacDonald, A.; Clos, J. A small heat shock protein is essential for thermotolerance and intracellular survival of Leishmania donovani. *J. Cell Sci.* **2014**, *127*, 4762–4773. [CrossRef]

48. Hombach-Barrigah, A.; Bartsch, K.; Smirlis, D.; Rosenqvist, H.; MacDonald, A.; Dingli, F.; Loew, D.; Spath, G.F.; Rachidi, N.; Wiese, M.; et al. Leishmania donovani 90 kD Heat Shock Protein—Impact of Phosphosites on Parasite Fitness, Infectivity and Casein Kinase Affinity. *Sci. Rep.* **2019**, *9*, 5074. [CrossRef]

49. Bifeld, E.; Tejera Nevado, P.; Bartsch, J.; Eick, J.; Clos, J. A versatile qPCR assay to quantify trypanosomatidic infections of host cells and tissues. *Med. Microbiol. Immunol.* **2016**, *205*, 449–458. [CrossRef]

50. Bifeld, E. Quantification of Intracellular Leishmania spp. Using Real–Time Quantitative PCR (qPCR). *Methods Mol. Biol.* **2019**, *1971*, 249–263.

51. Bifeld, E. Generation of Bone Marrow–Derived Macrophages for In Vitro Infection Experiments. *Methods Mol. Biol.* **2019**, *1971*, 237–247. [PubMed]

52. Peng, D.; Tarleton, R. EuPaGDT: a web tool tailored to design CRISPR guide RNAs for eukaryotic pathogens. *Microb. Genom.* **2015**, *1*, e000033. [CrossRef]

53. Hoyer, C.; Zander, D.; Fleischer, S.; Schilhabel, M.; Kroener, M.; Platzer, M.; Clos, J. A Leishmania donovani gene that confers accelerated recovery from stationary phase growth arrest. *Int. J. Parasitol.* **2004**, *34*, 803–811. [CrossRef] [PubMed]

54. Dickinson, D.J.; Ward, J.D.; Reiner, D.J.; Goldstein, B. Engineering the Caenorhabditis elegans genome using Cas9–triggered homologous recombination. *Nat. Methods* **2013**, *10*, 1028–1034. [CrossRef] [PubMed]

55. Bottcher, R.; Hollmann, M.; Merk, K.; Nitschko, V.; Obermaier, C.; Philippou-Massier, J.; Wieland, I.; Gaul, U.; Forstemann, K. Efficient chromosomal gene modification with CRISPR/cas9 and PCR–based homologous recombination donors in cultured Drosophila cells. *Nucleic Acids Res.* **2014**, *42*, e89. [CrossRef]

56. Hübel, A.; Krobitsch, S.; Horauf, A.; Clos, J. Leishmania major Hsp100 is required chiefly in the mammalian stage of the parasite. *Mol. Cell Biol.* **1997**, *17*, 5987–5995. [CrossRef]

57. Krobitsch, S.; Clos, J. A novel role for 100 kD heat shock proteins in the parasite *Leishmania donovani*. *Cell Stress Chaperones* **1999**, *4*, 191–198. [CrossRef]

58. Van Montfort, R.L.; Basha, E.; Friedrich, K.L.; Slingsby, C.; Vierling, E. Crystal structure and assembly of a eukaryotic small heat shock protein. *Nat. Struct. Biol.* **2001**, *8*, 1025–1030. [CrossRef]

59. Nuhs, A.; Schafer, C.; Zander, D.; Trube, L.; Tejera Nevado, P.; Schmidt, S.; Arevalo, J.; Adaui, V.; Maes, L.; Dujardin, J.C.; et al. A novel marker, ARM58, confers antimony resistance to Leishmania spp. *Int. J. Parasitol. Drugs Drug Resist.* **2014**, *4*, 37–47. [CrossRef]

60. Callahan, H.L.; Portal, I.F.; Bensinger, S.J.; Grogl, M. Leishmania spp: temperature sensitivity of promastigotes in vitro as a model for tropism in vivo. *Exp. Parasitol.* **1996**, *84*, 400–409. [CrossRef]

61. Piper, P.W. The heat shock and ethanol stress responses of yeast exhibit extensive similarity and functional overlap. *FEMS Microbiol. Lett.* **1995**, *134*, 121–127. [CrossRef] [PubMed]

62. Barak, E.; Amin-Spector, S.; Gerliak, E.; Goyard, S.; Holland, N.; Zilberstein, D. Differentiation of Leishmania donovani in host–free system: analysis of signal perception and response. *Mol. Biochem. Parasitol.* **2005**, *141*, 99–108. [CrossRef] [PubMed]

63. Cupolillo, E.; Brahim, L.R.; Toaldo, C.B.; de Oliveira-Neto, M.P.; de Brito, M.E.; Falqueto, A.; de Farias Naiff, M.; Grimaldi, G., Jr. Genetic polymorphism and molecular epidemiology of Leishmania (Viannia) braziliensis from different hosts and geographic areas in Brazil. *J. Clin. Microbiol.* **2003**, *41*, 3126–3132. [CrossRef] [PubMed]

64. Tobin, J.F.; Laban, A.; Wirth, D.F. Homologous recombination in Leishmania enriettii. *Proc. Natl. Acad. Sci. USA* **1991**, *88*, 864–868. [CrossRef]

65. Beverley, S.M. Protozomics: trypanosomatid parasite genetics comes of age. *Nat. Rev. Genet.* **2003**, *4*, 11–19. [CrossRef] [PubMed]

66. Wiese, M. A mitogen–activated protein (MAP) kinase homologue of Leishmania mexicana is essential for parasite survival in the infected host. *Embo. J.* **1998**, *17*, 2619–2628. [CrossRef]

67. Coelho, A.C.; Oliveira, J.C.; Espada, C.R.; Reimao, J.Q.; Trinconi, C.T.; Uliana, S.R. A Luciferase–Expressing Leishmania braziliensis Line That Leads to Sustained Skin Lesions in BALB/c Mice and Allows Monitoring of Miltefosine Treatment Outcome. *PLoS Negl. Trop. Dis.* **2016**, *10*, e0004660. [CrossRef]

68. Bastos, M.S.; Souza, L.A.; Onofre, T.S.; Silva, A.J.; Almeida, M.R.; Bressan, G.C.; Fietto, J.L. Achievement of constitutive fluorescent pLEXSY–egfp Leishmania braziliensis and its application as an alternative method for drug screening in vitro. *Mem. Inst. Oswaldo. Cruz.* **2017**, *112*, 155–159. [CrossRef]

69. Sharma, R.; Silveira-Mattos, P.S.; Ferreira, V.C.; Rangel, F.A.; Oliveira, L.B.; Celes, F.S.; Viana, S.M.; Wilson, M.E.; de Oliveira, C.I. Generation and Characterization of a Dual–Reporter Transgenic Leishmania braziliensis Line Expressing eGFP and Luciferase. *Front. Cell. Infect. Microbiol.* **2019**, *9*, 468. [CrossRef]

70. Andrade, J.M.; Murta, S.M. Functional analysis of cytosolic tryparedoxin peroxidase in antimony–resistant and –susceptible Leishmania braziliensis and Leishmania infantum lines. *Parasites Vectors* **2014**, *7*, 406. [CrossRef]

71. Andrade, J.M.; Baba, E.H.; Machado-de-Avila, R.A.; Chavez-Olortegui, C.; Demicheli, C.P.; Frezard, F.; Monte-Neto, R.L.; Murta, S.M. Silver and Nitrate Oppositely Modulate Antimony Susceptibility through Aquaglyceroporin 1 in Leishmania (Viannia) Species. *Antimicrob. Agents Chemother* **2016**, *60*, 4482–4489. [CrossRef] [PubMed]

72. Moreira, D.S.; Xavier, M.V.; Murta, S.M.F. Ascorbate peroxidase overexpression protects Leishmania braziliensis against trivalent antimony effects. *Mem. Inst. Oswaldo Cruz* **2018**, *113*, e180377. [CrossRef] [PubMed]

73. De Toledo, J.S.; Junqueira dos Santos, A.F.; Rodrigues de Moura, T.; Antoniazi, S.A.; Brodskyn, C.; Indiani de Oliveira, C.; Barral, A.; Cruz, A.K. Leishmania (Viannia) braziliensis transfectants overexpressing the miniexon gene lose virulence in vivo. *Parasitol. Int.* **2009**, *58*, 45–50. [CrossRef] [PubMed]

74. Smith, I.; Greenside, P.G.; Natoli, T.; Lahr, D.L.; Wadden, D.; Tirosh, I.; Narayan, R.; Root, D.E.; Golub, T.R.; Subramanian, A.; et al. Evaluation of RNAi and CRISPR technologies by large–scale gene expression profiling in the Connectivity Map. *PLoS Biol.* **2017**, *15*, e2003213. [CrossRef] [PubMed]

75. Adaui, V.; Lye, L.F.; Akopyants, N.S.; Zimic, M.; Llanos-Cuentas, A.; Garcia, L.; Maes, I.; De Doncker, S.; Dobson, D.E.; Arevalo, J.; et al. Association of the Endobiont Double–Stranded RNA Virus LRV1 With Treatment Failure for Human Leishmaniasis Caused by Leishmania braziliensis in Peru and Bolivia. *J. Infect. Dis.* **2016**, *213*, 112–121. [CrossRef] [PubMed]

76. Bourreau, E.; Ginouves, M.; Prevot, G.; Hartley, M.A.; Gangneux, J.P.; Robert-Gangneux, F.; Dufour, J.; Sainte-Marie, D.; Bertolotti, A.; Pratlong, F.; et al. Presence of Leishmania RNA Virus 1 in Leishmania guyanensis Increases the Risk of First–Line Treatment Failure and Symptomatic Relapse. *J. Infect. Dis.* **2016**, *213*, 105–111. [CrossRef]

77. Cantanhede, L.M.; Fernandes, F.G.; Ferreira, G.E.M.; Porrozzi, R.; Ferreira, R.G.M.; Cupolillo, E. New insights into the genetic diversity of Leishmania RNA Virus 1 and its species–specific relationship with Leishmania parasites. *PloS ONE* **2018**, *13*, e0198727. [CrossRef]

78. Ives, A.; Ronet, C.; Prevel, F.; Ruzzante, G.; Fuertes-Marraco, S.; Schutz, F.; Zangger, H.; Revaz-Breton, M.; Lye, L.F.; Hickerson, S.M.; et al. Leishmania RNA virus controls the severity of mucocutaneous leishmaniasis. *Science* **2011**, *331*, 775–778. [CrossRef]

79. Eren, R.O.; Reverte, M.; Rossi, M.; Hartley, M.A.; Castiglioni, P.; Prevel, F.; Martin, R.; Desponds, C.; Lye, L.F.; Drexler, S.K.; et al. Mammalian Innate Immune Response to a Leishmania–Resident RNA Virus Increases Macrophage Survival to Promote Parasite Persistence. *Cell. Host Microbe.* **2016**, *20*, 318–328. [CrossRef]

80. Doench, J.G.; Hartenian, E.; Graham, D.B.; Tothova, Z.; Hegde, M.; Smith, I.; Sullender, M.; Ebert, B.L.; Xavier, R.J.; Root, D.E. Rational design of highly active sgRNAs for CRISPR–Cas9–mediated gene inactivation. *Nat. Biotechnol.* **2014**, *32*, 1262–1267. [CrossRef]

81. Wong, N.; Liu, W.; Wang, X. WU–CRISPR: characteristics of functional guide RNAs for the CRISPR/Cas9 system. *Genome Biol.* **2015**, *16*, 218. [CrossRef] [PubMed]

82. Xu, H.; Xiao, T.; Chen, C.H.; Li, W.; Meyer, C.A.; Wu, Q.; Wu, D.; Cong, L.; Zhang, F.; Liu, J.S.; et al. Sequence determinants of improved CRISPR sgRNA design. *Genome Res.* **2015**, *25*, 1147–1157. [CrossRef] [PubMed]

83. Labuhn, M.; Adams, F.F.; Ng, M.; Knoess, S.; Schambach, A.; Charpentier, E.M.; Schwarzer, A.; Mateo, J.L.; Klusmann, J.H.; Heckl, D. Refined sgRNA efficacy prediction improves large– and small–scale CRISPR–Cas9 applications. *Nucleic Acids Res.* **2018**, *46*, 1375–1385. [CrossRef] [PubMed]

84. Graf, R.; Li, X.; Chu, V.T.; Rajewsky, K. sgRNA Sequence Motifs Blocking Efficient CRISPR/Cas9–Mediated Gene Editing. *Cell Rep.* **2019**, *26*, 1098–1103.e3. [CrossRef]
85. Yuen, G.; Khan, F.J.; Gao, S.; Stommel, J.M.; Batchelor, E.; Wu, X.; Luo, J. CRISPR/Cas9–mediated gene knockout is insensitive to target copy number but is dependent on guide RNA potency and Cas9/sgRNA threshold expression level. *Nucleic Acids Res.* **2017**, *45*, 12039–12053. [CrossRef]
86. Ng, H.; Dean, N. Dramatic Improvement of CRISPR/Cas9 Editing in Candida albicans by Increased Single Guide RNA Expression. *mSphere* **2017**, *2*. [CrossRef]
87. Jara, M.; Maes, I.; Imamura, H.; Domagalska, M.A.; Dujardin, J.C.; Arevalo, J. Tracking of quiescence in Leishmania by quantifying the expression of GFP in the ribosomal DNA locus. *Sci. Rep.* **2019**, *9*, 18951. [CrossRef]

Review

The Maze Pathway of Coevolution: A Critical Review over the *Leishmania* and Its Endosymbiotic History

Lilian Motta Cantanhêde, Carlos Mata-Somarribas, Khaled Chourabi, Gabriela Pereira da Silva, Bruna Dias das Chagas, Luiza de Oliveira R. Pereira, Mariana Côrtes Boité and Elisa Cupolillo *

Research on Leishmaniasis Laboratory, Oswaldo Cruz Institute, FIOCRUZ, Rio de Janeiro 21040360, Brazil; lilian.cantanhede@ioc.fiocruz.br (L.M.C.); carlos.somarribas@ioc.fiocruz.br (C.M.-S.); khaled.chourabi@ioc.fiocruz.br (K.C.); gabriela.silva@ioc.fiocruz.br (G.P.d.S.); bruna.chagas@ioc.fiocruz.br (B.D.d.C.); luizaper@ioc.fiocruz.br (L.d.O.R.P.); boitemc@ioc.fiocruz.br (M.C.B.)
* Correspondence: elisa.cupolillo@ioc.fiocruz.br; Tel.: +55-21-38658177

Abstract: The description of the genus *Leishmania* as the causative agent of leishmaniasis occurred in the modern age. However, evolutionary studies suggest that the origin of *Leishmania* can be traced back to the Mesozoic era. Subsequently, during its evolutionary process, it achieved worldwide dispersion predating the breakup of the Gondwana supercontinent. It is assumed that this parasite evolved from monoxenic Trypanosomatidae. Phylogenetic studies locate dixenous *Leishmania* in a well-supported clade, in the recently named subfamily Leishmaniinae, which also includes monoxenous trypanosomatids. Virus-like particles have been reported in many species of this family. To date, several *Leishmania* species have been reported to be infected by *Leishmania* RNA virus (LRV) and *Leishbunyavirus* (LBV). Since the first descriptions of LRVs decades ago, differences in their genomic structures have been highlighted, leading to the designation of LRV1 in *L.* (*Viannia*) species and LRV2 in *L.* (*Leishmania*) species. There are strong indications that viruses that infect *Leishmania* spp. have the ability to enhance parasitic survival in humans as well as in experimental infections, through highly complex and specialized mechanisms. Phylogenetic analyses of these viruses have shown that their genomic differences correlate with the parasite species infected, suggesting a coevolutionary process. Herein, we will explore what has been described in the literature regarding the relationship between *Leishmania* and endosymbiotic *Leishmania* viruses and what is known about this association that could contribute to discussions about the worldwide dispersion of *Leishmania*.

Keywords: *Leishmania*; *Leishmania* viruses; phylogeny; coevolution; endosymbiont protozoan viruses

check for
updates

Citation: Cantanhêde, L.M.; Mata-Somarribas, C.; Chourabi, K.; Pereira da Silva, G.; Dias das Chagas, B.; de Oliveira R. Pereira, L.; Côrtes Boité, M.; Cupolillo, E. The Maze Pathway of Coevolution: A Critical Review over the *Leishmania* and Its Endosymbiotic History. *Genes* **2021**, *12*, 657. https://doi.org/10.3390/genes12050657

Academic Editor: Jose M. Requena

Received: 15 January 2021
Accepted: 5 April 2021
Published: 27 April 2021

Publisher's Note: MDPI stays neutral with regard to jurisdictional claims in published maps and institutional affiliations.

1. Introduction

The origin of the *Leishmania* parasite dates back to the Mesozoic era, and models of its dispersion to the continents, still hypothetical, consider different scenarios [1]. The diversification of this group of dixenous parasites occurred on different continents, and currently, the *Leishmania* genus consists of dozens of different species worldwide, pathogenic to humans or not, which, by themselves, present complexities that are still not fully understood. There is some discussion on the taxonomy of *Leishmania*, and in this study, we will adopt the proposal of Kostygov et al. [2] and Espinosa et al. [3], naming four *Leishmania* subgenera: *L.* (*Leishmania*), *L.* (*Viannia*), *L.* (*Sauroleishmania*), and *L.* (*Mundinia*).

Despite efforts to unravel the mechanisms of *Leishmania* pathogenicity, to determine the risk of infection and to develop new treatments and vaccines against the parasite, there are still gaps in the state-of-the-art treatments to be explored. For example, one of the most well-defined aspects of the parasite, the *Leishmania* life cycle, has been updated by recent and important discoveries of factors that influence the parasite's dispersion ability [4]. An amazing field to be explored concerns the effects of endosymbiotic *Leishmania* virus presence, its relationship with *Leishmania* cells, and further clinical and epidemiological consequences. *Paraleishmania hertigi* and *Paraleishmania deanei*, formerly *Leishmania hertigi* and *Leishmania deanei* [2,5], respectively, were the first members of the subfamily Leishmaninae [6] identified as able to host virus-like particles [7]. Nevertheless, to date, no additional studies have been performed characterizing virus-like particles (VLPs) from these species. Only nine years later, *Leishmania (Viannia) guyanensis* and *Leishmania (Viannia) braziliensis*, which were both described as host viruses, were molecularly characterized [8], with the *L. guyanensis* virus named LR1 and the virus found in *L. braziliensis* named LR2. Both LR1 and LR2 were thought to contain single-stranded DNA [8], but soon after, they were demonstrated to contain circular double-stranded RNAs (dsRNAs) and were renamed *Leishmania* RNA virus (LRV) [9]. In the following years, LRV was described in 12 isolates of *L. braziliensis* and *L. guyanensis* from the Amazon region [10], and LRV was identified in one *Leishmania (Leishmania) major* isolate from a human patient in the former Soviet Union [11]. To the best of our knowledge, there are no studies that have searched for viruses in *L. (Sauroleishmania)* species, and only one species from *L. (Mundinia)*, *L. martiniquensis*, was recently found to harbour a virus named *Leishbunyavirus* (LBV).

Although not much attention has been given to *Leishmania* viruses classification, altogether, the literature classified the viruses infecting *Leishmania* species, LRV and LBV, into two virus families: Totiviridae and Leishbuviridae, respectively. Recent reports focused on *Leishmania* and the viral endosymbiont LRV first arose from questions not directly related to the virus but rather to Toll-like receptors and their association with variable immunological responses to *Leishmania* infection [12]. Important data have been gathered since then. Both viruses, LBV and LRV, influence the phenotypic expression of *Leishmania* infection in the vertebrate host, although biological aspects of *Leishmania*-harbouring viruses vs. virus-free *Leishmania* remain to be elucidated.

Thus, current data, in association with reports from decades ago, led us a step further in the understanding of this peculiar, dynamic, and million-year-old parasite. Some studies were recently published searching for and characterizing viruses in *Leishmania* parasites and in different members of the Trypanosomatidae family, suggesting endosymbiotic viruses as an ancient acquisition by these protozoans. Here, our main objective is to summarize previous and recent reports that characterize *Leishmania* viruses and the impact of this endosymbiosis and then to analyse their relationships with the parasite species that host them.

2. The *Leishmania* viruses

Virus-like particles (VLPs) in parasitic protozoans were first described in *Entamoeba histolytica* in the 1960s [13]. After that, several studies reported similar structures in many unicellular eukaryotes, such as *Giardia lamblia*, *Trichomonas vaginalis*, and members of the Trypanosomatidae family, including *Leishmania* spp. and *Trypanosoma* spp. For some additional protozoans, however, there are only studies reporting VLPs based on electron microscopy approaches but not by molecular methods. The International Committee on Taxonomy of Viruses (ICTV) recognized only the family Totiviridae gathering *Leishmania* viruses [14]. However, this, as well as families of viruses collected from other trypanosomatids, must be updated considering recent virus discovery and characterization [15].

Totiviridae consists of five genera: *Giardiavirus*, *Leishmaniavirus* (LRV), *Trichomonasvirus*, *Totivirus*, and *Victorivirus*. According to ICTV, *Leishmania* RNA virus 1 (LRV1) and *Leishmania* RNA virus 2 (LRV2) belong to the *Leishmaniavirus* genus. LRV was assumed to be capable of infecting *Leishmania* spp. only, with the two species identified as LRV1 and LRV2 (Figure 1), but recently, a member of this genus was also found in *Blechomonas* spp., a monoxenous trypanosomatid parasitizing fleas [15].

Recently, a new genus belonging to the Leishbuviridae family was proposed, *Leishbunyavirus* (LBV). The family Leishbuviridae includes the species *Leptomonas shilevirus*, which infects *Leptomonas moramango*, a monoxenic trypanosomatid [15], and LBV, which was found in *Leishmania martiniquensis* (Figure 1), a human pathogen that produces symptoms ranging from severe visceral disease to asymptomatic infections belonging to the subgenus *Leishmania (Mundinia)*. The virus was denominated *Leishmania martiniquensis leishbunyavirus* 1 (*Lmar*LBV1) and represents the only non-LRV virus found to infect a *Leishmania* species so far [16]. The subgenus *L.* (*Mundinia*) has been established recently and remains understudied. It consists of newly emerging, human-infecting *Leishmania* species and nonhuman pathogens distributed worldwide. It has been assumed that this subgenus represents the earliest diverging branch within *Leishmania*, possibly transmitted by a different vector [17].

Genus: Leishmaniavirus	**Genus: Leishbunyavirus**
Group III: Double strand RNA (dsRNA)	Group V: Negative strand RNA (ssRNA-)
Order: Ghabrivirales	Order: Bunyavirales
Family: Totiviridae	Family: Leishbuviridae
Species: ▲ *Leishmania RNA Virus* 1 (LRV1); ■ *Leishmania RNA Virus* 2 (LRV2).	Species: ● *Leptomonas shilevirus;* ◆ *Leishmania martiniquensis leishbunyavirus 1* (LmarLBV1).
Infected *Leishmania* species: ▲ *L. braziliensis, L. guyanensis, L. panamensis, L. shawi, L. naiffi, L. lainsoni;* ■ *L. major, L. tropica, L. infantum, L. aethiopica.*	Infected *Leishmania* species: ◆ *L. martiniquensis.*

Figure 1. Classification of *Leishmaniavirus* and *Leishbunyavirus* viruses and *Leishmania* species described so far harbouring each of these endosymbionts.

3. Exploiting Characteristics of *Leishmania*-Infecting Viruses

Leishmaniavirus species LRV1 and LRV2 were associated, respectively, with *Leishmania* (*Viannia*), found exclusively on the American continent, and with Old World *Leishmania* (*Leishmania*) species [18–22]. LBV, initially found in monoxenous trypanosomatids belonging to the subfamily Leishmaniinae and in the dixenous plant-parasitizing *Phytomonas* spp. [15], has also been detected in *Leishmania martiniquensis* [16] and possibly *Trypanosoma* spp. [15].

The Totiviridae family encompasses a broad range of viruses characterized by isometric virions, ranging from 30 to 40 nm in diameter, each containing a nonsegmented double-stranded RNA (dsRNA) genome, usually with two open reading frames (ORFs). LRV is a member of this family containing a ≅5.3 kb double-stranded RNA (dsRNA) genome [23,24] and organized into three ORFs. ORFs 2 and 3 encode a capsid protein (CP) and an RNA-dependent RNA polymerase (RdRP), respectively [25]. The first ORF is considered a predicted protein sequence and has shown no significant homology with known proteins [23]. Despite having small genomes, some totiviruses encode proteins in addition to RdRP and CP with known activity, such as the killer protein (KP4), produced by a fungal totivirus, which has proven antifungal activity [26]. Other totiviruses directly influence the expression of their host proteins, such as the virus that infects *Trichomonas vaginalis*, which, when present, is associated with an increase in the levels of proteins involved in the pathogenesis of the parasite [27]. Interestingly, although we do not know the protein encoded by ORF1 or its function, the viral capsid protein has endoribonuclease activity that precisely cleaves the transcript by ORF1 in both LRV1 [28] and LRV2 [29]. The two small RNA products resulting from the cleavage of their own endoribonuclease form a stable RNA/RNA complex, which can access host cell binding sites that are inaccessible to the transcript [30]. This configuration still requires further study. The classification of LRV in the Totiviridae family was due to its replication characteristics [31]. The low level of similarity (less than 40%) detected by comparing the nucleotide sequences from *L.* (*Viannia*) and *L. major* viruses enabled their classification into two different species, LRV1 and LRV2. Variation in the arrangement of the gene sequences is also observed between LRV1 and LRV2 [21,32]. LRV1 has an overlap between the regions encoding the viral capsid protein and the RNA polymerase, a particularity not observed for LRV2.

Leishbunyavirus belongs to the order Bunyavirales and is characterized as a virus exhibiting a negative-sense single-stranded RNA (ssRNA-) [33] organized in three genomic segments. The large segment encodes a viral RdRP, the medium segment encodes a surface glycoprotein precursor, and the small segment encodes a nucleoprotein [34]. Virions are usually 90 to 100 nm in diameter. The medium and small segments might present other ORFs involved in counteracting the host antiviral response, which may be present in both segments [35,36]. The infectivity and formation of viral particles in bunyaviruses depend on glycoproteins and type I transmembrane proteins that are proteolytically processed and glycosylated in the endoplasmic reticulum [36]. LmarLBV1 is a Bunyavirus and is the first non-LRV described infecting *Leishmania* [15,37].

Similar to other viruses, LRV and LBV require the resources of eukaryotic cells to sustain their metabolism. Furthermore, except for microRNAs (miRNAs) [38], dsRNA molecules are not produced by eukaryotic hosts, and eukaryotic cells have several mechanisms for detecting and inactivating these molecules [39,40]. The dsRNA viruses replicate within the capsid. Thus, the dsRNA genome is never exposed in the cytoplasm, which is an essential mechanism for evading host cell activation and antiviral action [41]. Transcription of the dsRNA genome by RdRP takes place within the virus [10]. The positive strand acts as messenger RNA (mRNA), giving rise to new viral particles, while the negative strand serves as a template for mRNA transcription [41].

4. A Brief History of the Detection and Dispersion of LRV1, LRV2 and LBV

LRV1 from the reference strain for *L. guyanensis* (MHOM/M4147) represents the first virus from kinetoplastids characterized by molecular approaches [8]. A few years later, the first study screening for the presence of LRV in *Leishmania* spp. strains from different geographical areas was conducted [42]. In this study, based on hybridization analysis, twelve LRV1 types (LRV1-1–LRV1-12) were defined, and it was shown for the first time that LRV1 could infect *L. braziliensis*, *L. guyanensis*, and various *Leishmania* strains from the Amazon Basin [42]. Comparative cDNA sequence analysis of LRV1-1 and LRV1-4 showed 77% identity, corroborating differences previously observed between these two types [32]. Furthermore, the comparison of two genomic regions from seven LRV types led to the description of two new types, LRV1-13 and LRV1-14, detected in *L. braziliensis* strains isolated from human patients from Bolivia [43].

In the early 1990s, parallel to the detection of LRV1 in two *L. (Viannia)* species, the discussion started as to whether the geographic distribution of *L. (Viannia)* spp. bearing LRV1 could be restricted to the Amazon Basin [42], despite widespread circulation of *L. braziliensis* in the American continent. Later, two other studies evaluated LRV1 in *L. braziliensis* from clinical samples and in *L. braziliensis* strains from south-eastern Brazil. All were negative [44,45], supporting the hypothesis of restricted circulation of *Leishmania* spp. bearing LRV1 to the Amazon Basin. Such findings exclude the possibility of a strict association between the presence of LRV1 and the severity of tegumentary leishmaniasis since there are also several leishmaniasis cases outside the Amazon Basin [20].

Recently, additional *L. (Viannia)* species were reported as infected by LRV1. Positive LRV1 samples were detected in tegumentary lesions from patients infected by *Leishmania (Viannia) lainsoni* and *Leishmania (Viannia) shawi* living in the western Brazilian Amazon region [20]. Later, LRV1 was demonstrated and characterized in the reference strain of *L. shawi* (MCEB/BR/1984/M8408), a strain isolated from a monkey [46]. A survey aiming to detect LRV in *Leishmania* strains deposited at the *Leishmania* collection of the Fundação Oswaldo Cruz-CLIOC. Available online: http://clioc.fiocruz.br (accessed on 6 April 2021) is underway and it was detected LRV1 in another *L. shawi* strain isolated from a human patient presenting CL in the Amazonas state (Table S1). An *Leishmania (Viannia) naiffi* strain from Amazonas state in Brazil was also reported to be *positive* for LRV1 [47] and 11 *L. naiffi* strains deposited at CLIOC were also positive, as well as one *L. lainsoni* (Table S1), the latter corroborating our previous study detecting LRV1 in clinical samples from patients infected by this species. All aforementioned results corroborated the assumption that LRV1 is restricted to *Leishmania* strains circulating in the Amazon Basin. However, we cannot rule out that the apparent narrow geographical distribution of LRV1 might be a result of biased surveys. Studying *Leishmania* spp. from Costa Rica, we detected an *L. guyanensis* strain positive for LRV1 (Table S1), reinforcing a recent finding indicating the circulation of LRV1 in this area [48].

In 1993, a virus was identified in an Old World *Leishmania* species, *L. major*, and was designated LRV2-1. It was described as immunologically distinct when compared to LRV1-1 and LRV1-4 [11]. The complete sequence of the virus found in *L. major* promastigotes MHOM/SU/1973/5-ASKH was published two years later, and it showed that the most relevant characteristic distinguishing the genomic structure of LRV2 from LRV1 and other totiviruses is the nonoverlapping capsid and RdRP genes [21].

LRV2 was detected in *L. major* [21], *Leishmania infantum* [22], *Leishmania aethiopica* [18,49], and *Leishmania tropica* [50]. Two studies conducted in Iran, in a zoonotic focus of cutaneous leishmaniasis (CL) and including visceral leishmaniasis (VL) patients, reported that the virus was detected in two different parasite specimens: one *L. infantum* strain derived from a VL patient unresponsive to treatment using meglumine antimoniate and one *L. major* strain from a great gerbil, *Rhombomys opimus* [22]. More recently, a survey was conducted in isolated promastigotes from 85 CL human patients from Iran. Eighty-three were identified as *L. major* and 2 as *L. tropica*. Fifty-nine (69.4%) presented LRV2, and one out of the two *L. tropica* isolates was also positive for LRV2 [50]. *L. tropica* was first demonstrated to be infected by LRV2 in a survey conducted in Turkey, in which 7 LRV2-positive *L. tropica* strains out of 24 were identified [51].

Recently, LRV2 was described in three (out of 3 examined) *L. major* strains in Turkey [51] and in two *L. major* strains isolated from CL patients from Uzbekistan. Sequence analysis indicated a high similarity between the two LRV2 isolates from Uzbekistan, which were closely related to the LRV2 isolate found in the *L. major* strain ASKH documented in Turkmenistan [21,49]. Thus, the presence of LRV2 in *L. major* is possibly frequent and widespread.

Recently, for the first time, a study demonstrated *L. (M.) martiniquensis* infected by endosymbiotic virus, a *Leishbunyavirus* (LBV). The molecular characterization revealed a genomic arrangement with three segments and sequences similar to those of LBV, which was first described infecting monoxenous *Crithidia* spp., a trypanosomatid member of the subfamily Leishmaniinae. However, to the best of our knowledge, the work published by Grybchuk and colleagues in 2018 [15] presented the most comprehensive study on LBV. In summary, thus far, LBV represents the most widespread and species-rich group of RNA viruses from trypanosomatids. This virus was found in *Crithidia* spp. from Ecuador, Ghana, and Russia and *L. moramango* from Madagascar, monoxenous trypanosomatid strains isolated from different hosts. Furthermore, using metatranscriptomic data from dipterans and horse leeches for viral and trypanosomatid surveys, they proposed this group of viruses associated with the subfamily Strigomonadinae and with *Trypanosoma* spp.

Regarding *Leishmania*, it is interesting that a geographically dispersed and multiple-host virus was detected in *L. (Mundinia)* species, the earliest branch within the genus *Leishmania*, which likely originated before Gondwana's breakup [52,53]. Another interesting feature of this group of parasites is concerned with its geographical dispersion and the diversity of vertebrates implicated as hosts, including humans. Interestingly, this group of parasites is probably not transmitted by sandflies. Comparative genomic analysis shows interesting differences in *L. (Mundinia)* from other *Leishmania* species [17].

5. LRV and LBV Modulating *Leishmania* spp. Phenotypes

Leishmania spp. infected with either Leishbunyaviridae or Totiviridae viruses show altered phenotypic expression. Several studies have been proposed to understand this impact, mainly involving the LRV1 endosymbiont, on the biology of different *L. (Viannia)* strains. The reason for this might be the enigmatic pathophysiology of CL and the intriguing hypothesis that LRV confers either a state of hypovirulence or hypervirulence on the host-parasite interaction [54].

Several groups have speculated on the influence of LRV1 on parasite virulence, and years have passed without major studies on the biological impact of LVR1 on *Leishmania* parasites [55]. Concern about LRV1 as a determinant of parasitic virulence reappeared in a 2011 study by Ives and colleagues using clones of *L. guyanensis* clinical isolates. Samples were classified due to their tendency to metastasize, ranging from highly metastatic (M+) to nonmetastatic (M−), using hamsters as the animal model. The authors found that a mucosal lesion-associated clone *L. guyanensis* carrying the virus (LgM+) increased the endogenous immune response in an unregulated manner, promoting an increase in inflammatory cytokines. These clones resulted in a phenotype of severe destruction of the nasopharyngeal mucosa when inoculated in mice, despite the significant reduction in the number of parasites. Macrophages infected with virus showed a phenotype similar to macrophages infected with parasites (LgM+), with increased expression levels of chemokines and cytokines such as CXCL10, CCL5, tumour necrosis factor-Alpha (TNF-α), and interleukin 6 (IL-6), also demonstrating that LRV1 alone induced the intensification of the inflammatory response to *Leishmania* antigens [12].

Thereafter, several studies explored the participation of LRV in the clinical evolution of the disease. Our group demonstrated that the relative risk of developing mucosal lesions in patients with Tegumentary leishmaniasis and LRV1 was three times higher than that in patients infected with parasites without LRV1 [20]. Moreover, the presence of LRV1 was associated with therapeutic failure cases in patients infected with *L. guyanensis* [56] and in patients infected with *L. braziliensis* [57]. However, other reports did not correlate LRV with distinct clinical phenotypes of TL [44,54] or treatment failure [58,59].

Assuming a mutualistic relationship between LRV and *Leishmania* spp., it is expected that *Leishmania* harbouring LRV1 could display better performance and fitness than virus-free strains facing certain environmental challenges. Routine evaluation of cultures maintained at CLIOC indicates two patterns of growth among *L. guyanensis* strains, and it was observed that LgLRV1+ survived longer and despite the environmental stress faced by parasites during in vitro cultivation, maintains viable parasites even in a nutrient-depleted environment without medium replacement (Figure S1) [60]. The reference strain for *L. guyanensis* (MHOM/BR/1975/M4147) is LRV1+, and a previous study demonstrated the detection of viable parasites until the end of the monitoring of the culture [61].

Studies have reported data on LRV+ and LRV− parasites under the same environment, for example, growing in the same culture medium [12,62], although there are apparently always fewer negative than positive parasites [63], suggesting that few LgLRV1− parasites may remain viable for a long time when cocultivated with LgLRV1+ parasites. We observed that experimentally mixed LgLRV1−/LgLRV1+ cultures presented a similar number of viable parasites at day 9 to that observed in single cultures for the LgLRV1+ strain (Figure S1), suggesting either (i) the counted parasites corresponded strictly to LgLRV+ cells or (ii) cocultivation enhances LVR− parasites' ability to survive. However, it is plausible that *Leishmania* spp., as described in *Trypanossoma brucei* [64], synthesize and secrete compounds in the shared environment, affecting population density and parasite behaviour, measured, for example, by growth rate in culture. It is possible that in addition to mechanisms such as cell-cell contact and secretion factors, exosome secretion, recently demonstrated for LRV1+ parasites, also contributes to this interaction [65–67].

Studies using mice infected by *L. guyanensis* LRV1+ demonstrated a higher parasite burden in lesions produced by these parasites than those produced by *L. guyanensis* LRV1− [12,68,69]. The immunization of mice with a vaccine produced from the LRV1 viral capsid protein decreases the burden of parasites in lesions after a new infection with *L. guyanensis* LRV1+ [68].

Little information is available concerning the influence of LRV on the biology and gene expression of *Leishmania* parasites when infected by these viruses. Teleologically, the viruses might influence the expression of many *Leishmania* genes, not only, but mainly those influenced by stressful conditions generated by parasite proliferation. Bearing in mind characteristics of infections caused by parasites containing LRV1, genes implicated on parasite proliferation and persistence are good target to be investigated also. Not less important are genes associated to therapeutic failure in infections caused by *Leishmania* parasites, pondering that cases of therapeutic failure have been associated with the presence of LRV1 in patients infected by *L. braziliensis* [57] and by *L. guyanensis* [56].

LRV is found in both stages of the *Leishmania* life cycle: promastigotes and intracellular amastigotes [8,9]. However, despite several studies exploring the effect of LRV in leishmaniasis pathogenesis, it is still unclear whether the virus effect is either the response of the vertebrate host to viral infection or if the virus affects the biology of its own host, *Leishmania* spp. [55,69]. A recent study evaluated the effect of LRV1 on the pathogenesis of TL using an isogenic, high viral load clone of *L. guyanensis* LRV− (from the M4147 strain). In doing so, it was possible to evaluate the effect of the virus in inducing the innate immune response. This study deciphered the mechanism by which LRV1 promotes parasitic persistence and disease progression and showed that this occurred due to the limited activation of inflammasomes in macrophages. Such an effect of LRV1 in modulating the immune response has also been demonstrated in human samples and was associated with mucosal leishmaniasis [70]. Additionally, as already mentioned, the presence of LRV1 and the viral load were identified as crucial factors in disease severity and pathology [62]. However, a question remains regarding the participation of LRV1 in modulating the immune response: it has been shown that the virus can be transported via exosomes [66], but at what point of infection is LRV1 exposed to the host cell, signalling the cascade that leads to the most severe phenotype of the disease?

Like LRV1, LRV2 present in *L. aethiopica* strains isolated from humans (LRV2-Lae) showed potential in modulating the immune response in macrophages, resulting in a hyperinflammatory and TLR3-dependent response [18]. In *Leishmania tropica*, LRV2 was detected in approximately 30% of the strains analysed [51]. *L. tropica* is an important aetiological agent of cutaneous leishmaniasis in the Old World, and there are several reports of this species in cases of mucosal leishmaniasis [71–74].

In Ethiopia and Brazil, a portion of patients with cutaneous lesions commonly progress to severe forms of the disease, such as mucosal leishmaniasis [75]. In those cases, the presence of LRV was associated with the development of the mucosal phenotype.

Despite the common influence of both LRV types on the immune response, other characteristics were not shared between them. For example, the LRV2 present in *L. major* isolates did not affect the therapeutic response [58], as already reported in infections by *L. guyanensis* and *L. braziliensis* LRV1+ [56,57]. However, a report of *Leishmania infantum* harbouring LRV2 described a patient with visceral leishmaniasis who had not responded to three cycles of systemic treatment. Therefore, not enough evidence is available to associate the presence of LRV2 with clinical phenotypes in VL caused by *L. infantum* [22].

The LBV detected in *L. martiniquensis* (LmarLBV1) also seems to influence parasite pathogenicity. Using an isogenic clone of *L. martiniquensis* without LBV (LmarLBV1-depleted), the influence of the virus on the biology of the parasite was evaluated, specifically concerning its ability to infect murine macrophages. The results showed that the LmarLBV1-depleted strain was less infective than the LmarLBV1 strain, indicating that LmarLBV1 facilitates parasite infectivity in vitro in the primary murine macrophage model [16].

6. The Maze Pathway of Coevolution of *Leishmania* spp. and Its Viruses

It is not yet fully known how *Leishmania* viruses are maintained and transmitted to *Leishmania* parasites. The most common mechanism for viral transmission in the Totiviridae family may be either vertical, horizontal (by cell fusion), or both, propagation [76]. Infection of non-LRV1-infected *Leishmania* parasites failed or was transitory when electroporation was attempted [77]. Mature viral particles of LRV could be transmitted to new parasites by cell division [11,41] or via exosomes [66]. More than 30% of exosomes produced by an *L. guyanensis* strain carry viral particles, and inside exosomes, LRV1 is able to resist inhospitable conditions until exosome-enveloped LRV1 infects other parasites [66]. Extracellular transmission of Totivirus in some protozoan parasites, such as *Giardia lamblia* [78], and in *L.*

guyanensis via exosomes [66] has been documented. Although this transmission is probably rare, virus-infected and noninfected parasites are still observed in the same culture [63]. It could not be ruled out that some parasites are resistant to virus infection, a hypothesis that remains to be tested.

The lack of a detectable infectious phase of LRV suggests a long-lasting relationship between the virus and the parasites, representing a symbiotic association. Indeed, studies have shown similar genetic intervals between *Leishmania* species and LRV1 and LRV2 [21,43]. Phylogenetic findings suggested that LRV acquisition by *Leishmania* parasites occurred prior to the divergence of Old and New World *Leishmania* parasites [43], but its interaction with the Trypanosomatid family was ancient, as indicated by the finding of LRV in *Blechomonas* [15].

Viral particles, LBV and LRV, were found in *Leishmania* species and in their closest phylogenetic clades *Endotrypanum* spp. and *Paraleishmania* spp. The loss and acquisition of both LBV and LRV probably occurred early in the family Trypanosomatidae, but additional research with different specimens from this family is necessary to make a proper inference for this hypothesis. Considering the knowledge gathered so far, the relationship between LBV and members of the Trypanosomatidae is older than that observed for LRV. LBV appears in *Trypanosoma* spp., regardless of whether *Blechomonas* is the first genus of the family harbouring LRV. LBV was detected in several members of the Trypanosomatidae family [15]. Although LRV, more specifically LRV3 and LRV4, was observed in *Blechomonas*, prior to the moment when Leishmaniinae split from other trypanosomatids, this virus emerged again in the *Leishmania* spp. branch. This could have coincided with the point in time when the dixenous life cycle emerged in Leishmaniinae, which could be supported by the identification of VLPs in *Paraleishmania* and *Endotrypanum* [2] as LRV, although a characterization of these particles is still required. Another possibility that can be assumed is the re-emergence of LRV before the time of *L. (Viannia)* and *L. (Leishmania)* diversification, considering VLPs found in *Paraleishmania* and *Endotrypanum* as non-LRVs. Comparative analyses of the *Leishmania* tree, based on random amplified polymorphic DNA (RAPD), and the LRV trees, obtained by sequence analysis of ORF3 or the 5′ untranslated region (5′-UTR), supported a long history of coevolution between LRV and the parasite-host strains, sustaining the hypothesis that LRV is an ancient virus of *Leishmania* spp. [43] and probably spread following host diversification (Figure 2).

Phylogenetic studies have shown that the transition from a monoxenous to a dixenous state occurred at least three times in the family Trypanosomatidae, giving rise to parasites of vertebrates, such as the *Trypanosoma* and *Leishmania* genera, and to *Phytomonas*, a dixenous phytopathogenic genus. Therefore, monoxenous parasites of invertebrates were ancestors of dixenous pathogens [79]. Considering the phylogenetic reconstruction of viruses found in many trypanosomatids using RdRP sequences, a well-supported clade for LBV was observed to be closely related to Phenuiviridae [15], a family including many viruses from insects, including the genus *Phlebovirus*, which is transmitted by sandfly species, the *Leishmania* vectors [80].

Assuming monophyly in the *Leishmania* clade and including their sister clades *Endotrypanum* spp. and *Paraleishmania* spp., two different points in time appear when the acquisition of these viruses could have occurred: first for the LBV in the subgenus *L. (Mundinia)* and then for the LRV in *L. (Viannia)* and *L. (Leishmania)*, with a later diversification into LRV1 and LRV2 at the same time that these two *Leishmania* subgenera split [18,43]. The challenge is now to uncover the points when gain and loss of the viruses appear in the process of diversification of the trypanosomatid taxa. Different strains from the same taxon can be found infected and noninfected by a specific virus, but it is still unknown whether the virus infection is an ancestral character or a derived one. The common ancestor for the *Leishmania* clade and their sister clades *Endotrypanum* spp. and *Paraleishmania* spp. could be virus-free, and independent viral acquisitions could have subsequently occurred. Different routes of both LBV and LRV acquisition and loss are possible in this protozoan group considering data gathered so far (Figure 2).

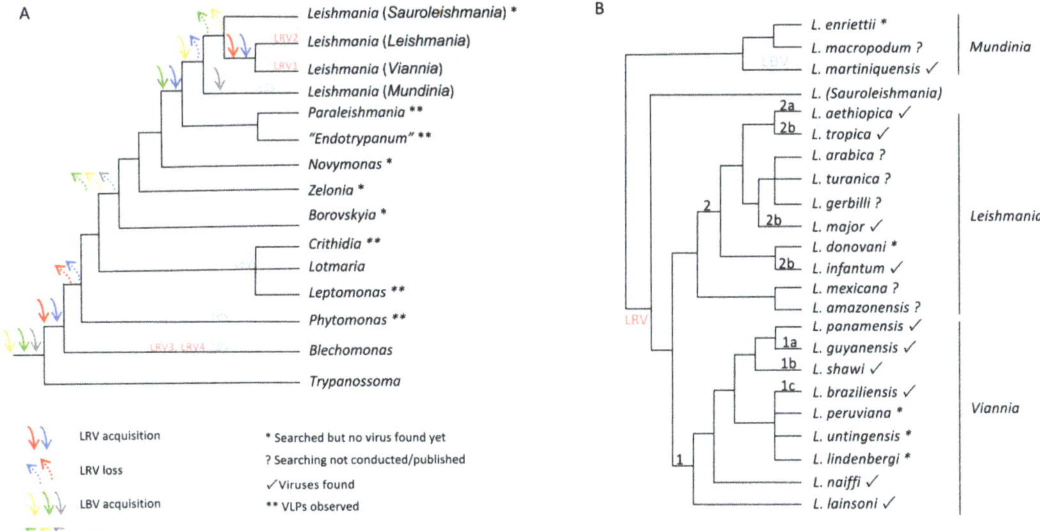

Figure 2. Schematic phylogenetic tree for the family Trypanosomatidae (**A**) and the genus *Leishmania* (**B**) based on published data [2,3,15] showing possibilities of *Leishbunyavirus* (LBV) and *Leishmania* RNA virus (LRV) acquisition by members of the family Trypanosomatidae and LRV dispersion across *Leishmania* species. Three scenarios are possible for LBV (green, yellow and grey arrows): green—ancient acquisition, with possible loss (dashed green arrow) in the first Leishmaniinae split and new acquisition in the clade containing *Leishmania*, *Paraleishmania*, and *"Endotrypanum"*, followed by loss in the split of *Leishmania* (*Mundinia*) from the other three *Leishmania* subgenera. This scenario assumes LBV not infecting *Novymonas*, *Zelonia* and *Borovskyia* and virus-like particles (VLPs) found in the clade containing Paraleishmania and "Endotrypanum" as LBV; yellow—the same as green, but with the last acquisition occurring in the split of *L.* (*Mundinia*) from the other *Leishmania* subgenera and subsequent loss in members of the other three *Leishmania* subgenera, assuming VLPs found in the clade Paraleishmania and "Endotrypanum" are not LBV; grey—ancient with possible loss (dashed grey arrow) in the first Leishmaniinae split and new acquisition when *L.* (*Mundinia*) split from the other *Leishmania* subgenera. Scenarios expected for LRV: blue—acquisition by a monoxenous trypanosomatid followed by sequential loss when another dixenous clade appears and acquisition in the clade containing *Leishmania*, Paraleishmania, and "Endotrypanum", followed by loss when *L.* (*Mundinia*) split from the other *Leishmania* subgenera and a new acquisition by clade *L.* (*Viannia*)/*L.* (*Leishmania*); this scenario assumes VLPs found in the clade containing Paraleishmania and "Endotrypanum" are LRV and the possibility of LRV infecting all *Leishmania* subgenera; red—acquisition by a monoxenous trypanosomatid followed by sequential loss when another dixenous clade appears and a new acquisition by clade *L.* (*Viannia*)/*L.* (*Leishmania*).

Alternatively, virus loss might have occurred independently and randomly. For strains from the same species, it is plausible that a given strain, or its ancestor, was infected, and during binary division, the virus was not equally transferred, resulting in both infected and noninfected descendants. This hypothesis also explains the observation of virus-infected and noninfected parasites in the same culture. To explore such an alternative, we consider LRV1 and *L.* (*Viannia*) as an example. LRV1 was detected in most of the *L.* (*Viannia*) species: *L. guyanensis*, *L. braziliensis*, *L. shawi*, *L. naiffi*, *L. lainsoni*, and *Leishmania panamensis*. Sequence analysis of LRV1 from *L. braziliensis*, *L. guyanensis*, and *L. shawi* showed clusters gathering according to the *Leishmania* species (Figure 2); the sole LRV1 sequence analysed from *L. shawi* was placed among two LRV1 sequences from *L. guyanensis* [46,63]. Curiously, the similarity between *L. shawi* and *L. guyanensis* was reported in many studies [81–83] and was also detected when LRV1-*L. guyanensis* and LRV1-*L. shawi* sequences were analysed [46]. Microsatellite analysis of *L.* (*Viannia*) spp. indicated that *L. guyanensis* is a distinct population within the *L.* (*Viannia*) subgenus (by microsatellite analysis), with no distinguishable subpopulations. However, differences in the reactivity profile with monoclonal antibodies were detected, overlapping the geographical distribution of the strains [84,85] and correlating with clusters formed after LRV1 *L. guyanensis* sequence analysis [46].

The case of *L. braziliensis* is especially interesting, as this species is widespread in the American continent, but so far, LRV1 has been detected only in strains isolated from the Amazon region. By microsatellite analyses, LRV(−) *L. braziliensis* strains belong to a distinct population from LRV1-infected *L. braziliensis* [86,87]. The intragroup diversity detected by the *L. braziliensis*-LRV1 sequence analysis is as high as the heterogeneity reported for this parasite species [88–90]. Two LRV1 clusters were demonstrated, corresponding to *L. braziliensis* from the western Amazon region (one from Bolivia and one from Brazil); an *L. braziliensis*-LRV1 sequence from French Guyana was placed in the middle, but with lower bootstrap support [46].

For *L. guyanensis*, *L. braziliensis*, other *L. (Viannia)* spp., and species infected by LRV2, infected and noninfected parasite cells were detected within the same strain. The same occurs for strains from the same regions. This assortment might be a significant determinant of coevolution [91] assuming that the degree of mixing, virus-free and virus-infected *Leishmania* spp. would increase *Leishmania* spp. exposure to viruses, therefore selecting for greater resistance and infectivity intervals. The characteristics of *Leishmania* and LRV could influence the probability of fluctuation in the direction of natural selection for a given phenotype over an evolutionary period of time (fluctuating selection dynamics—FSD). Furthermore, it could also be possible that *Leishmania* and their viruses are in combat, causing both to select adaptive characteristics, leading them to coevolve (arms race dynamics—ARD). The shift from FSD to ARD associated with population mixing is a possibility to be acknowledged [91]. Considering the infection by LRV in *Leishmania* species since the diversification of the subfamily Leishmaniinae, ARD could explain the lack of LRV-infected *L. braziliensis* outside of the Amazon Basin. If this is the case, *L. braziliensis* and LRV1 developed different resistance and infectivity (or strategies of infection), respectively. The raised hypothesis assumed the existence of *L. braziliensis* populations resistant to LRV infection. The methodology used to describe LRV transmission via exosomes [66] to uninfected *L. guyanensis* could be applied to test this assumption.

It seems that *L. braziliensis* parasites without LRV1 have been better adapted to the conditions encountered, especially in relation to the phlebotomine species, indicating that a bottleneck phenomenon occurred during the spread of *L. braziliensis*. Considering microsatellite analyses, there is one *L. (Viannia)* population in the Amazon region consisting of *L. braziliensis* strains and other *L. (Viannia)* species—*L. guyanensis* excluded. This diverse population is organized into subpopulations that match species identity [87]. In previous studies [20,43,47,64] and corroborated by the presented data, LRV1 infection was described in many *L. (Viannia)* species. It remains unsolved, however, whether LRV1 from these species is related to *L. braziliensis* LRV1. To address many of the points raised, it is important to conduct phylogenetic studies of LRV1, LRV2, LBV, and their hosts. It is noteworthy that the phylogenetic trees for LRV [43] and LBV [15] display congruence with those obtained for their hosts, suggesting coevolution and limitation of horizontal viral transmission.

7. Concluding Remarks

The hypothesis that parasites influence the population size or geographical dispersion of their host is opposed by a more acceptable hypothesis arguing that successful or well-adapted parasites evolve to be harmless to their host. Although virus-like particles and viruses were first detected in *Leishmania* parasites some decades ago, the impact of this interaction and the diversity of these endosymbionts have recently drawn considerable attention, mainly due to the virulence trade-off experimentally demonstrated in the context of *Leishmania (Viannia) guyanensis* and LRV1. However, the theory regarding the evolution of interactions among different endosymbiotic viruses and *Leishmania* spp. is still in its emerging stages. Recent studies have reported the discovery of several viruses in trypanosomatids, indicating the existence of unknown viral diversity, which needs to be further investigated and can provide important evolutionary information. At least two virus families have already been described as *Leishmania* spp. endosymbionts, but we still do not know if these viruses occur only in *Leishmania* spp. or if they can be detected elsewhere, such as in the invertebrate host of *Leishmania* spp. It is plausible to assume a dynamic symbiotic relationship in this long-term interaction between LRV or LBV and *Leishmania* spp., but the influence of either LRV or LBV on *Leishmania* biology is not yet clear. At least in some circumstances, it seems that this interaction causes a stressful condition, promoting increased tolerance of *Leishmania* spp. to some environments and augmenting its replication rate. It is a fact that both viruses influence leishmaniasis pathogenesis, but it is still unclear whether this is a consequence of the vertebrate host response to the virus living in *Leishmania* spp. cytoplasm or a biological response of *Leishmania* spp. to the endosymbiotic viruses. The impact of a "parasitized parasite" in the initial moments of a natural infection is also an aspect that deserves attention. The phenotype of higher pathogenicity linked to *Leishmania* spp. bearing viruses might be linked to an increased evolutionary fitness might be considered to signal that viral acquisition was beneficial to the parasite. However, there is also the possibility of better fitness for those organisms that are less pathogenic, which could have the chance to produce asymptomatic infections, to be maintained longer in the vertebrate host, and to be dispersed to new environments.

The screening of viruses in *Leishmania* spp. is still limited to a few studies, but so far, the evidence has indicated that LRV1 is restricted to the American continent and associated with *Leishmania* (*Viannia*) species and that LRV2 is linked to the Old World and hosted by *Leishmania* (*Leishmania*). LBV was detected only in *L. martiniquensis*, a species belonging to a subgenus not closely related to *L.* (*Viannia*) or *L.* (*Leishmania*). For both LRV1 and LRV2, there were different genotypes and correlations with the parasitized *Leishmania* species. The consequence of *Leishmania*-LRV or *Leishmania*-LBV coevolution was probably dependent on coevolutionary dynamics, involving (i) fluctuating selection affecting the frequency of some genotypes, especially those linked to resistance and infectivity [92] or fluctuations in the ranges of resistance and infectivity [93], and (ii) antagonist coevolution turning towards either increasing infectivity, resistance, or both. There is an important imbroglio of evolution and ecology linked to the relationship between *Leishmania* spp. and LRV or LBV, these interactions providing a direct impact on the evolutionary route.

Supplementary Materials: The following are available online at https://www.mdpi.com/article/10.3390/genes12050657/s1, Supplementary Data 1: LRV1 detection in *Leishmania* spp. Table S1: Strains of *L.* (*Viannia*) positive for LRV1. Supplementary Data 2: Growth curve of *Leishmania guyanensis* strains infected or not with the *Leishmania* RNA virus in single or mixed culture. Figure S1: Growth curve of *Leishmania guyanensis* strains infected or not with the *Leishmania* RNA virus in single or mixed culture.

Author Contributions: L.M.C. and E.C., conceptualization, writing and design of the figures; M.C.B. and E.C., scientific critical overview and constant writing revision; C.M.-S., K.C., G.P.d.S., and B.D.d.C., contributed with unpublished data; C.M.-S. and L.d.O.R.P., writing—review and editing. All authors contributed to the final critical review of the manuscript. All authors have read and agreed to the published version of the manuscript.

Funding: This study was supported by the Instituto Nacional de Ciência e Tecnologia, INCT-EpiAmo. Coordenação de Aperfeiçoamento de Pessoal de Nível Superior–Brasil (CAPES)–Finance Code 001; Programa PRINT FIOCRUZ-CAPES. CNPq (Researcher Fellow, 302622/2017-9). Faperj (CNE, E26-202.569/2019). IOC-Fiotec (PAEF LPL) Oswaldo Cruz Foundation.

Institutional Review Board Statement: Not applicable.

Informed Consent Statement: Not applicable.

Data Availability Statement: Not applicable.

Conflicts of Interest: The authors declare no conflict of interest.

References

1. Steverding, D. The history of leishmaniasis. *Parasites Vectors* **2017**, *10*, 1–10. [CrossRef]
2. Kostygov, A.Y.; Yurchenko, V. Revised classification of the subfamily Leishmaniinae (Trypanosomatidae). *Folia Parasitol.* **2017**, *64*, 1–5. [CrossRef] [PubMed]
3. Espinosa, O.A.; Serrano, M.G.; Camargo, E.P.; Teixeira, M.M.G.; Shaw, J.J. An appraisal of the taxonomy and nomenclature of trypanosomatids presently classified as Leishmania and Endotrypanum. *Parasitology* **2018**, *145*, 430–442. [CrossRef]
4. Serafim, T.D.; Coutinho-Abreu, I.V.; Oliveira, F.; Meneses, C.; Kamhawi, S.; Valenzuela, J.G. Sequential blood meals promote Leishmania replication and reverse metacyclogenesis augmenting vector infectivity. *Nat. Microbiol.* **2018**, *3*, 548–555. [CrossRef]
5. Cupolillo, E.; Medina-Acosta, E.; Noyes, H.; Momen, H.; Grimaldi, G. A revised classification for Leishmania and Endotrypanum. *Parasitol. Today* **2000**, *16*, 142–144. [CrossRef]
6. Jirků, M.; Yurchenko, V.Y.; Lukeš, J.; Maslov, D.A. New species of insect trypanosomatids from costa rica and the proposal for a new subfamily within the Trypanosomatidae. *J. Eukaryot. Microbiol.* **2012**, *59*, 537–547. [CrossRef] [PubMed]
7. Croft, S.L.; Molyneux, D.H. Studies on the ultrastructure, virus-like particles and infectivity of Leishmania hertigi. *Ann. Trop. Med. Parasitol.* **1979**, *73*, 213–226. [CrossRef]
8. Tarr, P.I.; Aline, R.F.; Smiley, B.L.; Scholler, J.; Keithly, J.; Stuart, K. LR1: A candidate RNA virus of Leishmania. *Proc. Natl. Acad. Sci. USA* **1988**, *85*, 9572–9575. [CrossRef]
9. Widmer, G.; Comeau, A.M.; Furlong, D.B.; Wirth, D.F.; Patterson, J.L. Characterization of a RNA virus from the parasite Leishmania. *Proc. Natl. Acad. Sci. USA* **1989**, *86*, 5979–5982. [CrossRef]
10. Weeks, R.S.; Patterson, J.L.; Stuart, K.; Widmer, G. Transcribing and replicating particles in a double-stranded RNA virus from Leishmania. *Mol. Biochem. Parasitol.* **1992**, *52*, 207–213. [CrossRef]
11. Cadd, T.L.; MacBeth, K.; Furlong, D.; Patterson, J.L. Mutational analysis of the capsid protein of Leishmania RNA virus LRV1-4. *J. Virol.* **1994**, *68*, 7738–7745. [CrossRef] [PubMed]
12. Ives, A.; Ronet, C.; Prevel, F.; Ruzzante, G.; Fuertes-Marraco, S.; Schutz, F.; Zangger, H.; Revaz-Breton, M.; Lye, L.-F.; Hickerson, S.M.; et al. Leishmania RNA virus controls the severity of mucocutaneous leishmaniasis. *Science* **2011**, *331*, 775–778. [CrossRef] [PubMed]
13. Miller, J.H.; Swartzwelder, J.C. Virus-like Particles in an Entamoeba histolytica Trophozoite. *J. Parasitol.* **1960**, *46*, 523. [CrossRef]
14. Walker, P.J.; Siddell, S.G.; Lefkowitz, E.J.; Mushegian, A.R.; Dempsey, D.M.; Dutilh, B.E.; Harrach, B.; Harrison, R.L.; Hendrickson, R.C.; Junglen, S.; et al. Changes to virus taxonomy and the International Code of Virus Classification and Nomenclature ratified by the International Committee on Taxonomy of Viruses (2019). *Arch. Virol.* **2019**, *164*, 2417–2429. [CrossRef] [PubMed]
15. Grybchuk, D.; Akopyants, N.S.; Kostygov, A.Y.; Konovalovas, A.; Lye, L.F.; Dobson, D.E.; Zangger, H.; Fasel, N.; Butenko, A.; Frolov, A.O.; et al. Viral discovery and diversity in trypanosomatid protozoa with a focus on relatives of the human parasite Leishmania. *Proc. Natl. Acad. Sci. USA* **2018**, *115*, E506–E515. [CrossRef]
16. Grybchuk, D.; MacEdo, D.H.; Kleschenko, Y.; Kraeva, N.; Lukashev, A.N.; Bates, P.A.; Kulich, P.; Leštinová, T.; Volf, P.; Kostygov, A.Y.; et al. The first Non-LRV RNA virus in leishmania. *Viruses* **2020**, *12*. [CrossRef]
17. Butenko, A.; Kostygov, A.Y.; Sádlová, J.; Kleschenko, Y.; Bečvář, T.; Podešvová, L.; MacEdo, D.H.; Žihala, D.; Lukeš, J.; Bates, P.A.; et al. Comparative genomics of Leishmania (Mundinia). *Bmc Genom.* **2019**, *20*, 1–12. [CrossRef]
18. Zangger, H.; Hailu, A.; Desponds, C.; Lye, L.-F.; Akopyants, N.S.; Dobson, D.E.; Ronet, C.; Ghalib, H.; Beverley, S.M.; Fasel, N. Leishmania aethiopica field isolates bearing an endosymbiontic dsRNA virus induce pro-inflammatory cytokine response. *PLoS Negl. Trop. Dis.* **2014**, *8*, e2836. [CrossRef]
19. Salinas, G.; Zamora, M.; Stuart, K.; Saravia, N. Leishmania RNA viruses in Leishmania of the Viannia subgenus. *Am. J. Trop. Med. Hyg.* **1996**, *54*, 425–429. [CrossRef]
20. Cantanhêde, L.M.; da Silva Júnior, C.F.; Ito, M.M.; Felipin, K.P.; Nicolete, R.; Salcedo, J.M.V.; Porrozzi, R.; Cupolillo, E.; de Godoi Mattos Ferreira, R. Further Evidence of an Association between the Presence of Leishmania RNA Virus 1 and the Mucosal Manifestations in Tegumentary Leishmaniasis Patients. *Plos Negl. Trop. Dis.* **2015**, *9*, e0004079. [CrossRef]
21. Scheffter, S.M.; Ro, Y.T.; Chung, I.K.; Patterson, J.L. The complete sequence of Leishmania RNA virus LRV2-1, a virus of an Old World parasite strain. *Virology* **1995**, *212*, 84–90. [CrossRef]
22. Hajjaran, H.; Mahdi, M.; Mohebali, M.; Samimi-Rad, K.; Ataei-Pirkooh, A.; Kazemi-Rad, E.; Naddaf, S.R.; Raoofian, R. Detection and molecular identification of leishmania RNA virus (LRV) in Iranian Leishmania species. *Arch. Virol.* **2016**, *161*, 3385–3390. [CrossRef]
23. Stuart, K.D.; Weeks, R.; Guilbride, L.; Myler, P.J. Molecular organization of Leishmania RNA virus 1. *Proc. Natl. Acad. Sci. USA* **1992**, *89*, 8596–8600. [CrossRef]
24. Widmer, G.; Patterson, J.L. Genomic structure and RNA polymerase activity in Leishmania virus. *J. Virol.* **1991**, *65*, 4211–4215. [CrossRef]
25. Ghabrial, S.A. University of K. Totivirus. In *Encyclopedia of Virology*; Elsevier Ltd.: Lexington, KY, USA, 2008; pp. 163–174.
26. Allen, A.; Islamovic, E.; Kaur, J.; Gold, S.; Shah, D.; Smith, T.J. Transgenic maize plants expressing the Totivirus antifungal protein, KP4, are highly resistant to corn smut. *Plant Biotechnol. J.* **2011**, *9*, 857–864. [CrossRef] [PubMed]
27. Provenzano, D.; Khoshnan, A.; Alderete, J.F. Involvement of dsRNA virus in the protein composition and growth kinetics of host Trichomonas vaginalis. *Arch. Virol.* **1997**, *142*, 939–952. [CrossRef] [PubMed]

28. Ro, Y.; Patterson, J.L. Identification of the Minimal Essential RNA Sequences Responsible for Site-Specific Targeting of the Leishmania RNA Virus 1-4 Capsid Endoribonuclease Identification of the Minimal Essential RNA Sequences Responsible for Site-Specific Targeting of the Leis. *J. Virol.* **2000**. [CrossRef] [PubMed]

29. MacBeth, K.J.; Ro, Y.T.; Gehrke, L.; Patterson, J.L. Cleavage site mapping and substrate-specificity of Leishmaniavirus 2-1 capsid endoribonuclease activity. *J. Biochem.* **1997**, *122*, 193–200. [CrossRef] [PubMed]

30. Saiz, M.; Ro, Y.T.; Wirth, D.F.; Patterson, J.L. Host cell proteins bind specifically to the capsid-cleaved 5′ end of Leishmaniavirus RNA. *J. Biochem.* **1999**, *126*, 538–544. [CrossRef] [PubMed]

31. Patterson, J.L. Viruses of protozoan parasites. *Exp. Parasitol.* **1990**, *70*, 111–113. [CrossRef]

32. Scheffter, S.; Widmer, G.; Patterson, J.L. Complete sequence of Leishmania RNA virus 1-4 and identification of conserved sequences. *Virology* **1994**, *199*, 479–483. [CrossRef]

33. Lefkowitz, E.J.; Dempsey, D.M.; Hendrickson, R.C.; Orton, R.J.; Siddell, S.G.; Smith, D.B. Virus taxonomy: The database of the International Committee on Taxonomy of Viruses (ICTV). *Nucleic Acids Res.* **2018**, *46*, D708–D717. [CrossRef]

34. Wichgers Schreur, P.J.; Kormelink, R.; Kortekaas, J. Genome packaging of the Bunyavirales. *Curr. Opin. Virol.* **2018**, *33*, 151–155. [CrossRef]

35. Sun, Y.; Li, J.; Gao, G.F.; Tien, P.; Liu, W. Bunyavirales ribonucleoproteins: The viral replication and transcription machinery. *Crit. Rev. Microbiol.* **2018**, *44*, 522–540. [CrossRef] [PubMed]

36. Elliott, R.M. Molecular biology of the Bunyaviridae. *J. Gen. Virol.* **1990**, *71*, 501–522. [CrossRef] [PubMed]

37. Akopyants, N.S.; Lye, L.F.; Dobson, D.E.; Lukeš, J.; Beverley, S.M. A novel bunyavirus-like virus of trypanosomatid protist parasites. *Genome Announc.* **2016**, *4*, 4–5. [CrossRef] [PubMed]

38. Bartel, D.P. MicroRNAs: Genomics, biogenesis, mechanism, and function. *Cell* **2004**, *116*, 281–297. [CrossRef]

39. Carthew, R.W.; Sontheimer, E.J. Origins and Mechanisms of miRNAs and siRNAs. *Cell* **2009**, *136*, 642–655. [CrossRef] [PubMed]

40. Gitlin, L.; Andino, R. Nucleic Acid-Based Immune System: The Antiviral Potential of Mammalian RNA Silencing. *J. Virol.* **2003**, *77*, 7159–7165. [CrossRef]

41. Maga, J.A.; Widmer, G.; LeBowitz, J.H. Leishmania RNA virus 1-mediated cap-independent translation. *Mol. Cell. Biol.* **1995**, *15*, 4884–4889. [CrossRef]

42. Guilbride, L.; Myler, P.J.; Stuart, K. Distribution and sequence divergence of LRV1 viruses among different Leishmania species. *Mol. Biochem. Parasitol.* **1992**, *54*, 101–104. [CrossRef]

43. Widmer, G.; Dooley, S. Phylogenetic analysis of Leishmania RNA virus and leishmania suggests ancient virus-parasite association. *Nucleic Acids Res.* **1995**, *23*, 2300–2304. [CrossRef] [PubMed]

44. Pereira, L.D.O.R.; Maretti-Mira, A.C.; Rodrigues, K.M.; Lima, R.B.; de Oliveira-Neto, M.P.; Cupolillo, E.; Pirmez, C.; de Oliveira, M.P. Severity of tegumentary leishmaniasis is not exclusively associated with Leishmania RNA virus 1 infection in Brazil. *Memórias Do Inst. Oswaldo Cruz* **2013**, *108*, 665–667. [CrossRef]

45. Macedo, D.H.; Menezes-Neto, A.; Rugani, J.M.; Rocha, A.C.; Silva, S.O.; Melo, M.N.; Lye, L.F.; Beverley, S.M.; Gontijo, C.M.; Soares, R.P. Low frequency of LRV1 in Leishmania braziliensis strains isolated from typical and atypical lesions in the State of Minas Gerais, Brazil. *Mol. Biochem. Parasitol.* **2016**, *210*, 50–54. [CrossRef] [PubMed]

46. Cantanhêde, L.M.; Fernandes, F.G.; Eduardo Melim Ferreira, G.; Porrozzi, R.; De Godoi Mattos Ferreira, R.; Cupolillo, E. New insights into the genetic diversity of Leishmania RNA Virus 1 and its species-specific relationship with Leishmania parasites. *PLoS ONE* **2018**, *13*, 1–16. [CrossRef]

47. Vieira-Gonçalves, R.; Fagundes-Silva, G.A.; Heringer, J.F.; Fantinatti, M.; Da-Cruz, A.M.; Oliveira-Neto, M.P.; Guerra, J.A.O.; Gomes-Silva, A. First report of treatment failure in a patient with cutaneous leishmaniasis infected by Leishmania (Viannia) naiffi carrying Leishmania RNA virus: A fortuitous combination? *Rev. Da Soc. Bras. De Med. Trop.* **2019**, *52*, 10–12. [CrossRef]

48. Kariyawasam, R.; Mukkala, A.N.; Lau, R.; Valencia, B.M.; Llanos-Cuentas, A.; Boggild, A.K. Virulence factor RNA transcript expression in the Leishmania Viannia subgenus: Influence of species, isolate source, and Leishmania RNA virus-1. *Trop. Med. Health* **2019**, *47*, 1–9. [CrossRef] [PubMed]

49. Kleschenko, Y.; Grybchuk, D.; Matveeva, N.S.; Macedo, D.H.; Ponirovsky, E.N.; Lukashev, A.N.; Yurchenko, V. Molecular Characterization of Leishmania RNA virus 2 in Leishmaniamajor from Uzbekistan. *Genes* **2019**, *10*, 830. [CrossRef] [PubMed]

50. Saberi, R.; Fakhar, M.; Hajjaran, H.; Ataei-Pirkooh, A.; Mohebali, M.; Taghipour, N.; Ziaei Hezarjaribi, H.; Moghadam, Y.; Bagheri, A. Presence and diversity of Leishmania RNA virus in an old zoonotic cutaneous leishmaniasis focus, northeastern Iran: Haplotype and phylogenetic based approach. *Int. J. Infect. Dis.* **2020**, *101*, 6–13. [CrossRef]

51. Nalçacı, M.; Karakuş, M.; Yılmaz, B.; Demir, S.; Özbilgin, A.; Özbel, Y.; Töz, S. Detection of Leishmania RNA virus 2 in Leishmania species from Turkey. *Trans. R. Soc. Trop. Med. Hyg.* **2019**, *113*, 410–417. [CrossRef]

52. Lukeš, J.; Butenko, A.; Hashimi, H.; Maslov, D.A.; Votýpka, J.; Yurchenko, V. Trypanosomatids Are Much More than Just Trypanosomes: Clues from the Expanded Family Tree. *Trends Parasitol.* **2018**, *34*, 466–480. [CrossRef]

53. Harkins, K.M.; Schwartz, R.S.; Cartwright, R.A.; Stone, A.C. Phylogenomic reconstruction supports supercontinent origins for Leishmania. *Infect. Genet. Evol.* **2016**, *38*, 101–109. [CrossRef]

54. Ogg, M.M.; Carrion, R.; Botelho, A.C.D.C.; Mayrink, W.; Correa-Oliveira, R.; Patterson, J.L. Short report: Quantification of leishmaniavirus RNA in clinical samples and its possible role in pathogenesis. *Am. J. Trop. Med. Hyg.* **2003**, *69*, 309–313. [CrossRef] [PubMed]

55. Hartley, M.-A.; Ronet, C.; Zangger, H.; Beverley, S.M.; Fasel, N. Leishmania RNA virus: When the host pays the toll. *Front. Cell. Infect. Microbiol.* **2012**, *2*, 99. [CrossRef] [PubMed]

56. Bourreau, E.; Ginouves, M.; Prévot, G.; Hartley, M.-A.; Gangneux, J.-P.; Robert-Gangneux, F.; Dufour, J.; Sainte-Marie, D.; Bertolotti, A.; Pratlong, F.; et al. Leishmania-RNA virus presence in L. guyanensis parasites increases the risk of first-line treatment failure and symptomatic relapse. *J. Infect. Dis.* 2015; 1–28. [CrossRef]

57. Adaui, V.; Lye, L.; Akopyants, N.S.; Zimic, M.; Llanos-cuentas, A.; Garcia, L.; Maes, I.; Doncker, S. De Association of the Endobiont Double-Stranded RNA Virus LRV1 With Treatment Failure for Human Leishmaniasis Caused by Leishmania braziliensis in Peru and Bolivia. *J. Infect. Dis.* **2016**, *213*. [CrossRef]

58. Abtahi, M.; Eslami, G.; Cavallero, S.; Vakili, M.; Hosseini, S.S.; Ahmadian, S.; Boozhmehrani, M.J.; Khamesipour, A. Relationship of Leishmania RNA Virus (LRV) and treatment failure in clinical isolates of Leishmania major. *BMC Res. Notes* **2020**, *13*, 1–6. [CrossRef]

59. Ginouvès, M.; Couppié, P.; Simon, S.; Bourreau, E.; Rogier, S.; Brousse, P.; Travers, P.; Pommier de Santi, V.; Demar, M.; Briolant, S.; et al. Leishmania virus genetic diversity is not related to leishmaniasis treatment failure. *Clin. Microbiol. Infect.* **2020**. [CrossRef]

60. Choisy, M.; Hide, M.; Bañuls, A.-L.; Guégan, J.-F. Rocking the curve. *Trends Microbiol.* **2004**, *12*, 534–536. [CrossRef] [PubMed]

61. Mendes, B.P.; Da Silva, I.A.; Damata, J.P.; Castro-Gomes, T.; Vieira, L.Q.; Ribeiro-Dias, F.; Horta, M.F. Metacyclogenesis of Leishmania (Viannia) guyanensis: A comprehensive study of the main transformation features in axenic culture and purification of metacyclic promastigotes by negative selection with Bauhinia purpurea lectin. *Parasitology* **2019**, 716–727. [CrossRef] [PubMed]

62. Zangger, H.; Ronet, C.; Desponds, C.; Kuhlmann, F.M.; Robinson, J.; Hartley, M.-A.; Prevel, F.; Castiglioni, P.; Pratlong, F.; Bastien, P.; et al. Detection of Leishmania RNA virus in Leishmania parasites. *PLoS Negl. Trop. Dis.* **2013**, *7*, e2006. [CrossRef]

63. Tirera, S.; Ginouves, M.; Donato, D.; Caballero, I.S.; Bouchier, C.; Lavergne, A.; Bourreau, E.; Mosnier, E.; Vantilcke, V.; Couppié, P.; et al. Unraveling the genetic diversity and phylogeny of Leishmania RNA virus 1 strains of infected Leishmania isolates circulating in French Guiana. *PLoS Negl. Trop. Dis.* **2017**, *11*, 1–20. [CrossRef]

64. Silvester, E.; Young, J.; Ivens, A.; Matthews, K.R. Europe PMC Funders Group Interspecies quorum-sensing in co-infections can manipulate trypanosome transmission potential. *Interspecies* **2018**, *2*, 1471–1479. [CrossRef]

65. Schuh, C.M.A.P.; Aguayo, S.; Zavala, G.; Khoury, M. Exosome-like vesicles in Apis mellifera bee pollen, honey and royal jelly contribute to their antibacterial and pro-regenerative activity. *J. Exp. Biol.* **2019**, *222*. [CrossRef] [PubMed]

66. Atayde, V.D.; da Silva Lira Filho, A.; Chaparro, V.; Zimmermann, A.; Martel, C.; Jaramillo, M.; Olivier, M. Exploitation of the Leishmania exosomal pathway by Leishmania RNA virus 1. *Nat. Microbiol.* **2019**, *4*, 714–723. [CrossRef] [PubMed]

67. Atayde, V.D.; Aslan, H.; Townsend, S.; Hassani, K.; Kamhawi, S.; Olivier, M. Exosome Secretion by the Parasitic Protozoan Leishmania within the Sand Fly Midgut. *Cell Rep.* **2015**, *13*, 957–967. [CrossRef] [PubMed]

68. Castiglioni, P.; Hartley, M.-A.; Rossi, M.; Prevel, F.; Desponds, C.; Utzschneider, D.T.; Eren, R.-O.; Zangger, H.; Brunner, L.; Collin, N.; et al. Exacerbated Leishmaniasis Caused by a Viral Endosymbiont can be Prevented by Immunization with Its Viral Capsid. *PLoS Negl. Trop. Dis.* **2017**, *11*, e0005240. [CrossRef] [PubMed]

69. Ronet, C.; Beverley, S.M.; Fasel, N. Muco-cutaneous leishmaniasis in the New World: The ultimate subversion. *Virulence* **2011**, *2*, 547–552. [CrossRef]

70. de Carvalho, R.V.H.; Lima-Junior, D.S.; da Silva, M.V.G.; Dilucca, M.; Rodrigues, T.S.; Horta, C.V.; Silva, A.L.N.; da Silva, P.F.; Frantz, F.G.; Lorenzon, L.B.; et al. Leishmania RNA virus exacerbates Leishmaniasis by subverting innate immunity via TLR3-mediated NLRP3 inflammasome inhibition. *Nat. Commun.* **2019**, *10*. [CrossRef]

71. Baneth, G.; Zivotofsky, D.; Nachum-Biala, Y.; Yasur-Landau, D.; Botero, A.M. Mucocutaneous Leishmania tropica infection in a dog from a human cutaneous leishmaniasis focus. *Parasites Vectors* **2014**, *7*, 1–5. [CrossRef]

72. Shirian, S.; Oryan, A.; Hatam, G.R.; Daneshbod, Y. Mixed mucosal leishmaniasis infection caused by Leishmania tropica and leishmania major. *J. Clin. Microbiol.* **2012**, *50*, 3805–3808. [CrossRef]

73. Morsy, T.A.; Khalil, N.M.; Salama, M.M.; Hamdi, K.N.; al Shamrany, Y.A.; Abdalla, K.F. Mucosal leishmaniasis caused by Leishmania tropica in Saudi Arabia. *J. Egypt. Soc. Parasitol.* **1995**, *25*, 73–79. [PubMed]

74. Strazzulla, A.; Cocuzza, S.; Pinzone, M.R.; Postorino, M.C.; Cosentino, S.; Serra, A.; Cacopardo, B.; Nunnari, G. Mucosal leishmaniasis: An underestimated presentation of a neglected disease. *Biomed Res. Int.* **2013**, *2013*. [CrossRef] [PubMed]

75. Schönian, G.; Akuffo, H.; Lewin, S.; Maasho, K.; Nylén, S.; Pratlong, F.; Eisenberger, C.L.; Schnur, L.F.; Presber, W. Genetic variability within the species Leishmania aethiopica does not correlate with clinical variations of cutaneous leishmaniasis. *Mol. Biochem. Parasitol.* **2000**, *106*, 239–248. [CrossRef]

76. Dalzoto, P.R.; Glienke-Blanco, C.; Kava-Cordeiro, V.; Ribeiro, J.Z.; Kitajima, E.W.; Azevedo, J.L. Horizontal transfer and hypovirulence associated with double-stranded RNA in Beauveria bassiana. *Mycol. Res.* **2006**, *110*, 1475–1481. [CrossRef]

77. Armstrong, T.C.; Keenan, M.C.; Widmer, G.; Patterson, J.L. Successful transient introduction of Leishmania RNA virus into a virally infected and an uninfected strain of Leishmania. *Proc. Natl. Acad. Sci. USA* **1993**, *90*, 1736–1740. [CrossRef]

78. Ro, Y.; Scheffter, S.M.; Patterson, J.L. Specific in vitro cleavage of a Leishmania virus capsid-RNA-dependent RNA polymerase polyprotein by a host cysteine-like protease. *J. Virol.* **1997**, *71*, 8983–8990. [CrossRef]

79. Maslov, D.A.; Votýpka, J.; Yurchenko, V.; Lukeš, J. Diversity and phylogeny of insect trypanosomatids: All that is hidden shall be revealed. *Trends Parasitol.* **2013**, *29*, 43–52. [CrossRef]

80. Moriconi, M.; Rugna, G.; Calzolari, M.; Bellini, R.; Albieri, A.; Angelini, P.; Cagarelli, R.; Landini, M.P.; Charrel, R.N.; Varani, S. Phlebotomine sand fly–borne pathogens in the Mediterranean Basin: Human leishmaniasis and phlebovirus infections. *Plos Negl. Trop. Dis.* **2017**, *11*, 1–19. [CrossRef]

81. da Silva, L.A.; de Sousa, C.d.S.; da Graça, G.C.; Porrozzi, R.; Cupolillo, E. Sequence analysis and PCR-RFLP profiling of the hsp70 gene as a valuable tool for identifying Leishmania species associated with human leishmaniasis in Brazil. *Infect. Genet. Evol.* **2010**, *10*, 77–83. [CrossRef]

82. Boité, M.C.; Mauricio, I.L.; Miles, M.A.; Cupolillo, E. New Insights on Taxonomy, Phylogeny and Population Genetics of Leishmania (Viannia) Parasites Based on Multilocus Sequence Analysis. *PLoS Negl. Trop. Dis.* **2012**, *6*. [CrossRef]

83. Cupolillo, E.; Grimaldi, G.; Momen, H. A General Classification of New World Leishmania Using Numerical Zymotaxonomy. *Am. J. Trop. Med. Hyg.* **1994**, *50*, 296–311. [CrossRef] [PubMed]

84. Romero, G.A.S.; Ishikawa, E.; Cupolillo, E.; Toaldo, C.B.; Guerra, M.V.; Vinitius de Farias Guerra, M.; Gomes Paes, M.; de Oliveira Macêdo, M.V.; Shaw, J.J. Identification of antigenically distinct populations of Leishmania (Viannia) guyanensis from Manaus, Brazil, using monoclonal antibodies. *Acta Trop.* **2002**, *82*, 25–29. [CrossRef]

85. Grimaldi, G.; Momen, H.; Naiff, R.D.; McMahon-Pratt, D.; Barrett, T.V. Characterization and classification of leishmanial parasites from humans, wild mammals, and sand flies in the Amazon region of Brazil. *Am. J. Trop. Med. Hyg.* **1991**, *44*, 645–661. [CrossRef] [PubMed]

86. Oddone, R.; Schweynoch, C.; Schönian, G.; De Sousa, C.D.S.; Cupolillo, E.; Espinosa, D.; Arevalo, J.; Noyes, H.; Mauricio, I.; Kuhls, K. Development of a multilocus microsatellite typing approach for discriminating strains of Leishmania (Viannia) species. *J. Clin. Microbiol.* **2009**, *47*, 2818–2825. [CrossRef]

87. Kuhls, K.; Cupolillo, E.; Silva, S.O.; Schweynoch, C.; Boité, M.C.; Mello, M.N.; Mauricio, I.; Miles, M.; Wirth, T.; Schönian, G. Population Structure and Evidence for Both Clonality and Recombination among Brazilian Strains of the Subgenus Leishmania (Viannia). *PLoS Negl. Trop. Dis.* **2013**, *7*. [CrossRef] [PubMed]

88. Gomes, R.F.; Macedo, A.M.; Pena, S.D.J.; Melo, M.N. Leishmania (Viannia) braziliensis: Genetic Relationships between Strains Isolated from Different Areas of Brazil as Revealed by DNA Fingerprinting and RAPD. *Exp. Parasitol.* **1995**, *80*, 681–687. [CrossRef]

89. Rougeron, V.; De Meeûs, T.; Hide, M.; Waleckx, E.; Bermudez, H.; Arevalo, J.; Llanos-Cuentas, A.; Dujardin, J.C.; De Doncker, S.; Le Ray, D.; et al. Extreme inbreeding in Leishmania braziliensis. *Proc. Natl. Acad. Sci. USA* **2009**, *106*, 10224–10229. [CrossRef]

90. Odiwuor, S.; Veland, N.; Maes, I.; Arévalo, J.; Dujardin, J.C.; Van der Auwera, G. Evolution of the Leishmania braziliensis species complex from amplified fragment length polymorphisms, and clinical implications. *Infect. Genet. Evol.* **2012**, *12*, 1994–2002. [CrossRef]

91. Gomez, P.; Ashby, B.; Buckling, A. Population mixing promotes arms race host-parasite coevolution. *Proc. R. Soc. B Biol. Sci.* **2014**, *282*, 20142297. [CrossRef]

92. Gandon, S.; Buckling, A.; Decaestecker, E.; Day, T. Host-parasite coevolution and patterns of adaptation across time and space. *J. Evol. Biol.* **2008**, *21*, 1861–1866. [CrossRef]

93. Sasaki, A. Host-parasite coevolution in a multilocus gene-for-gene system. *Proc. R. Soc. B Biol. Sci.* **2000**, *267*, 2183–2188. [CrossRef] [PubMed]

genes

Article

De Novo Transcriptome Meta-Assembly of the Mixotrophic Freshwater Microalga *Euglena gracilis*

Javier Cordoba [1], Emilie Perez [1,2], Mick Van Vlierberghe [2], Amandine R. Bertrand [2], Valérian Lupo [2], Pierre Cardol [1] and Denis Baurain [2,*]

[1] InBioS—PhytoSYSTEMS, Laboratoire de Génétique et Physiologie des Microalgues, ULiège, B-4000 Liège, Belgium; j.cordoba@outlook.es (J.C.); emilie.perez@alumni.uliege.be (E.P.); pierre.cardol@uliege.be (P.C.)

[2] InBioS—PhytoSYSTEMS, Unit of Eukaryotic Phylogenomics, ULiège, B-4000 Liège, Belgium; mvanvlierberghe@doct.uliege.be (M.V.V.); amandine.bertrand@doct.uliege.be (A.R.B.); valerian.lupo@doct.uliege.be (V.L.)

* Correspondence: denis.baurain@uliege.be; Tel.: +32-4-366-3864

Abstract: *Euglena gracilis* is a well-known photosynthetic microeukaryote considered as the product of a secondary endosymbiosis between a green alga and a phagotrophic unicellular belonging to the same eukaryotic phylum as the parasitic trypanosomatids. As its nuclear genome has proven difficult to sequence, reliable transcriptomes are important for functional studies. In this work, we assembled a new consensus transcriptome by combining sequencing reads from five independent studies. Based on a detailed comparison with two previously released transcriptomes, our consensus transcriptome appears to be the most complete so far. Remapping the reads on it allowed us to compare the expression of the transcripts across multiple culture conditions at once and to infer a functionally annotated network of co-expressed genes. Although the emergence of meaningful gene clusters indicates that some biological signal lies in gene expression levels, our analyses confirm that gene regulation in euglenozoans is not primarily controlled at the transcriptional level. Regarding the origin of *E. gracilis*, we observe a heavily mixed gene ancestry, as previously reported, and rule out sequence contamination as a possible explanation for these observations. Instead, they indicate that this complex alga has evolved through a convoluted process involving much more than two partners.

Keywords: transcriptome assembly; gene expression; transcriptional regulation; ontology network; co-expression network; taxonomic analysis; database contamination; kleptoplastidy

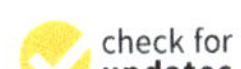
check for **updates**

Citation: Cordoba, J.; Perez, E.; Van Vlierberghe, M.; Bertrand, A.R.; Lupo, V.; Cardol, P.; Baurain, D. *De Novo* Transcriptome Meta-Assembly of the Mixotrophic Freshwater Microalga *Euglena gracilis*. *Genes* **2021**, *12*, 842. https://doi.org/10.3390/genes12060842

Academic Editor: Jose M. Requena

Received: 8 April 2021
Accepted: 27 May 2021
Published: 29 May 2021

Publisher's Note: MDPI stays neutral with regard to jurisdictional claims in published maps and institutional affiliations.

1. Introduction

Euglena gracilis is a secondary green alga that can grow in a wide variety of environments. *E. gracilis* belongs to the euglenids, a monophyletic group of free-living, single-celled flagellates that inhabit aquatic ecosystems. Euglenids are distinguished mainly by their unique type of cell covering, the pellicle. The latter is a complex structure composed of proteinaceous strips covered by a cell membrane and underlain by the microtubule system and the cisternae of the endoplasmic reticulum [1]. Together, euglenids, symbiontids (free-living flagellates living in low-oxygen marine sediments), diplonemids (free-living marine flagellates) and kinetoplastids (free-living and parasitic flagellates, e.g., *Trypanosoma*) form the monophyletic group of Euglenozoa [2–5]. Euglenids are early diverged members of the *Euglenozoa* and distant relatives to the kinetoplastids [6]. Thus, analysing *E. gracilis* genomic information is a way to approach the evolution of parasitism, due to their common ancestry with kinetoplastids [7,8]. For example, it has been shown that many additional subunits of the mitochondrial respiratory chain previously considered exclusive to kinetoplastids are shared with *E. gracilis*, and therefore cannot be associated with the parasitic lifestyle [9]. Yet, it is worth mentioning that free-living bodonids (e.g., *Bodo saltans*) are better comparators for parasitism [10,11]. The relationship between euglenids and kinetoplastids has been first proposed by T. Cavalier-Smith based on ultrastructural similarities (e.g., "mitochondrial cristae shaped like a flattened disc with a narrow neck") [12], then supported by other lines of evidence, such as alignments of nuclear rRNA [13], the addition of a leader sequence to nuclear pre-mRNAs [14] and the presence of trypanothione reductase in *E. gracilis*, previously found only in kinetoplastids [15].

E. gracilis bears a complex plastid [16], derived from a green alga belonging to *Pyramimonadales*, and acquired by a free living phagotrophic eukaryovorous euglenid ancestor [17–19]. As the result of a so-called "secondary" endosymbiosis, this chloroplast is bound by three membranes, whereas primary plastids only have two membranes [20,21]. Whatever the specific event, endosymbiosis is accompanied by massive gene loss and gene transfer from the genome of the symbiont to the nuclear genome of the host (Endosymbiotic Gene Transfer or EGT) [22]. Moreover, there can be gene transfers from sources other than the symbiont giving rise to the observed plastid [Horizontal (or Lateral) Gene Transfer or HGT/LGT], for example, over (more or less cryptic) transient endosymbioses (e.g., "shopping bag" [23–25] and "red carpet" [26] hypotheses). Alternatively, HGT can occur in a, possibly ulterior, "non-endosymbiotic context" [27,28] (e.g., "limited transfer window" hypothesis" [29]), because it may be easier to duplicate or recruit a foreign gene for servicing the nascent plastid than to get it from the symbiont itself [30]. In any case, both EGT and HGT have shaped the nuclear genome of photosynthetic euglenids, leading to heavy genetic mosaicism (e.g., [7,31,32]).

Due to its great metabolic flexibility, a large number of culture media and growing conditions have been used to study *E. gracilis* over the past 60 years [33–37]. Commonly, the mineral composition remains similar from one medium to another, but three parameters vary greatly: the pH (which can be acidic or neutral), the source of organic carbon (e.g., acetate, ethanol, and succinate) and the concentration of the carbon source (from 10 mM to more than 150 mM). *E. gracilis* can therefore exploit a variety of organic carbon sources, as well in the dark (heterotrophic conditions) as in the light (mixotrophic conditions), where a high concentration of organic carbon leads to a decrease in photosynthesis by repressing chlorophyll biosynthesis, reflecting the fact that this organism switches between nutritional modes and combines them readily [38–40]. *E. gracilis* is also known for its atypical metabolic pathways, some of them producing compounds of commercial interest. In photosynthetic euglenoids, carbon reserves are stored in the cytoplasm in the form of paramylon (β-1,3-glucan), in place of the starch (α-1,4 and α-1,6-glucan) typical of the green line [41,42]. Paramylon can be used to produce bioplastics [43] and, similarly to other β-glucans, has been reported to display some anti-tumoural activity [44]. In anoxic (fermentative) conditions, *E. gracilis* has the unique ability among microalgae to convert paramylon into wax ester compounds suitable for drop-in jet biofuels conversion because of their low freezing point [45–47]. *E. gracilis* is also used as a source of dietary supplements (e.g., the most bioactive form of vitamin E, α-tocopherol, is present in *E. gracilis* biomass in a relatively high amount) [48].

Due to its evolutionary and biotechnological interests, *E. gracilis* is the best studied member of the euglenids. Its chloroplast genome (143 kb) was among the first plastid genomes ever sequenced [49], while its tiny mitochondrial genome has been recently resolved [50,51]. To date, few studies have used high throughput sequencing technologies to publish Omics information on *E. gracilis* [7,52,53]. In this respect, attempts to sequence its nuclear genome are also very recent (initially estimated between 1 Gb to 9 Gb; see [54] for a review). These efforts have culminated with the release of a very large (500 Mb) and highly fragmented draft genome, as authors recalled, due to gapped contigs or unknown base representation in half of the genome [7].

In this work, we have assembled a consensus transcriptome taking advantage of the raw read data publicly available, including newly generated transcriptomic libraries, for a total of five different data sources. Our assembly protocol was very thorough, with a special emphasis on potential contaminant sequences, resulting in the most complete transcriptome released to date for *E. gracilis*, according to a systematic comparison with the two other public transcriptomes [7,53]. After functional and taxonomic annotation of the predicted coding sequences, we performed a comparative study of their expression level across a range of culture conditions and studies, which allowed us to build an information-rich network of co-expressed genes. However, these results confirm that transcriptional control is not the primary level of genetic regulation in euglenozoans, while our taxonomic analyses point to highly mixed gene ancestry, compatible with a kleptoplastidic phase of plastid acquisition.

2. Materials and Methods

2.1. Data Collection

2.1.1. Public Repositories

Searching for public RNA-Seq data for *E. gracilis* in the International Nucleotide Sequence Database Collaboration (INSDC) returned eight studies. We further recovered an additional dataset, produced and submitted to the European Nucleotide Archive (ENA) repositories by ourselves (see Section 2.1.2 for details). Of these nine studies, only five short read datasets (5 experiments/23 samples) that used Illumina technology to analyse whole transcriptomes were exploitable. Among the discarded experiments, PRJEB4713 contained 454 GS FLX Titanium long reads, a size that is difficult to handle by the chosen assembler, while PRJEB21674 only included a single euglenid sample (among 1179), yet labelled as "Euglena sp.", PRJNA294935 primarily contained mitochondrial sequences, and PRJNA12797 (built out of ESTs) was not accessible from public repositories. At last, PRJDB4781 was not included because our meta-assemblies had been completed by the date of its release (October 2019). The data files from the five retained experiments were downloaded using fastq-dump utility from the SRA Toolkit with -I and –split-file arguments to divide files into forward and reverse paired reads. We also collected the two transcriptome assemblies hitherto available, GEFR01 and GDJR01. The former was encoded under study accession PRJNA298469, which corresponds to experiments B and C, and the latter, which corresponds to experiment D, was encoded as study PRJNA289402. For further details on experimental design or/and samples, see Table 1.

2.1.2. In-House Experiments, Cell Culture and Sequencing

The strain of *E. gracilis* (1224-5/25) was obtained from SAG (Sammlung von Algenkulturen Göttingen, Germany). Cells were cultured in liquid mineral medium tris-minimum-phosphate (TMP) at pH 7.0 and 25 °C, supplemented with a mixture of vitamins (vitamin B1 2·10-2 mM, vitamin B8 10-4 mM and vitamin B12 10-4 mM). In three samples, acetate (60 mM) was added as a carbon source, under different photosynthetic photon flux densities (PPFD, T8 fluorescent neon tubes) (in the dark, at low PPFD ($50 \ \mu E \ m^{-2} \ s^{-1}$) or at medium PPFD ($200 \ \mu E \ m^{-2} \ s^{-1}$), while in a fourth sample, acetate was not supplied and light was set to low PPFD ($50 \ \mu E \ m^{-2} \ s^{-1}$). For each sample, the cells in the exponential phase ($1–2 \times 10^{-6}$ cells/mL) were recovered by centrifugation, 10 min at 500 g. Total RNA was extracted with the protocol outlined in [55], then fragmented and retro-transcribed before standardization using the Duplex-Specific Nuclease kit (Evrogen, Russia). Each library was prepared using the Illumina total mRNA kit (Illumina, San Diego, CA, USA) and quantified by qPCR using the KAPA Library Quantification Kit (Roche, Switzerland). Subsequently, samples were sequenced in both reading directions (paired-end 2×100 nt) on four separate tracks of a high-speed sequencer Illumina HiSeq 2000, yielding on average ca. 235 million reads per sample. Library preparation, DSN normalization and high-throughput sequencing by Illumina technology were carried out by the GIGA genomics platform (https://www.gigagenomics.uliege.be (accessed on 23 July 2014)). Raw reads have been deposited at the ENA database under the study accession number PRJEB38787 (Table 1).

2.2. Data Assembly

A schematic representation of the de novo transcriptome reconstruction and analysis pipeline is given in Figure 1. All computations were performed on a grid computer.

Table 1. Representation of the collected data and overview of the experimental design. Exp. Code: letter assigned to each experiment (one letter per study). Study Acc.: public accession number of the BioProject. Sample Code: first letter corresponds to the experiment, first digit to experimental conditions of the samples, and second digit (if any) to the replicates. Run Acc.: public accession number of read FASTQ files. Temp.: estimated Celsius degrees of cell culture temperature. Medium: type of cell culture medium, rich (R) or mineral (M) plus carbon source (+C). Light: estimated light experimental conditions, darkness (D), low-light (LL) and high-light (HL). Shaking: rpm of shaker incubator. Cult. Cond.: trophic regime, fermentative (F), heterotrophic (H), phototrophic (P) or mixotrophic (M). Harvest Phase: development stage of the culture when collected, exponential phase (Exp) or stationary phase (Stat).

Exp. Code	Study Acc.	Sample Code	Run Acc.	Temp.	Medium	Light	Shaking	Cult Cond.	Harvest Phase	Reference
A	PRJNA310762	A.1.1	SRR3159774	25	R + C	D	0	H	Exp	[7]
		A.1.2	SRR3159775	25	R + C	D	0	H	Exp	
		A.1.3	SRR3159776	25	R + C	D	0	H	Exp	
		A.2.1	SRR3159777	25	R + C	LL	0	M	Exp	
		A.2.2	SRR3159778	25	R + C	LL	0	M	Exp	
		A.2.3	SRR3159779	25	R + C	LL	0	M	Exp	
B	PRJEB10085	B.1	ERR974915	21	M + C	LL	0	M	Stat	[52]
		B.2	ERR974916	30	R+C	D	200	H	Stat	
C	PRJNA298469	C.0	SRR2628535	25	M	LL	0	M	Stat	[7]
D	PRJNA289402	D.0	SRR3195326	26	R+C	HL	120	M	Stat	[53]
		D.1.1	SRR3195327	26	R+C	HL	120	M	Stat	
		D.1.2	SRR3195329	26	R+C	HL	120	M	Stat	
		D.1.3	SRR3195331	26	R+C	HL	120	M	Stat	
		D.2.1	SRR3195332	26	R+C	HL	120	F	Stat	
		D.2.2	SRR3195334	26	R+C	HL	120	F	Stat	
		D.2.3	SRR3195335	26	R+C	HL	120	F	Stat	
		D.3.1	SRR3195338	26	R+C	HL	120	F	Stat	
		D.3.2	SRR3195339	26	R+C	HL	120	F	Stat	
		D.3.3	SRR3195340	26	R+C	HL	120	F	Stat	
E	PRJEB38787	E.1	ERR4227585	25	M	LL	100	P	Exp	This study
		E.2	ERR4227586	25	M+C	D	100	H	Exp	
		E.3	ERR4227587	25	M+C	LL	100	M	Exp	
		E.4	ERR4227588	25	M+C	HL	100	M	Exp	

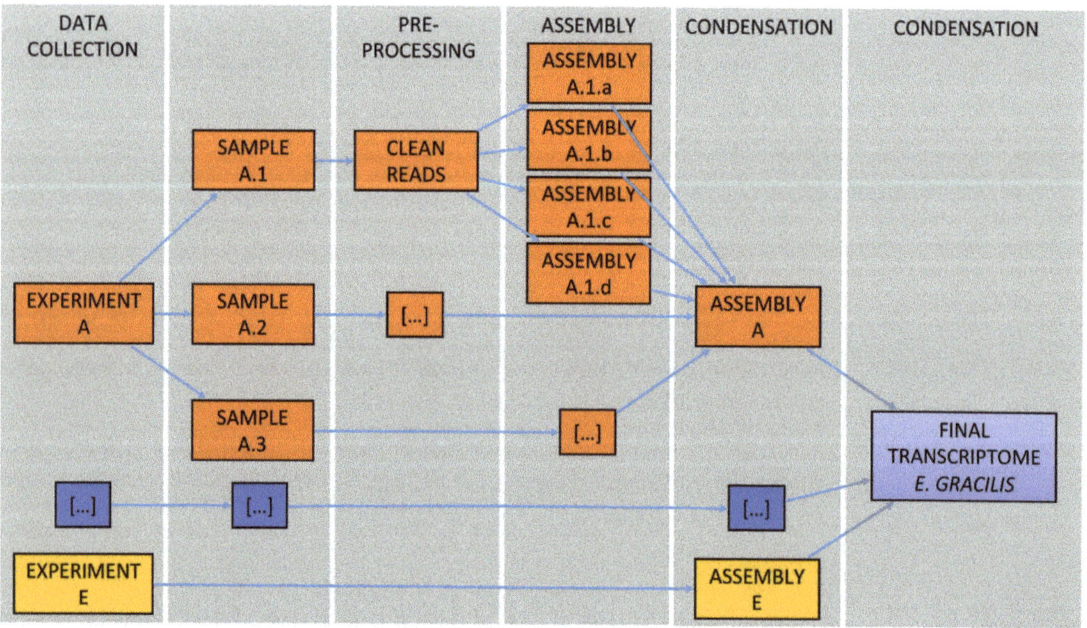

Figure 1. Schematic representation of our de novo transcriptome meta-assembly pipeline.

2.2.1. Data Pre-Processing

Every raw read file (run accessions SRR/ERR) was treated as one sample, even if two or more files were replicates of the same experimental condition. Once collected and transformed into fastq files, all samples were treated separately. Raw reads were analysed with FastQC v0.11.6 to assess the quality of the data [56]. PRINSEQ-lite.pl v0.20.4 was used to remove reads that contained more than one ambiguous nucleotide [57]. Then, Trimmomatic v0.32 was used with the following parameters (ILLUMINACLIP: TruSeq3-PE.fa:2:30:10 SLIDINGWINDOW: 4:25 LEADING: 3 TRAILING: 3 MINLEN: 25) to truncate the low quality regions of certain sequences and cut adapters and other Illumina-specific sequences from the reads [58]. Output data was sorted into three different batches as paired, unpaired and singleton reads. Finally, read quality was re-assessed using FastQC, and the resulting plots visually compared to those obtained in the beginning to check the effect of the filtering procedure.

2.2.2. Transcriptome Assembly

Pre-processed reads (paired, unpaired and singleton reads) were assembled per experiment in two steps to yield five transcriptomes, one per experiment. We used Trinity v2.4.0 software [59] for de novo transcriptome assembly. During the first step, samples of each experiment were assembled four times, combining values (one/two) of minimum count for k-mers to be assembled (–min_kmer_cov) with normalization turned off (–no_normalize_reads) or on (default) to provide maximal sensitivity for reconstructing lowly expressed transcripts. In all cases, we used the default parameters with a minimum contig length (–min_contig_length) of 100 nt. Second, to reconstruct one single transcriptome per experiment, the four assembled transcriptome replicates were pooled together with the tr2aacds.pl script (using default parameters) from the EvidentialGene v2016.07.11 software package [60,61].

2.2.3. Transcriptome Decontamination

To ensure the purity of the five transcriptomes, we determined the guanine-cytosine (GC) content distribution across reconstructed transcripts. Furthermore, we explored the potential contamination of the five transcriptomes individually by comparing their transcripts against the NCBI nucleotide database (nt) using BLASTN v2.2.28 [62,63]. We used a conservative approach with an E-value threshold of 1×10^{-50} and an identity threshold of 90% to maximize the identification of true matches. The best hit for each query was selected, and the organism name (sscinames) of these top matches were collected, tabulated and quantified. Abundant organisms other than *Euglena* were flagged as putative contaminants. To obtain uncontaminated transcriptomes, the original reads were first aligned to the corresponding genomes (downloaded from Ensembl [64] using Bowtie 2 v2.2.6 in local mode (–local –no-unal)) [65,66]. Reads for which the alignment score exceeded the default minimal value of 20 + 8.0 * ln(L), where L is the read length, were removed. Then, the remaining (i.e., unaligned) reads were assembled again following the procedure described in Section 2.2.2.

2.2.4. Generation of a Consensus Transcriptome

The five resulting transcriptomes (one per experiment) were further combined and analysed with the tr2aacds.pl and evgmrna2tsa2.pl (-onlypubset) scripts from EvidentialGene to select the overall best candidate transcripts. The remaining reconstructed transcripts were discarded because they were classified either as redundant, fragmented or uninformative coding sequences, based on untranslated region (UTR) length, gaps, amino acid quality, and stop and start codon presence. After reducing redundancy, EvidentialGene clustered the best transcripts by groups of likely isoforms using CD-HIT v4.6.8 [67,68] and a similarity threshold of 90% on the amino-acid sequences. Sequences were considered as true isoforms (i.e., representing the same gene) when sharing high-identity ($\geq$98%) exon-sized fragments, as determined with BLASTN v2.2.28 (E-value cut off of 1×10^{-19}).

Transcripts proposed by EvidentialGene as the most representative isoform for each gene were selected for annotation (see Sections 2.4 and 2.5) and for studying gene expression (Sections 2.6–2.8).

2.3. Assessment of Transcriptome Quality

Additional analyses were performed to determine the quality of the assembled transcripts. The same set of analyses was also performed on the two other transcriptomes publicly available (GEFR01 [7] and GDJR01 [53]) for comparison with the present study. First, basic statistics based on the length of transcripts and the number of ORFs were computed. Read representation was determined by mapping back the cleaned reads (see Section 2.2.1) to each of the three transcriptomes with the aligner Bowtie 2 v2.2.6 (–local, –no-unal) as described in [65]. Note that unpaired and singleton reads were excluded from all quality statistics. In parallel, we used two evaluation tools, Detonate v1.11 [69] and TransRate v1.0.3 [70], to get reference-free quality scores for the three transcriptomes.

To check the presence of the spliced leader (SL) sequence [14] in the three public transcriptomes, we used wordmatch from the EMBOSS software package [71] and three length thresholds (12, 14 and 24 nt) found in the literature [52,53]. Matches were only considered when falling at the $5'$-end of a transcript, whether in forward or reverse orientation, as transcripts are not oriented in the transcriptomes. More precisely, each transcript was first reverse-complemented, and both versions (forward and reverse) were truncated at 40 nt before running wordmatch. Besides, transcripts actually corresponding to rRNA sequences were identified by combining RNAmmer v1.2 [72] and MegaBLAST v2.2.28 [62] searches (E-value cut-off of 1×10^{-50}, the latter using accessions X12890.1 (*E. gracilis* rrnC operon), M12677.1 (SSU rRNA 18S) and X53361.2 (LSU rRNA 28S) as queries. Regarding coding sequences, we estimated the numbers of putative genes with GeneMarkS-T (beta version) [73] and measured transcriptome completeness with BUSCO v.3.0.1 [74,75] using both "Eukaryota" and "Protists *ensembl*" datasets.

Lastly, we used CD-HIT-2D v4.6.8 [67,68] to identify similar predicted protein sequences between transcriptomes with our transcriptome as a reference. We explored different word sizes (2 to 5) at several thresholds of sequence identity (ranging from 0.5 to 0.9). Sequences from the other two public transcriptomes that could not be clustered with sequences of our consensus transcriptome were tentatively aligned using BLASTP v2.2.28 instead [62]. We further calculated the expression of presumably "missing" sequences in GDJR01 (D) and GEFR01 (B-C), respectively, following the procedure described in Section 2.5. The sequence was deemed invalid and not considered missing if its expression was below one transcript per kilobase million (TPM) in the transcriptome from which it had been identified. In a complementary analysis, highly similar nucleotide sequences from the three transcriptomes were clustered all together at once using CD-HIT-EST (identity threshold of 0.9, word size of 8, coverage of the shorter sequence of 0.9). Within each cluster, transcripts were pooled per transcriptome and their properties used to compare the three transcriptomes over all clusters, in terms of redundancy, length and identity. Analyses were performed either on all clusters or only on clusters shared across the three transcriptomes.

2.4. Transcript Annotation

The annotation procedure was carried out in three steps. First, assembled transcripts (i.e., the EvidentialGene representative isoforms) were annotated with EggNOG-mapper v1 [76,77]. We used HMMER to compare our data with the eukaryotic database of EggNOG, prioritizing coverage. Second, we annotated our transcripts by similarity using PSI-BLAST v2.2.28 searches [62] (E-value cut-off of 0.001) against Swiss-Prot [78]. Third, we aligned the assembled transcripts to the NCBI protein (*nr*) database [63] using TBLASTN v2.2.28 [62] (same E-value cut-off). We recovered Gene Ontology terms (GO) [79] and Kyoto Encyclopedia of Genes and Genomes Orthologs terms (KO) [80] of each transcript for further term enrichment analysis and network representation (see Section 2.7 for details). For that purpose, EggNOG features were assigned when possible to a transcript; if annotation was missing, PSI-BLAST v2.2.28 annotation was provided instead, or even TBLASTN v2.2.28 features whenever the two first previous methods failed. For mitochondrion and plastid-specific analyses, the components of the photosynthetic and respiratory electron transport chains were identified by BLASTP v2.2.28 searches [62] (E-value cut-off of 0.001) against reference proteins described in the literature. Hence, respiratory subunits were taken from [9,81,82], whereas subunits of photosystem I, photosystem II, cytochrome b6f complex, cF1Fo ATP-synthase were sourced from [83], and LHC polyproteins from [84].

2.5. Taxonomic Analyses

Taxonomic affinities were determined based on BLASTX v2.2.28 [62] searches against a broadly sampled proteome database, composed of 73 manually selected eukaryotes [85] and 19,802 representative prokaryotes subsampled from a curated database of 27,762 genomes [86]. For each assembled transcript, a last common ancestor (LCA) was computed based on their closest relatives (best hits, if any) in the database, provided they had a bit-score $\geq$80 and were within 95% of the bit-score of the first hit (MEGAN-like algorithm [86,87]). Organellar (plastid and mitochondrion) encoded proteins were distinguished from nuclear-encoded proteins by querying (BLASTP) two *E. gracilis* organelle databases assembled from the NCBI RefSeq "Proteins" portal [63]. To identify with certainty an organelle-encoded protein, only hits with a minimum percentage identity of 99% and a strictly identical length were considered. Such organelle-encoded sequences were expected at least from our own reads, which were generated in the absence of poly-A selection.

In parallel, tetranucleotide frequencies (TNFs) were computed for individual transcripts using the default settings of compseq from the EMBOSS software package [71]. Then, assembled transcripts for which a taxonomic affiliation had been obtained were ranked following their GC content and split into four partitions of equal size in terms of number of transcripts. Finally, ten principal component analyses (PCAs) were computed on TNFs, each one based on 1000 randomly chosen transcripts, using the prcomp function of the STATS v3.4.3 R base package [88]. For each PCA, two different colour schemes were applied on data points: the broad taxonomic affiliation of the transcript LCA (divided into four groups: Viridiplantae, Kinetoplastida, other Eukaryota and Bacteria), and the GC-content partition of the transcript.

2.6. Expression Quantification

The abundance of assembled transcripts was estimated by using RSEM v1.2.31 [89] and Bowtie2 v2.2.6 aligner [65,66]. Specifically, we used the align_and_estimate_abundance.pl Perl script wrapped in the Trinity v2.4.0 software package [59]. Data was then processed with abundance_estimates_to_matrix.pl Perl script without normalization parameters to generate the final expression matrix. Expression values are provided in transcripts per kilobase million (TPM) and pooled per gene (i.e., gene-level counts) [90].

Each count value was log2-transformed and converted to a Z-score to make samples comparable (sample mean was subtracted from each sample observation and divided by sample standard deviation). Batch effects were tentatively removed with the help of the SVA v.3.26.0 R package [91], so as to adjust data for unwanted sources of variation. However, such correction proved to be ineffective and thus abandoned (see Results and Discussion). For downstream analyses, only the 2500 most variable genes were retained (based on their expression variance across the 23 samples).

2.7. Gene Clustering Based on Expression Profiles

The 2500 most variable genes were clustered using the Partitioning around medoids (PAM) algorithm (from the CLUSTER v.2.0.7 R package) [92], which creates a fixed number of clusters (k) by minimizing the sum of the dissimilarities of the observations to their closest representative object (medoid). To capture both positive and negative relationships between gene pairs, we used a dissimilarity matrix of expression based on the squared Pearson correlation ($d = 1 - r^2$). The optimal cluster segregation was selected by cycling through the number of potential solutions, ranging from k = 5 to 75. In each solution, an average of maximal absolute correlations within-cluster (w-k cor$_{max}$) and an average of minimum absolute correlations between-cluster medoids (b-k cor$_{min}$) were computed. To intercept the point where optimal cluster segregation occurred, a reinterpretation of the Dunn index was used, and we computed the b-k cor$_{min}$ and w-k cor$_{max}$ ratio, choosing the solution with the minimal ratio value. At this optimal point, decreasing or increasing the number of cluster solutions would not better explain the data [93]. Heat map and hierarchical clustering analyses (correlation was used as the distance and centroid linkage clustering as the method) of expression data were carried out using the pheatmap function from the pheatmap v1.0.12 R package [94] and, when necessary, row-wise data (gene expression of the transcripts) was aggregated using k-means clustering to facilitate visual inspection of expression across conditions.

2.8. Gene Ontology (Enrichment) Analyses

The clusters based on the 2500 most variable genes were further analysed to visualize overrepresented biological terms using the whole GO and KEGG term space from Section 2.4 as a background. We explored enriched pathways within the expression clusters using ClueGo v2.5.0 tool [95], a visualization plug-in implemented in the Cytoscape v3.6.0 environment [96]. Term overrepresentation was estimated by an enrichment test based on the hypergeometric distribution followed by Benjamini–Hochberg adjustment for multiple testing. An annotation network was built with the ClueGo plug-in from kappa scores, which reflect the associations between genes and GO and KEGG terms. Network specificity was set between 3 and 12 GO hierarchy levels, and term selection was set to a minimum of 3% genes per cluster. Kappa score threshold was set to 0.3, and we allowed GO parent-child term fusion. Moreover, we explored the network with the MCODE algorithm [97], implemented as a Cytoscape plug-in, to detect densely connected regions or hubs in the network. Those hubs were found in the network establishing a degree cut-off of 2 for network scoring criteria, without including loops. Option Fluff was selected and parameters for Cluster Finding panel were set at 0.1 and 0.2 for node density and node score cut-off, respectively, a minimum of 2 edges per node of cluster cores (K-Core) and a maximum depth of 100.

3. Results and Discussion

3.1. Data Collection/Datasets

Out of the eight datasets publicly available for *E. gracilis*, only four [PRJNA310762 (A), PRJEB10085 (B), PRJNA298469 (C), PRJNA289402 (D)], were retained to assemble our consensus transcriptome, along with our own experiment PRJEB38787 (E; Table 1), which used Duplex-Specific thermostable nuclease (DSN) normalization to avoid poly-A selection. These five datasets totalled circa 2.6 billion raw Illumina reads (100-nt long), of which 70% belong to our experiment. After quality treatment, between 5 and 7% of reads were lost in experiments PRJNA310762 (A), PRJNA298469 (C) and PRJNA289402 (D), whereas the rejection of reads was more important in experiments PRJEB10085 (B) and PRJEB38787 (E). In PRJEB10085 (B), 19% of reads were truncated as a consequence of low-quality regions, whereas in PRJEB38787 (E), 50% of reads were discarded because of the high number of ambiguous nucleotides, especially in reverse reads. Hence, we got 57.8 million of good quality reads out of 62 after pre-processing of experiment PRJNA310762 (A) [7], 310 million reads out of 383 for experiment PRJEB10085 (B) [52], and 267.7 million from experiment PRJNA289402 (D). In the latter case, we used all samples as input, whereas Yoshida et al. (2016) only used the reads from cells grown in mixotrophic conditions to build their assembly [53]. Finally, Ebenezer et al. (2019) used 410 million reads as input for their transcriptome assembly, probably as the result of combining reads from PRJEB10085 (B) and PRJNA298469 [7].

After quality filtering, ca. 1.5 billion reads were retained, pre-processed read files of each individual experiment were assembled in four replicates using Trinity and then condensed into one individual transcriptome per experiment using EvidentialGene, which served as the basis for creating the consensus transcriptome (see Materials and Methods for details). Overall, PRJEB38787 (E), PRJEB10085 (B), PRJNA289402 (D), PRJNA310762 (A) and PRJNA298469 (C) experiments accounted for 55, 20, 17, 4, and 2% of the pre-processed reads used for the individual assemblies, respectively.

3.2. De Novo Assembly Evaluation

3.2.1. Individual Assemblies

The presence of sequences within a data set that originate from sources other than the sequenced sample is a known limitation of RNA-Seq experiments (e.g., [98,99] in human datasets). For some studies, such as large-scale phylogenomics, contaminants can be very problematic and must be dealt with using an array of different approaches [100]. Thus, before combining the individual five transcriptomes into a final consensus transcriptome, all assembled sequences were BLASTed against the NCBI nucleotide (*nt*) database [63] to identify possible contaminants. Using stringent thresholds, we found in the five transcriptomes only 948 unique hits of reconstructed transcripts that matched organisms other than *E. gracilis*. These organisms were considered as possible contaminants. Among them, we selected the five organisms whose abundance was the greatest (*Homo sapiens*, *Saccharomyces cerevisiae*, *Escherichia coli*, *Ovis aries* and *Caenorhabditis elegans*). It is noteworthy that sheep (and cow) DNA is commonly sequenced on our genomic platform. By mapping all pre-processed reads to the nuclear genome of these five species, we found that contaminants were less than 0.01% of the reads matching one of the contaminant genomes. In comparison, it has been shown that 0.13% of contaminant reads were present on average in a subset of 150 sequencing data files from the 1000 Genomes Project [101]. In the case of PRJNA298469 (C), we flagged as contaminants 68 reads per million reads (RPM), a larger proportion compared to the other experiments, which varied between 2 and 29 RPM (Table 2). Contaminant reads were removed and new assemblies of each experiment were generated anew from decontaminated reads, following the same procedure as above (see Section 2.2.2 for details). Afterwards, a new BLAST analysis was performed to quantify whether the contamination level was reduced. As expected, hits matching to *C. elegans*, *Escherichia coli*, *H. sapiens*, *O. aries* and *Saccharomyces cerevisiae* decreased, while hits matching to *Euglena* remained similar (Supplementary Figure S1). Besides, we traced the non-*Euglena* sequences that persisted in the final consensus transcriptome presented just below (see Section 3.2.2). Overall, from 716 unique hits of non-*Euglena* sequences identified with the latter BLAST analysis, only 64 were still present in the final consensus transcriptome (see Section 3.3.2 for details on the contamination sources). As a case in point, the complex genetic makeup of *E. gracilis* (e.g., [52]) makes it difficult to determine when a sequence, even if very peculiar, has been acquired from a very distantly related species or whether it can be a contaminant (see also Section 3.3.2 for an attempt to differentiate the two cases). For example, the glyoxylate cycle is localized within the mitochondria in *E. gracilis* and isocitrate lyase and malate synthase form only one bifunctional enzyme, called EgGCE [102,103]. A bifunctional enzyme for the glyoxylate cycle is also found in the worm *C. elegans* (opisthokonts), revealing an independent acquisition of the bifunctional enzyme by convergent evolution in these two organisms [104].

The five decontaminated individual transcriptomes were then evaluated with TransRate to check their uniformity. Four transcriptomes yielded ca. 42,342 ($\pm$6159) transcripts on average, whilst the number of reconstructed sequences in experiment PRJEB10085 (B) was more than twice the average, 95,490 sequences (Table 2). In addition, the computed GC content was 58% for experiment PRJEB10085 (B), a lower percentage compared to the other assembled transcriptomes, which was around 64%. Finally, we discovered a high frequency of sequences under 500 nt and characterized by a lower GC content (Supplementary Figure S2). After those small sequences were removed (representing 62% of the transcripts), TransRate statistics were recomputed and yielded values more in line with other experiments, both in terms of number of sequences (36,287) and GC content (62%).

We could not determine what the removed sequences were by similarity searches. They might represent some sort of artefact, contamination, or even be the result of a specific feature of experiment PRJEB10085 (B), for example the sequencing of a different strain, i.e., *E. gracilis* var. saccharophila Klebs (SAG 1224/7a) [52], whereas the other four experiments all used the Z strain (SAG 1224-5/25).

Table 2. Basic statistics based on transcript properties of reconstructed transcriptomes from collected data. ACC: study accession, REF: bibliographic reference, RAW: number of downloaded reads, PRE: number of good reads after pre-processing, CNT: number of reads removed after pre-processing considered as contamination (reads per million; rpm), SEQ: number of transcripts, MIN: minimal sequence length, MAX: maximal sequence length, MEAN: mean sequence length, TOTAL: combined sequence length, SEQ < 200: number of transcripts under 200 n, SEQ > 1 k: number of transcripts over 1000 nt, SEQ > 10 k: number of transcripts over 10,000 nt, ORF: number of sequences with a predicted open reading frame, ORF (%): for contigs with an ORF, the mean % of the contig covered by the ORF, N[z]: minimum contig length needed to cover [z]% of the transcriptome. GC (%): percentage of guanine-cytosine content, PART and PART (%): number and percentage of sequences contributed to the final consensus transcriptome (see below). In PRJEB10085 (B) (filtered), sequences <500 nt were further discarded (see text).

Statistic	A	B	B (Filtered)	C	D	E
ACC	PRJNA310762	PRJEB10085	PRJEB10085	PRJNA298469	PRJNA289402	PRJEB38787
REF	[7,52,53]					This study
RAW	61,531,862	383,416,636	383,416,636	27,096,926	285,148,782	1,902,226,200
PRE	57,862,467	310,302,570	310,302,570	25,244,887	267,779,751	875,299,135
CNT	740 (12 rpm)	9080 (29 rpm)	9080 (29 rpm)	1750 (68 rpm)	1191 (4 rpm)	2403 (2 rpm)
SEQ	38,559	95,490	36,287	42,363	37,425	51,021
MIN	101	101	500	101	101	101
MAX	13,929	21,744	21,744	11,354	26,839	10,795
MEAN	1043	647	1312	810	1120	610
TOTAL	40,861,413	64,426,688	47,615,807	34,438,742	42,382,170	31,671,589
SEQ < 200	4330	17,074	0	782	3051	3989
SEQ > 1 k	16,289	18,638	18,638	10,932	17,048	7104
SEQ > 10 k	4	15	15	1	13	1
ORF	24,757	29,060	27,842	27,063	24,817	26,882
ORF (%)	88%	82%	83%	89%	87%	93%
N90	576	347	654	419	606	367
N70	1140	667	1101	686	1187	528
N50	1607	1282	1574	1014	1658	753
N30	2257	2033	2243	1452	2318	1090
N10	3600	4026	3707	2358	3812	1850
GC (%)	64%	58%	62%	64%	64%	64%
PART	22,234	-	27,730	10,129	19,663	11,602
PART (%)	24.3%	-	30.3%	11.1%	21.5%	12.7%

3.2.2. Final Consensus Transcriptome

To obtain our final transcriptome, we combined the individual five decontaminated transcriptomes into a consensus transcriptome. Regardless of the aforementioned differences in the amount of pre-processed reads per dataset, the contribution of transcripts from each study in the final consensus transcriptome was rather balanced, where PRJEB10085 (B), PRJNA310762 (A), PRJNA289402 (D), PRJEB38787 (E), and PRJNA298469 (C) accounted for 30.3%, 24.3%, 21.5%, 12.7%, and 11.1%, respectively (Table 2). The resulting transcripts were classified into non-redundant protein-encoding genes, and one representative isoform was selected for each gene. Our new transcriptome was then compared with the other two publicly available transcriptomes, GDJR01 (D) [53] and GEFR01 (B-C) [7] (Table 3). Ebenezer et al. (2019) [7] used a combination of in-house generated sequences (PRJNA298469 (C)) and publicly available data from O'Neill et al. (2015) [52] (PRJEB10085 (B)) to assemble a transcriptome. Assembly transcriptome statistics were computed with TransRate. The overall number of sequences reported in the present work is 91,040, with N50 of 1432 nt, whereas in GDJR01 (D), it was 113,152 (N50 1604), and 72,506 (N50 1242) in GEFR01 (B-C).

The mean length of our transcripts was 1096 nt, a value closer to GDJR01 (D) than GEFR01 (B-C), which was ca. 200 nt smaller. The number of protein coding regions predicted by GeneMarkS-T (58,542) and the number of open reading frames (ORF) found with TransRate (62,287) are slightly smaller than in GDJR01 (D), but about twice greater than in GEFR01 (B-C). Our own sequences were classified into 49,922 predicted non-redundant protein-encoding genes, which is comparable to GDJR01 (D), but almost eighteen thousand genes more than in GEFR01 (B-C). As expected, these recomputed numbers are similar to those reported in the original publications of Yoshida et al. (2016) [53] and Ebenezer et al. (2019) [7]. Additionally, O'Neill et al. (2015) [52] found over 32,000 unique components for their *E. gracilis* transcriptome. The total size of our consensus transcriptome is 100 Mb, whilst the size of GDJR01 (D) is 122 Mb, 63 Mb for GEFR01 (B-C) and 38.4 Mb for O'Neill et al. (2015) [52] transcriptome. Overall, the genome size of *E. gracilis* has been estimated from total DNA content to range between 1 Gbp to 9 Gbp [54]. In contrast, the most recent estimation based on high throughput sequencing data was 332–500 Mb in size for the whole haploid genome [7] but, because half of the genome is gapped or has unknown base representation, the authors pointed out that this latter estimation was likely to be approximate.

Table 3. Basic statistics of transcript properties computed for the three public transcriptome assemblies, including the consensus transcriptome generated in the present work, and completed with data retrieved from the publications of Ebenezer et al. (2019) [7] and Yoshida et al. (2016) [53]. Row titles are as in Table 2, except for CDS: number of unique coding sequences (i.e., ORFs or UNIGENEs), GMS-T and GMS-T (%): number and percentage of predicted protein coding regions calculated by GeneMarkS-T.

Statistic	GEFR01	GDJR01	HBDM01
REF	[7,53]		This study
SEQ	72,506	113,152	91,040 [1]
MIN	202	201	201
MAX	25,763	21,553	26,839
MEAN	869	1087	1096
TOTAL	63,049,595	122,976,775	100,187,451
SEQ < 200	0 [1]	0 [1]	0 [1]
SEQ > 1 k	19,740	49,277	37,294
SEQ > 10 k	25	27	24
ORF [2]	30,467	65,943	62,287
ORF (%)	79%	73%	85%
N90	374	523	545
N70	704	1130	965
N50	1242	1604	1432
N30	1916	2181	2049
N10	3344	3347	3410
GC (%)	61%	63%	63%
CDS	32,128	49,826	49,922
GMS-T	35,929	63,432	58,542
GMS-T (%)	49%	56%	64%

[1] Submission tools for sequence repositories do not accept transcripts ≤ 200 nt. Hence, the number of sequences in the public version of HBDM01 is lower than reported elsewhere in this work. [2] ORFs were determined with TransDecoder, whereas CDS were determined with EvidentialGene (or a similar tool, depending on the study).

The pre-processed reads from the five experiments were aligned back to the three public transcriptomes as a metric of completeness. In most cases, the percentage of mapping was over 80%, reaching even more than 90%, with the exception of reads produced by ourselves PRJEB38787 (E), which had a representation of ~75% and ~50% in GEFR01 (B-C) and GDJR01 (D), respectively (Table 4). It is probable that our reads have a lower mapping percentage because they were generated from DSN-normalized total RNA samples, for which analyses of a preliminary sequencing lane revealed many reads corresponding to non-mRNA sequences (e.g., rRNA). However, the specifically low mapping to GDJR01

(D) cannot be explained easily because "transcripts" matching to rRNA sequences were identified in all three public transcriptomes (Supplementary Archive File S1).

Table 4. Mapping fraction of pre-processed reads from each collected dataset (rows) to the three public transcriptome assemblies (columns), GEFR01 [7], GDJR01 [53] and HBDM01 (this study).

Code	Accession	Reference	GEFR01	GDJR01	HBDM01
A	PRJNA310762	[7]	87.40%	92.51%	93.38%
B	PRJEB10085	[52]	84.68%	90.13%	91.49%
C	PRJNA298469	[7]	80.26%	91.66%	90.39%
D	PRJNA289402	[53]	85.25%	95.04%	94.28%
E	PRJEB38787	This study	75.28%	51.39%	80.76%

Using BUSCO on our predicted proteins, we found that the consensus transcriptome contained 84.8% of complete eukaryotic orthologs and half of them were duplicated, while 10.6% were missing (Figure 2). In comparison, we estimated the completeness of GDJR01 (D) at 80.8% of complete orthologs, of which a fifth were duplicated, and completeness of GEFR01 (B-C) at 76.9%, with only 4% of them duplicated. Moreover, we observed that lower percentages of complete orthologs were accompanied by higher numbers of fragmented and missed sequences. Overall, our consensus transcriptome appears to be the most complete, GEFR01 (B-C) being the least. Ebenezer et al. (2019) [7] also determined BUSCO completeness in GDJR01 (D) and GEFR01 (B-C) transcriptomes in addition to the original transcriptome presented by O'Neill et al. (2015) [52] and similarly concluded that GEFR01 (B-C) was the least complete transcriptome. Beyond transcripts missing due to low expression, discrepancies in the number of complete orthologs predicted by the different studies may also be due to the use of different tools for protein prediction. Whereas we used cdna_bestorf.pl script from EvidentialGene, the other studies used TransDecoder [59], which, reportedly, tends to predict larger amounts of proteins, but performs worse for true transcripts [105]. Despite these differences, the general representation scores of the reads in the assembled transcripts were similar across the three public transcriptomes, even if depending on the exact evaluation software used (Table 5).

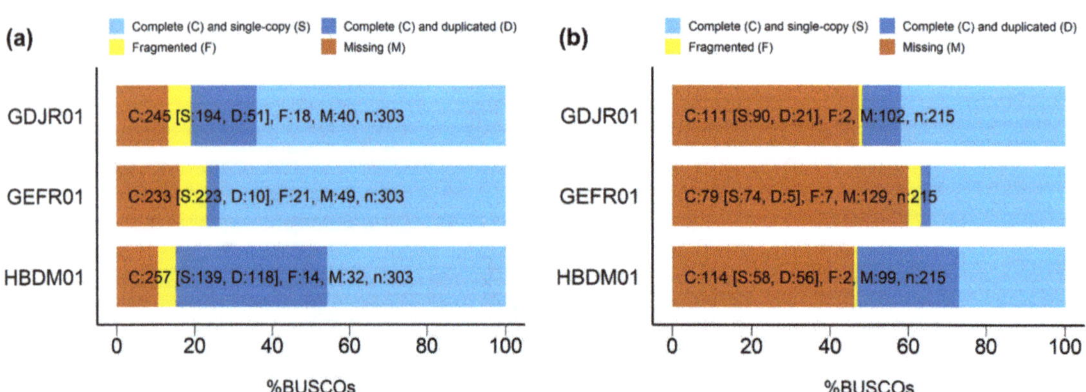

Figure 2. BUSCO-generated charts showing the relative completeness of the three public transcriptome assemblies, GEFR01 [7], GDJR01 [53] and HBDM01 (this study). BUSCO datasets were based on odb9. (**a**) "Eukaryota" (303 BUSCOs); (**b**) "Protists *ensembl*" (215 BUSCOs).

Table 5. TransRate and Detonate assembly scores for the three public transcriptome assemblies, GEFR01 [7], GDJR01 [53] and HBDM01 (this study). Scores indicate how well transcripts are supported by the RNA-Seq data.

Assembly Score	GEFR01	GDJR01	HBDM01
TransRate Score	0.1789	0.0304	0.0430
TransRate Optimal Score	0.2051	0.1729	0.0764
Detonate Score	$-97{,}461 \times 10^6$	$-97{,}561 \times 10^6$	$-97{,}459 \times 10^6$

As already mentioned, one evidence supporting the evolutionary relationship between trypanosomatids and euglenids are trans-splicing mechanisms [14]. We found that the SL-sequence was present in no more than 10.8% of transcripts in our transcriptome, far from the approximately 53–60% prevalence reported before [14,53], and closer to the 16% found by [52]. However, when performing the exact same analysis on the other two public transcriptomes, we find contrasting results, with SL-sequence matches recovered in at most of 2% and 30.3% of GEFR01 and GDJR01, respectively (Table 6). This indicates that the transcriptome of Yoshida et al. (2016) [53] has the most complete transcripts in 5-end, even though our own assembly includes 200 transcripts with a full-length perfect match to the 24-nt SL-sequence (vs. 45 and 5 for GEFR01 and GDJR01, respectively). Comparison of the mapping coverage for the three public transcriptomes shows that partial matches (12–14 nt) are much more numerous than full-length matches, as expected, but that the former are concentrated at the very beginning of the transcripts, which suggests that they are genuine SL-sequences (Supplementary Figure S3).

Table 6. SL-sequence related statistics for the three public transcriptome assemblies, GEFR01 [7], GDJR01 [53] and HBDM01 (this study). These correspond to exact matches limited to the first 40 nucleotides of each transcript.

Threshold (nt)	Statistic	GEFR01	GDJR01	HBDM01
24	Forward matches	24	5	86
	Reverse matches	21	0	114
	Total matches	45	5	200
	Average length (nt)	24.00	24.00	24.00
14	Forward matches	176	16,580	3370
	Reverse matches	200	12,999	3265
	Total matches	376	29,579	6635
	Average length (nt)	16.28	15.57	15.59
12	Forward matches	749	18,322	4403
	Reverse matches	766	16,016	5397
	Total matches	1515	34,338	9800
	Average length (nt)	13.37	15.19	14.68

Finally, we determined whether sequences of the other two available transcriptomes were present in our consensus transcriptome through two complementary approaches: one pairwise, sensitive and based on protein sequences, and one global, conservative and based on nucleotide sequences (Supplementary Table S1b). First, when using CD-HIT-2D with our transcriptome as a reference, a word size of 2 and an identity threshold of 0.4, 26.1% (34,490) of total sequences from GDJR01 (D) were missing and 37.6% (28,552) of total sequences from GEFR01 (B-C). Missing sequences were BLASTed (TBLASTN E-value cut-off of 0.001) against our transcriptome, and 20.5% (27,152) of total sequences of GDJR01 (D) were recaptured and 24.8% (18,870) of GEFR01 (B-C) (Supplementary Table S1a). After computing TPM values using the pre-processed reads generated in this study, we found that only 518 missing sequences of GDJR01 (D) were expressed above 1 TPM and 1595 in GEFR01 (B-C), which means that potentially 0.5% and 2% of the truly expressed sequences from GDJR01 (D) and GEFR01 (B-C), respectively, are missing from our consensus transcriptome. Hence, these sensitive analyses suggest that we captured more than 98% of the sequences produced in the other transcriptomes hitherto published. Second, CD-HIT-EST was used to compute clusters of related transcripts at an identity threshold of 90%. We recovered 121,851 clusters, in which the three transcriptomes had very similar patterns of presence and representation (Supplementary Table S1b). Hence, each transcriptome had at least one transcript in 60,220 to 66,041 clusters, whereas they each provided the representative (longest) sequence in 39,434 to 41,610 clusters. Singleton cluster statistics were slightly different, with GEFR01 having 29,997 specific clusters, followed by GDJR01 (27,058) and then our own transcriptome (19,028). When focusing on the 24,164 clusters shared between the three transcriptomes, we see that our transcriptome contributes the highest number of representative sequences, which confirms that they are generally longer than their homologues in the other two transcriptomes. This is also visible in a direct comparison of the mean an maximum transcript length across the three transcriptomes, whether on the 121,851 or the 24,164 clusters (Supplementary Figure S4). In contrast, comparison of the median and max identity between transcripts of the three datasets reveals that GEFR01 sequences are the most similar on average to the sequences from the two other transcriptomes. They are also the less redundant, with the lowest number of transcripts per cluster.

Altogether, these comparative analyses indicate that the three publicly available transcriptomes each have a distinct edge on the other two: Ebenezer et al. (2019) [7] assembled a compact set of sequences nonetheless providing a large fraction of unique transcripts, whereas Yoshida et al. (2016) [53] obtained a more redundant transcriptome, but with many transcripts complete at their 5-end, as evidenced by the detection of SL-sequences, and for our part, we generated the longest transcripts on average, including a few hundred featuring a full-length SL-sequence, with moderate redundancy.

3.3. Global (Transcriptome) Annotation

3.3.1. Functional Annotation of Transcripts

The combination of annotation strategies in our 49,922 predicted non-redundant protein-encoding genes yielded 9916 sequences with GO terms, 7775 KEGG orthologs, 13,298 sequences with a functional annotation and 13,850 with a taxonomic affiliation (Supplementary Table S2; see also Section 3.3.2). In the same way, O'Neill et al. (2015) [52] found 14,389 proteins with annotated functions out of the 32,128 predicted proteins of their transcriptome, whereas out of the 49,826 unique components reported by Yoshida et al. (2016) [53], approximately 11,314 were functionally annotated. Ebenezer et al. (2019) [7] annotated over 19,000 sequences, but without discerning what kind of attributes were associated in each case.

In comparison to the annotation performed in the other transcriptomes, we were able to find all the enzymes of the mevalonate pathway, including the diphosphomevalonate decarboxylase (EC 4.1.1.33), which was missing in the work of O'Neill et al. (2015) [52], thereby revealing that the last reaction is catalysed by a canonical enzyme. Regarding the carbohydrate-active enzymes, we found results similar to those outlined by O'Neill et al. (2015) [52]. Hence, we identified a great number of glycosyltransferases (311) and glycoside hydrolases (80), of which a quarter (19) were different types of glucanases (Supplementary Table S3). Corroborating the results of Yoshida et al. (2016) [53], we found two transcripts encoding glucan synthases, but could not identify transcripts encoding a 1,3-β-D-glucan phosphorylase, despite that such an enzyme has been previously characterised biochemically [106,107].

In *E. gracilis*, the photoreceptor is considered by some authors to be a rhodopsin-like protein where the retinal chromophore is a carotenoid [108]. We found five enzymes involved in retinol metabolism (EC 2.3.1.76; EC 3.1.1.64, EC 2.3.1.135; EC 1.1.1.105, EC 1.3.99.23) but, in line with Ebenezer et al.'s (2019) [7] findings, we could not find any rhodopsin-like protein candidates. Instead, we found 47 genes involved in visual perception processes (GO:0007601) and, more broadly, 333 genes related to photoresponse (Supplementary Table S4), including 13 cAMP/cGMP phosphodiesterases involved in

amplification of luminous signal, 15 GTPase regulators, nine arrestins, which are important for regulating signal transduction at G protein-coupled receptors, eight cryptochromes, and three cyclic nucleotide-gated channels of rod photoreceptors. In addition, we found 13 proteins of the paraflagellar rod, a structure observed in euglenids, kinetoplastids and dinoflagellates [109–111]. Such a structure is associated with the paraflagellar body (also called paraxonemal body, PAB) in *E. gracilis* [112]. We also found 49 transcripts coding for photoactivated adenylate cyclases (PAC), which are light-sensitive proteins of PAB [113]. Of these, 43 clearly show a bacterial affinity in our analyses, whereas two are highly similar two trypanosomatid sequences [114].

To better understand the general functionality of the consensus transcriptome, we reported the GO annotation results as high-level terms of the three ontologies without the detail of the specific fine-grained terms. For such a task, we used the generic GO Slim Mapper tool of The Saccharomyces Genome Database [115], and the list of summarized GO terms (GO slim) can be found in Supplementary Table S5. As we used a compendium of culture conditions, we expected to capture the sum of functionalities represented by the studies individually. We found a total number of 164 GO terms after GO slim analysis, represented by core metabolism (41), transport (13), cell organization (15) and maintenance (25), nucleotide metabolism (35) and protein synthesis (17), vesicle or cilium organization (15) among others. The annotation from O'Neill et al. (2015) [52] was classified into 157 GO categories while Yoshida et al. (2016) [53] determined, under mixotrophic conditions, that the main functional categories were genetic information processing (399 components), translation (291 components), and energy metabolism (239 components). Besides, genes belonging to the latter three categories were generally down-regulated during anaerobic treatment [53]. In the same way, Ebenezer et al. (2019) [7] indicated that major categories were dominated by core metabolic, structural and informational process supergroups, consistent with the current work and previous studies [52,53].

3.3.2. Taxonomic Annotation of Transcripts

As a complex alga resulting from a secondary endosymbiosis between a euglenozoan host and a chlorophyte alga, *E. gracilis* bears genes from multiple origins [16,25]. In terms of sequence similarity (and depending on the current sampling in reference organisms), its nuclear genome is expected to be composed of four main gene classes: (i) *Euglena*-specific genes, (ii) kinetoplastid-specific genes, (iii) eukaryotic genes (i.e., widespread in other eukaryotes), and (iv) (green) genes acquired during the secondary endosymbiosis [31]. Over the last fifteen years, this issue has been extensively studied, both using similarity [52,53] and phylogenetic [7,9,31,32,116–119] approaches, either at small (i.e., targeted subsets) [9,116–118] or larger (i.e., transcriptomic) scales and, when at larger scale, either by focusing on the chloroplast [119] or by surveying "unbiased" transcript collections [7,31,32,52,53]. All these studies have revealed that *E. gracilis* display sequence similarities to a panel of organisms that is larger than predicted by a simple theory of secondary symbiogenesis [120,121]. Unsurprisingly, our large-scale similarity analyses of the consensus transcriptome confirm the results of these previous works (Figure 3). A first observation is that only 28% of the predicted non-redundant protein-encoding genes (13,850 out of 49,922) bear any exploitable similarity with sequences in reference databases. Among those, 937 (7%) correspond to organisms to which we could not assign a specific taxon, whereas 4054 (29%) were only identified as "Eukaryota". The remaining gene similarities are distributed among kinetoplastids (1364, 10%), green plants (977, 7%) and other subgroups of eukaryotes, whether photosynthetic, such as cryptophytes (530, 4%) and haptophytes (468, 3%), or not, e.g., opisthokonts (947, 7%). Bacterial groups account for 1690 transcripts (12%), among which the most prominent are proteobacteria (34% of bacteria) and cyanobacteria (212, 13%). Only 40 (2%) and 15 (0.9%) transcripts are affiliated to the PVC group or Chlamydiae, respectively [122]. As expected [31], focusing on 119 nuclear-encoded genes involved in mitochondrial and photosynthetic electron transfer chains increases the similarity signal in favour of kinetoplastids (20 out of 86, 22%) and green plants (20 out of 33, 58%), respectively (Supplementary Figure S5; see also HTML Supplementary Files S2 and S3).

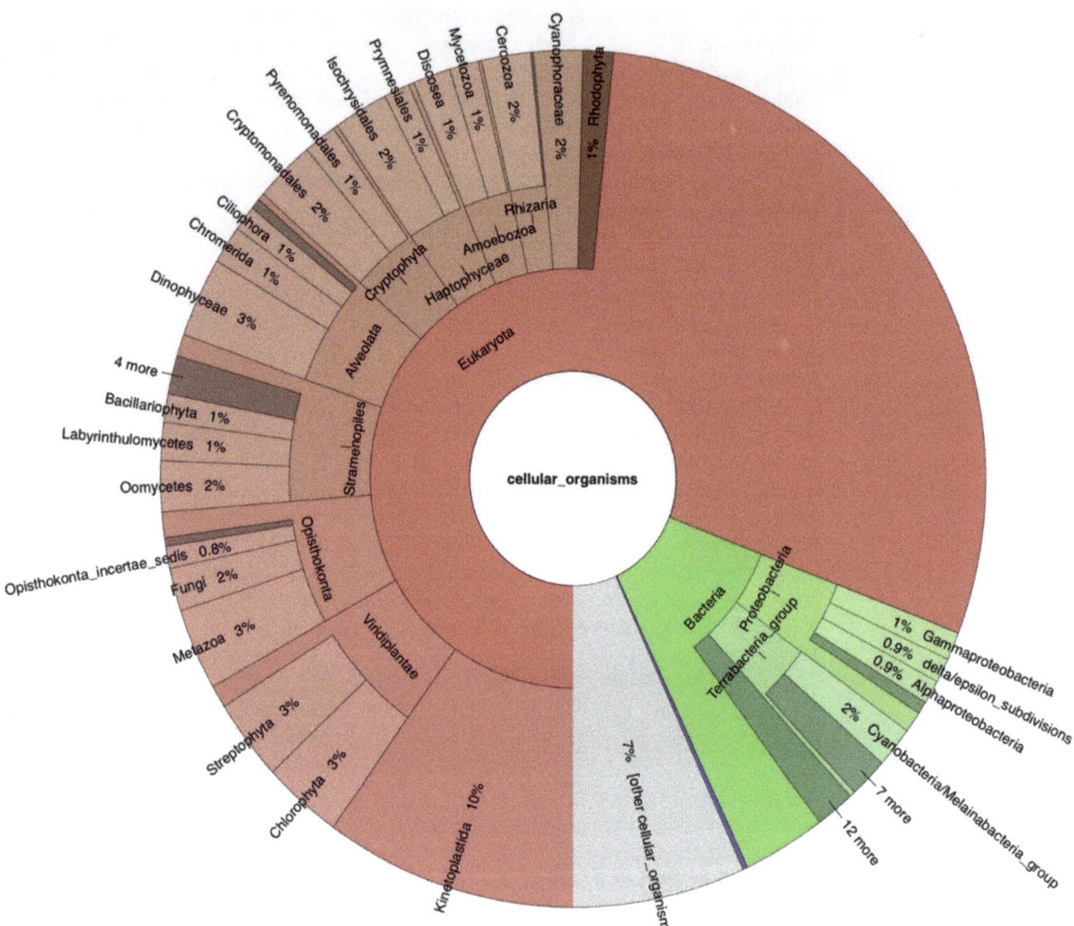

Figure 3. Taxonomic analysis of reconstructed transcripts (BLASTX MEGAN-like affiliations). The Krona chart is a zoom on the 13,850 transcripts to which a taxonomy could be associated, i.e., 28% of the 49,922 reconstructed transcripts. Among this classified fraction, 937 (7%) correspond to organisms to which we cannot assign a specific taxon ("other cellular organisms"). The thin blue slice is labelled "Archaea" (0.2%). The interactive chart is available as HTML Supplementary File S1.

Similarly to other complex algae (e.g., cryptophytes and chlorarachniophytes [123], ochrophytes and haptophytes [124,125]), *E. gracilis* transcriptomes show a heavily mixed ancestry in terms of gene donor lineages. However, it is a known (yet somewhat neglected) issue that publicly available transcriptomes can be contaminated by foreign sequences because of ecology (e.g., predator–prey, host–parasite or symbiotic relationships), or due to cross-contamination (either in the lab or on sequencing platforms) (see [126] and references therein). That is why we exerted special care to avoid including non-*Euglena* transcripts when assembling the five individual transcriptomes (see Section 3.2.1). In our final consensus transcriptome, we still identified 64 sequences as contaminants, of which 23 are false positives, owing to strong sequence similarity with different kinetoplastids (9 transcripts), green plants or algae (7), or non-green microalgae (7). Since the transcriptome had already been publicly released at the time, the other 41 remaining sequences were retained in subsequent analyses, but tagged as contaminants (Supplementary Table S6). Moreover, we used the taxonomic annotation of the 13,850 annotated transcripts to determine whether contaminants could be identified by their base composition pattern (see [127] and references therein). To this end, PCA plots were computed based on transcript tetranucleotide frequencies. Two types of colour annotation were then applied: one following a scale of GC-content and one following the taxonomy (Supplementary Figure S6). It appears that the taxonomic signal is mixed throughout these PCAs, whereas GC-content clearly corresponds to the PC1 axis. Thus, it was not possible in our case to identify and sort out contaminated transcripts (if any) from *Euglena* transcripts with this approach.

3.4. Systematic Functional Annotation of Top Differentially Expressed Genes

To better understand the functional organization of the most relevant *E. gracilis* genes under the assayed culture conditions, we computed a network of ontologies, based on transcript expression levels across all samples and studies (Supplementary Table S7). For this purpose, we only selected GO and KEEG terms that corresponded to the 2500 most variable genes (in terms of expression) to determine which biological functions were represented and how they were related to each other. The resulting organized network contained 119 nodes, with an average of nine neighbours per node, and 436 genes from the initial 2500 genes were retained (some genes being part of multiple hubs). We then used the MCODE algorithm to find evidence of higher order organization (Figure 4). The network was composed of nine modules (or hubs), each defined by one ontological category (Supplementary Table S8). Hub number 1 (72 transcripts) reflects "regulation of DNA damage checkpoint", with transcripts involved in apoptosis, control of transcription and other developmental processes. Unlike hub number 7 (see below), hub 1 has a stress response component. Hub 2 (191 transcripts) is the largest hub, and comprises genes involved in translational initiation and termination, or protein targeting to a membrane, and is thus defined by "ribosome" terms. Hub 2 is connected to hubs 3, 5 and 6 in the network. Categorized as a "thylakoid" hub, hub 3 (133 transcripts) is the second largest hub. It mainly comprises photosynthetic electron transport chain transcripts and other components that respond to light stimuli. According to taxonomic annotation, the majority of the genes represented in this hub come from green organisms. Transcripts involved in protein kinase activity were found in Hub 4 (23 transcripts), defined as "cyclin-dependent protein serine/threonine kinase regulator activity". Hub 5 (25 transcripts) corresponded mainly to processes involved in genetic information processing, such as spliceosome, exosome, chromosome-associated proteins, or chaperones. Hub 6 (79 transcripts) is defined by several categories related to mitochondrial protein complexes and mitochondria transport, and has a central position in the network (connections to hubs 1, 2, 3 and 8). Hub 7 (46 transcripts) was defined by "DNA integrity checkpoint" ontology terms and consisted of cell cycle processes, such as transition from G1 phase to S or the previously mentioned DNA integrity checkpoint. Hub 8 (53 transcripts) was categorized as "response to temperature stimulus" and was composed mainly of transcripts that encode heat shock proteins. Components of hub 9 (22 transcripts) were related to "negative regulation of translation". Overall, our 2500 most relevant genes appear to be distributed around the central role of the mitochondrion, whose origin traces back to the euglenozoan host cell [31]. In this respect, our taxonomic analysis specifically revealed that more than 10% of genes are related to kinetoplastids (the closest available proxy for the host cell) in all hubs, except for hub 3, categorized as "thylakoid" (Supplementary Table S9).

3.5. Cluster Annotation Enrichment Analysis and Gene Co-Expression

From the same top 2500 variable genes, we identified positive and negative relationships between pairs of genes based on gene expression. We tried to capture genes that behave conjointly across the various experimental conditions and group them into clusters. According to our expectations where a gene would be binary regulated (up or down), the optimal k solution should range between 2^5 (32) and 2^{13} (8192) (accounting for 5 to 13 distinct experimental conditions with a total sample number of 23; see Table 1). We computed the optimal number of clusters and determined that 36 clusters was the most suitable solution for the selected genes (Supplementary Figure S7). To better understand the underlying biological processes inside the clusters, ontologies that were overrepresented were extracted and analysed. Only five out of the 36 clusters were characterized by significantly overrepresented ontological terms (Supplementary Table S10). In total, those five clusters were composed of 631 transcripts out of the 2500 initially used for clustering, and 52% of them had at least one annotation attribute. Their expression can be visualized in hierarchically clustered heat maps (Figure 5).

Results from the enrichment tests revealed that "nucleosome category" was overrepresented in cluster 1, which contains transcripts of the "DNA damage checkpoint" and "ribosome" hubs of the ontological network, hub 1 and 2, respectively (see above). These transcripts encode histones, and core components of "nucleosome", that participate in wrapping and compacting DNA into chromatin. The observation that DNA packaging, transcription and translation shared the same gene expression pattern may be relevant because in euglenids, as well as in dinoflagellates, chromosomes are permanently condensed [128]. Furthermore, transcripts encoding different components of the chloroplast reaction centres of hub 3 were also found in this cluster. This cluster was characterized by a larger down-regulated expression in PRJEB38787 (E), while other experiments were slightly over and under zero. Cluster 4 was enriched in "photosynthetic electron transport" and "DNA damage checkpoint" related terms mainly present in hub 3, with several transcripts encoding ATP synthase subunits in the former and cell cycle and apoptosis regulator proteins in the latter. Gene expression in cluster 4 was homogeneous with values ranging between one or minus one, except for a group of genes greatly down-regulated in studies PRJNA310762 (A), PRJEB10085 (B), PRJNA298469 (C), and likely to be not expressed in such experiments. About a third of the transcripts from cluster 19 encode different types of serine/threonine proteins and are ontologically typified by "cyclin-dependent protein serine/threonine kinase regulator activity", which are processes closely related to cell cycle regulation. Their expression was slightly negative in the experiment PRJEB38787 (E) and positive in PRJEB10085 (B) while it remained unaltered in the rest of the experiments. "Neuroblast proliferation" and "neuroblast division" categories illustrated cluster 24, which, considering the unicellular nature of *E. gracilis*, was more likely to be related to cytoskeletal structure of eukaryotic cells formed during cell division or cell polarity than regulation of neurogenesis. In study PRJNA289402 (D), ABC transporters, fatty acid and polyketide synthesis were more down-regulated than in the remaining studies. Lastly, cluster 25 was enriched in "positive regulation of mitochondria organization" due to the presence of putative mitochondrial heat shock proteins that were co-regulated across studies. Besides, expression of cluster 25 was disparate for PRJNA289402 (D), compared with the other studies. A main difference was a group of transcripts largely downregulated in the PRJNA289402 (D) experiment, while they were upregulated in the remaining studies. Those transcripts putatively encode different components of the nitrogen metabolism, some chloroplastic electron transport chain components and ATP-dependent RNA helicase. A few transcripts related to cell cycle and translation, present in the annotation network, were found in cluster 25.

The cluster patterns reported above show that expression is driven by study rather than experimental conditions of the studies. Even if disappointing, these findings were similar after the tentative SVA correction of the batch effect present in the studies (Supplementary Figure S8). Presumably, our approach was not able to properly capture the batch effect, maybe due to an unbalanced batch-group design of the studies [129]. Nonetheless, we observed that a selection of 133 genes, coding for the components of the photosynthetic and respiratory electron transport chains, were grouped together. This subset of genes, located in the chloroplast and in the mitochondrion, respectively, was selected because most of the experimental conditions (light/dark, presence or absence of acetate in the medium, oxic/anoxic environment) of the studies were expected to affect respiration and photosynthesis. As illustrated in Figure 6, the expression of these genes is also driven by the study rather than by the reported physico-chemical parameters of each experiment. Yet, most components of the mitochondrial electron transport chain among the 133 selected genes were grouped together after hierarchical clustering of their expression, while chloroplastic components exploded into different subgroups. Concretely, genes coding for light-harvesting complexes grouped together distantly from other chloroplastic components. These transcripts are nuclear-encoded and showed a taxonomic affinity to Streptophyta (Supplementary Table S11).

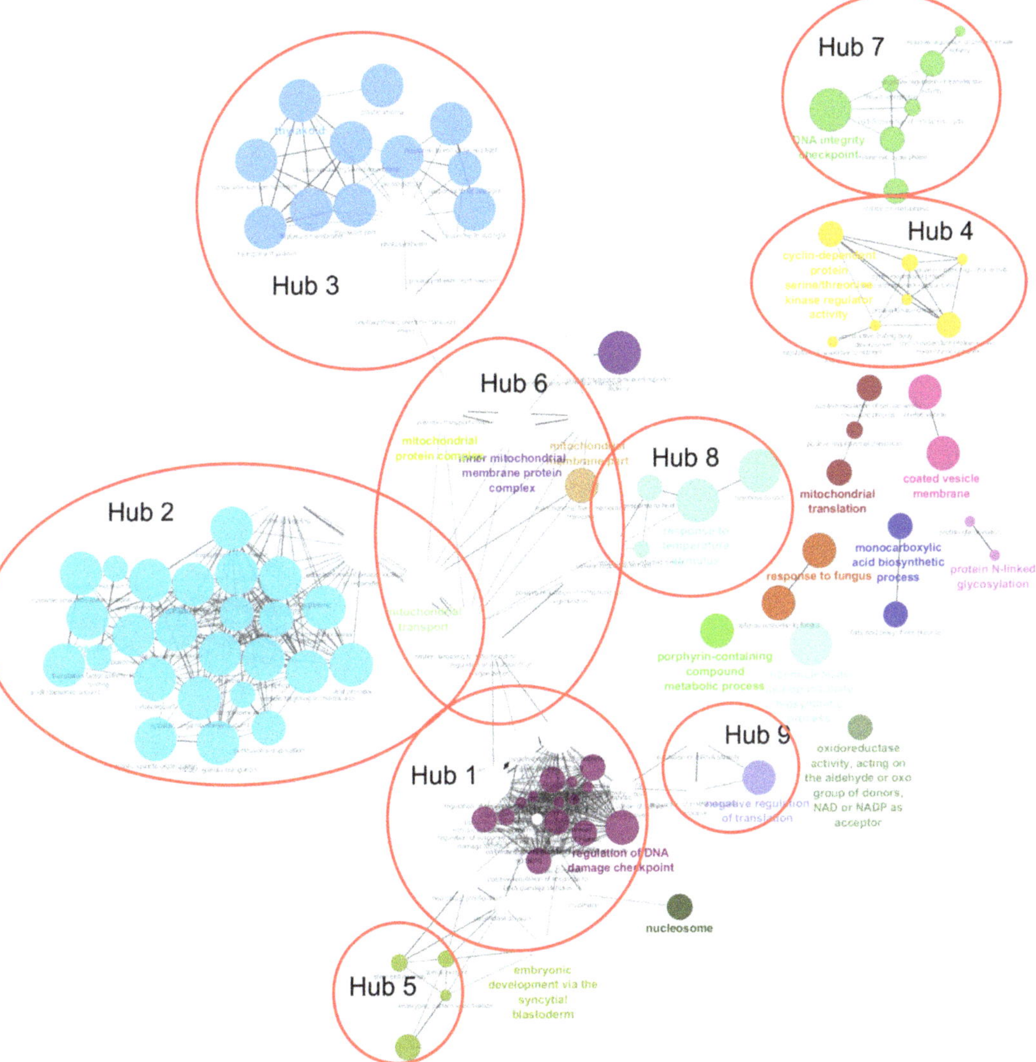

Figure 4. Annotation network of ontological terms showing the functional organization and relationships between the 2500 most variable genes. GO and KEGG terms were considered as a large pool in which the genes could be associated with 0 to N terms. Such associations served as the basis to infer the network (see text). Colours correspond to ontological terms (or groups of related ontological terms).

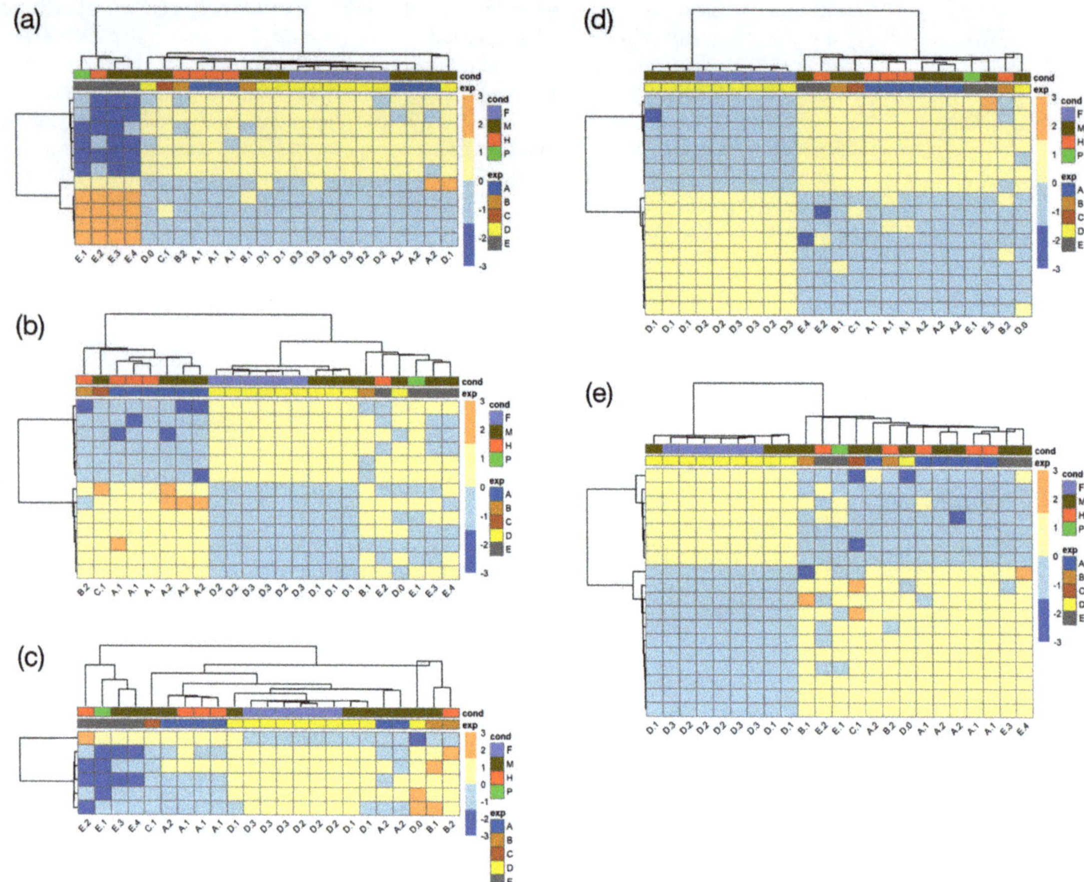

Figure 5. Selected co-expression clusters computed on the 2500 most variable genes. Only the five clusters characterized by significantly overrepresented ontological terms (featuring 631 transcripts) are shown. Heat maps and trees regroup samples behaving similarly across genes on the horizontal axis and genes behaving similarly across samples on the vertical axis; gene expression is vertically clustered to facilitate visualization (see text). Samples are colour-coded both by condition (F = fermentative, M = mixotrophic, H = heterotrophic, P = phototrophic) and by study (A = PRJNA310762, B = PRJEB10085, C = PRJNA298469, D = PRJNA289402, E = PRJEB38787). (**a**) Cluster 1; (**b**) cluster 4; (**c**) cluster 19; (**d**) cluster 24; (**e**) cluster 25.

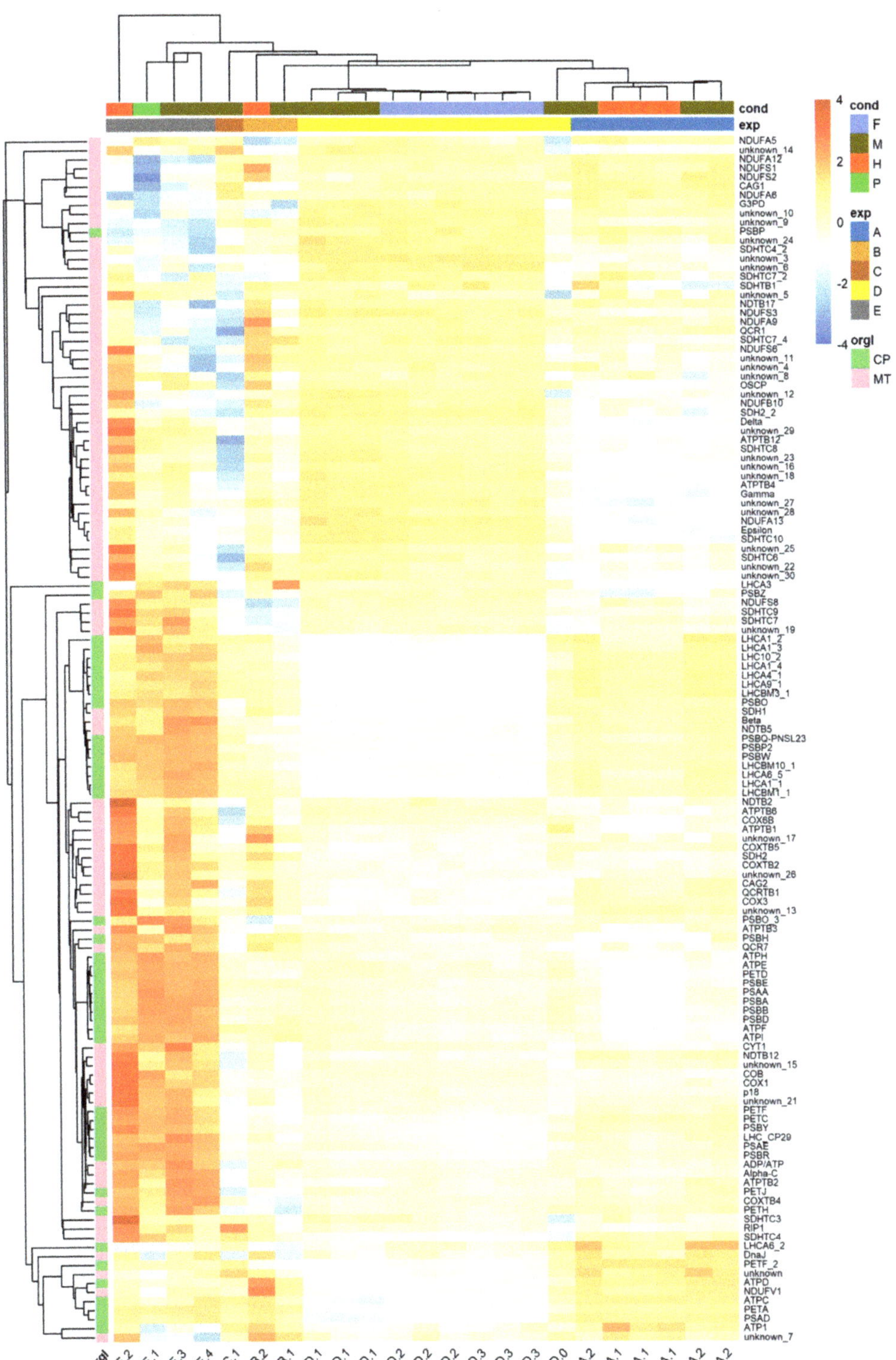

Figure 6. Expression heat map of 133 genes involved in electron transport chains. Heat maps and trees regroup samples behaving similarly across genes on the horizontal axis and genes behaving similarly across samples on the vertical axis (see

text). Samples are colour-coded both by condition (F = fermentative, M = mixotrophic, H = heterotrophic, P = phototrophic) and by study (A = PRJNA310762, B = PRJEB10085, C = PRJNA298469, D = PRJNA289402, E = PRJEB38787). Genes are colour-coded by organelle (CP = chloroplast; MT = mitochondrion).

Overall, our last analysis indicates that genes that share common metabolic functions are packed together, as would be expected, even though the expression is driven by study rather than culture condition. Beyond the technical issues that may have contributed to a loss of exploitable signal (e.g., heterogeneous experimental "design", see Table 1, uncorrected batch effects), these negative results can also be interpreted as additional evidence for the idea that, similar to what is known in trypanosomatids, nuclear gene expression in *E. gracilis* is not primarily regulated at the transcriptional level. In these parasites, gene regulation mostly occurs at the post-transcriptional level, through stabilization/degradation of mRNA molecules and control of mRNA translation (see [8] for a recent review of the issue). While the former mechanism should in principle change transcript abundance, the latter one might not be visible in comparative transcriptomics. For example, Yoshida et al. (2016) observed little change at the transcriptomic level following anaerobic treatment. Moreover, these changes in gene expression were inconsistent with respect to the activation of paramylon degradation and wax ester production [53]. In a more systematic investigation, Ebenezer et al. (2019) reported a striking lack of correlation between transcriptomic and proteomic data when comparing light and dark conditions [7]. As already mentioned, the raw transcriptomic data from these two studies were included in the present work (along with those of O'Neill et al. (2015) [52] and our own data), which allowed us to compare gene expression across a wider range of culture conditions at once. A few meaningful clusters of genes (i.e., following functional term enrichment) could be identified based on shared expression patterns across samples, which suggests that there is some biological signal in transcript abundance. However, the dominance of batch effects on these levels further questions the usefulness of transcriptomics for functional studies in *E. gracilis*.

4. Conclusions

Owing to its singular evolutionary origin, a merger between a chlorophyte alga and a phagotrophic unicellular belonging to a non-model eukaryotic group [20], *E. gracilis* is a fascinating, multifaceted chimeric organism, whose significance is constantly growing in domains as varied as the production of bio-based products [43], the treatment of wastewater ([130]), the provision of food supplements for space exploration [131], or the elucidation of mechanisms it shares with its parasitic trypanosome cousins [8,9,15] (see also the other articles of the present Special Issue).

By building a consolidated transcriptome of this photosynthetic eukaryote, we aimed at providing a solid resource to the community, taking into account previous work [7,52,53], yet enriched with unreleased data (obtained back in 2012–2014; Supplementary Figure S9) [132]. Our final consensus transcriptome comprises 91,040 unique transcripts and 49,922 predicted non-redundant protein-encoding genes. It appears to be the most complete up-to-date, at least according to sequence metrics, the number of universal orthologs found, read percentages supporting the assembly, and the fact that most of the *E. gracilis* sequences available to date have been included. Hence, we have been able to capture more than 98% of the sequences produced in the other transcriptomes hitherto published, while the number of predicted genes is in the same range [7,53]. This suggests that there was still some room for improvement, contrary to expectations for the opposite [7], and it might be related to the inclusion of reads obtained without poly-A selection, but following DSN normalization.

Annotating these transcripts, whether from a functional or taxonomic point of view, remains a challenge, notably because of the lack of well-characterized closely related organisms, the trypanosomes being relatively derived parasites [133]. This results in a mere 26–27% of our predicted genes annotated by sequence similarity, above the 23% of Yoshida et al. (2016) [53], but below the 45% of O'Neill et al. 2015 [52] and the 52–55% of

Ebenezer et al. (2019) [7], who further considered orthogroup sharing as annotation. In principle, this should encourage more large-scale studies, e.g., comparative transcriptomics performed in a wide range of culture conditions and stresses, in order to build a reliable gene expression network from co-expression data, and thereby provide alternative means for annotating genes of unknown function. Alas, as it now appears quite clearly, gene expression is mostly controlled at the post-transcriptional level in euglenozoans [7,8], including the regulation of chloroplast development in photosynthetic euglenids [134]. This implies that functional studies in *E. gracilis* have to be carried out through proteomics rather than transcriptomic approaches (e.g., [119,135]). This is fully possible considering the availability of several high-quality transcriptome assemblies to feed reference databases for proteomic fragment identification, including the one presented in this work. In this respect, the unfortunate lack of a complete genome beyond the draft level, even if frustrating, is not an insuperable issue [7].

Regarding the highly mixed taxonomic affinities of *Euglena* transcripts, our similarity searches yielded proportions in line with previous studies, even when those studies were based on more reliable phylogenetic approaches [136], such as the comprehensive work of Ebenezer et al. (2019) [7]. Altogether, the current knowledge points to the "shopping bag" [23–25] (or "red-carpet" [26]) model for the evolutionary origin of *Euglena*, i.e., transient endosymbioses during which multiple rounds of HGT/EGT have progressively shaped the plastid proteome. Yet, it is noteworthy that such a gene mixture would also be compatible with a kleptoplastidic origin for photosynthetic euglenids, in which the transient "endosymbioses" would actually imply stolen plastids and not intact symbionts. Moreover, some predatory euglenids, such as *Peranema trichophorum*, can feed either by phagocytosis of whole cells or by drilling a hole in their prey and then sucking up its cellular contents [137], a process known as myzocytosis [138]. Beyond providing a selective force for transferring genes to the host nucleus to service the ingested plastids, as in the recently characterized ARS (Antarctic Ross Sea) dinoflagellate bearing haptophyte-derived kleptoplastids [139], a kleptoplastidic model would also better fit the three membranes of the euglenid chloroplasts [20,140] and the presence of kleptoplastids acquired by myzocytosis in the early branching *Rapaza viridis* [141].

Supplementary Materials: The following are available online at https://www.mdpi.com/article/ 10.3390/genes12060842/s1. Figure S1: Taxonomic distribution of best BLAST hits before and after decontamination. Figure S2: GC-content distribution across reconstructed transcripts and in function of transcript length. Figure S3: Mapping coverage analysis for the 24-nt SL-sequence on the 5-end of the transcripts. Figure S4: Comparison of transcript count, length and identity over clusters of highly similar transcripts. Figure S5: Taxonomic analysis of reconstructed transcripts corresponding to mitochondrial and photosynthetic electron transfer chains. Figure S6: PCA plots computed on the tetranucleotide frequencies of taxonomically annotated reconstructed transcripts. Figure S7: Correlation values for a range of cluster solutions. Figure S8: PCA plots computed on gene expression before and after SVA batch effect correction. Figure S9: Quality-control of the total RNA prepared in our lab. Table S1a: Pairwise overlap between the new consensus transcriptome and two publicly available transcriptomes. Table S1b: Global overlap between the three public transcriptomes. Table S2: Annotation of the 49,922 predicted non-redundant protein-encoding genes. Table S3: List of 392 genes corresponding to carbohydrate-active enzymes. Table S4: List of 380 genes involved in visual perception processes and photoresponse. Table S5: List of 164 GO slim terms generated by the Slim Mapper tool. Table S6: List of 64 possibly contaminant transcripts persisting in the final consensus transcriptome. Table S7: Expression values in transcripts per kilobase million (TMP) for the 49,922 genes. Table S8: Composition of the 9 hubs in the ontology network. Table S9: Taxonomic analysis of the 9 hubs in the ontology network. Table S10: Composition of the 5 clusters in the gene co-expression network. Table S11: Expression values (in TPM) of 133 genes involved in photosynthetic and respiratory electron transfer chains. Archive file S1: RNAmmer and MegeBLAST reports for rRNA sequences. HTML file S1: Interactive Krona chart for the taxonomic affiliations of the 49,922 genes. HTML file S1: Krona chart for the nuclear genes involved in the mitochondrial electron transfer chain. HTML file S3: Krona chart for the nuclear genes involved in the photosynthetic electron transfer chain.

Author Contributions: Conceptualization, P.C. and D.B.; methodology, J.C. and D.B.; software, J.C., M.V.V., A.R.B., V.L. and D.B.; formal analysis, J.C., M.V.V., A.R.B. and V.L.; investigation, J.C. and E.P.; resources, P.C. and D.B.; data curation, J.C., M.V.V., V.L. and D.B.; writing—original draft preparation, J.C. and D.B.; writing—review and editing, J.C., P.C. and D.B.; visualization, J.C., E.P., M.V.V. and A.R.B.; supervision, P.C. and D.B.; project administration, J.C., P.C. and D.B.; funding acquisition, P.C. and D.B. All authors have read and agreed to the published version of the manuscript.

Funding: This research was funded by the Belgian FRS-FNRS (Fonds de la Recherche Scientifique), grant number PDR T.0032, and the European Research Council (ERC), H2020-EU BEAL project 682580, both awarded to P.C. Computational resources were provided through two grants to D.B. (University of Liège "Crédit de démarrage 2012" SFRD-12/04 and FRS-FNRS "Crédit de recherche 2014" CDR J.0080.15). E.P., M.V.V., A.R.B. and V.L. were (or still are) FRIA fellows (Fonds pour Formation à la Recherche dans l'Industrie et dans l'Agriculture) of the FRS-FNRS. P.C. is a Senior Research Associate from the FRS-FNRS.

Institutional Review Board Statement: Not applicable.

Informed Consent Statement: Not applicable.

Data Availability Statement: Publicly available datasets were generated in this study. These can be found at the European Nucleotide Archive (ENA), under study accession PRJEB38787 and TSA project accession HBDM01000000. Most custom analysis scripts have been deposited on GitHub (https://github.com/microalgues/clustering (accessed on 22 August 2019)).

Acknowledgments: We are grateful to Patrick E. Meyer for useful advice on network analyses.

Conflicts of Interest: The authors declare no conflict of interest.

References

1. Leander, B.S.; Farmer, M.A. Comparative morphology of the euglenid pellicle. II. Diversity of strip substructure. *J. Eukaryot. Microbiol.* **2001**, *48*, 202–217. [CrossRef]
2. Adl, S.M.; Simpson, A.G.; Lane, C.E.; Lukes, J.; Bass, D.; Bowser, S.S.; Brown, M.W.; Burki, F.; Dunthorn, M.; Hampl, V.; et al. The revised classification of eukaryotes. *J. Eukaryot. Microbiol.* **2012**, *59*, 429–493. [CrossRef]
3. Breglia, S.A.; Yubuki, N.; Hoppenrath, M.; Leander, B.S. Ultrastructure and molecular phylogenetic position of a novel euglenozoan with extrusive episymbiotic bacteria: Bihospites bacati n. gen. et sp. (Symbiontida). *BMC Microbiol.* **2010**, *10*, 145. [CrossRef]
4. Burki, F. The Eukaryotic Tree of Life from a Global Phylogenomic Perspective. *Cold Spring Harb. Perspect. Biol.* **2014**, *6*, a016147. [CrossRef] [PubMed]
5. Zakrys, B.; Milanowski, R.; Karnkowska, A. Evolutionary Origin of Euglena. *Adv. Exp. Med. Biol.* **2017**, *979*, 3–17. [CrossRef]
6. Butenko, A.; Opperdoes, F.R.; Flegontova, O.; Horak, A.; Hampl, V.; Keeling, P.; Gawryluk, R.M.R.; Tikhonenkov, D.; Flegontov, P.; Lukes, J. Evolution of metabolic capabilities and molecular features of diplonemids, kinetoplastids, and euglenids. *BMC Biol.* **2020**, *18*, 23. [CrossRef]
7. Ebenezer, T.E.; Zoltner, M.; Burrell, A.; Nenarokova, A.; Novak Vanclova, A.M.G.; Prasad, B.; Soukal, P.; Santana-Molina, C.; O'Neill, E.; Nankissoor, N.N.; et al. Transcriptome, proteome and draft genome of Euglena gracilis. *BMC Biol.* **2019**, *17*, 11. [CrossRef]
8. Vesteg, M.; Hadariova, L.; Horvath, A.; Estrano, C.E.; Schwartzbach, S.D.; Krajcovic, J. Comparative molecular cell biology of phototrophic euglenids and parasitic trypanosomatids sheds light on the ancestor of Euglenozoa. *Biol. Rev. Camb. Philos. Soc.* **2019**, *94*, 1701–1721. [CrossRef] [PubMed]
9. Perez, E.; Lapaille, M.; Degand, H.; Cilibrasi, L.; Villavicencio-Queijeiro, A.; Morsomme, P.; Gonzalez-Halphen, D.; Field, M.C.; Remacle, C.; Baurain, D.; et al. The mitochondrial respiratory chain of the secondary green alga Euglena gracilis shares many additional subunits with parasitic Trypanosomatidae. *Mitochondrion* **2014**, *19 Pt B*, 338–349. [CrossRef] [PubMed]
10. Jackson, A.P.; Quail, M.A.; Berriman, M. Insights into the genome sequence of a free-living Kinetoplastid: Bodo saltans (Kinetoplastida: Euglenozoa). *BMC Genom.* **2008**, *9*, 594. [CrossRef] [PubMed]
11. Deschamps, P.; Lara, E.; Marande, W.; Lopez-Garcia, P.; Ekelund, F.; Moreira, D. Phylogenomic analysis of kinetoplastids supports that trypanosomatids arose from within bodonids. *Mol. Biol. Evol.* **2011**, *28*, 53–58. [CrossRef]
12. Cavalier-Smith, T. Eukaryote kingdoms: Seven or nine? *Biosystems* **1981**, *14*, 461–481. [CrossRef]
13. Sogin, M.L.; Elwood, H.J.; Gunderson, J.H. Evolutionary diversity of eukaryotic small-subunit rRNA genes. *Proc. Natl. Acad. Sci. USA* **1986**, *83*, 1383–1387. [CrossRef] [PubMed]
14. Tessier, L.H.; Keller, M.; Chan, R.L.; Fournier, R.; Weil, J.H.; Imbault, P. Short leader sequences may be transferred from small RNAs to pre-mature mRNAs by trans-splicing in Euglena. *EMBO J.* **1991**, *10*, 2621–2625. [CrossRef] [PubMed]
15. Montrichard, F.; Le Guen, F.; Laval-Martin, D.L.; Davioud-Charvet, E. Evidence for the co-existence of glutathione reductase and trypanothione reductase in the non-trypanosomatid Euglenozoa: Euglena gracilis Z. *FEBS Lett.* **1999**, *442*, 29–33. [CrossRef]

16. Sibbald, S.J.; Archibald, J.M. Genomic Insights into Plastid Evolution. *Genome Biol. Evol.* **2020**, *12*, 978–990. [CrossRef] [PubMed]
17. Rogers, M.B.; Gilson, P.R.; Su, V.; McFadden, G.I.; Keeling, P.J. The complete chloroplast genome of the chlorarachniophyte Bigelowiella natans: Evidence for independent origins of chlorarachniophyte and euglenid secondary endosymbionts. *Mol. Biol. Evol.* **2007**, *24*, 54–62. [CrossRef] [PubMed]
18. Turmel, M.; Gagnon, M.C.; O'Kelly, C.J.; Otis, C.; Lemieux, C. The chloroplast genomes of the green algae Pyramimonas, Monomastix, and Pycnococcus shed new light on the evolutionary history of prasinophytes and the origin of the secondary chloroplasts of euglenids. *Mol. Biol. Evol.* **2009**, *26*, 631–648. [CrossRef]
19. Jackson, C.; Knoll, A.H.; Chan, C.X.; Verbruggen, H. Plastid phylogenomics with broad taxon sampling further elucidates the distinct evolutionary origins and timing of secondary green plastids. *Sci. Rep.* **2018**, *8*, 1523. [CrossRef]
20. Gibbs, S.P. The chloroplasts of Euglena may have evolved from symbiotic green algae. *Can. J. Bot.* **1978**, *56*, 2883–2889. [CrossRef]
21. Cavalier-Smith, T. Membrane heredity and early chloroplast evolution. *Trends Plant Sci.* **2000**, *5*, 174–182. [CrossRef]
22. Timmis, J.N.; Ayliffe, M.A.; Huang, C.Y.; Martin, W. Endosymbiotic gene transfer: Organelle genomes forge eukaryotic chromosomes. *Nat. Rev. Genet.* **2004**, *5*, 123–135. [CrossRef]
23. Larkum, A.W.; Lockhart, P.J.; Howe, C.J. Shopping for plastids. *Trends Plant. Sci.* **2007**, *12*, 189–195. [CrossRef] [PubMed]
24. Howe, C.J.; Barbrook, A.C.; Nisbet, R.E.; Lockhart, P.J.; Larkum, A.W. The origin of plastids. *Philos. Trans. R. Soc. Lond. B Biol. Sci.* **2008**, *363*, 2675–2685. [CrossRef] [PubMed]
25. Keeling, P.J. The number, speed, and impact of plastid endosymbioses in eukaryotic evolution. *Annu. Rev. Plant Biol.* **2013**, *64*, 583–607. [CrossRef]
26. Ponce-Toledo, R.I.; Lopez-Garcia, P.; Moreira, D. Horizontal and endosymbiotic gene transfer in early plastid evolution. *New Phytol.* **2019**, *224*, 618–624. [CrossRef]
27. Teich, R.; Zauner, S.; Baurain, D.; Brinkmann, H.; Petersen, J. Origin and distribution of Calvin cycle fructose and sedoheptulose bisphosphatases in plantae and complex algae: A single secondary origin of complex red plastids and subsequent propagation via tertiary endosymbioses. *Protist* **2007**, *158*, 263–276. [CrossRef] [PubMed]
28. Petersen, J.; Ludewig, A.K.; Michael, V.; Bunk, B.; Jarek, M.; Baurain, D.; Brinkmann, H. Chromera velia, endosymbioses and the rhodoplex hypothesis–plastid evolution in cryptophytes, alveolates, stramenopiles, and haptophytes (CASH lineages). *Genome Biol. Evol.* **2014**, *6*, 666–684. [CrossRef]
29. Barbrook, A.C.; Howe, C.J.; Purton, S. Why are plastid genomes retained in non-photosynthetic organisms? *Trends Plant Sci.* **2006**, *11*, 101–108. [CrossRef]
30. Singer, A.; Poschmann, G.; Muhlich, C.; Valadez-Cano, C.; Hansch, S.; Huren, V.; Rensing, S.A.; Stuhler, K.; Nowack, E.C.M. Massive Protein Import into the Early-Evolutionary-Stage Photosynthetic Organelle of the Amoeba Paulinella chromatophora. *Curr. Biol.* **2017**, *27*, 2763–2773. [CrossRef] [PubMed]
31. Ahmadinejad, N.; Dagan, T.; Martin, W. Genome history in the symbiotic hybrid Euglena gracilis. *Gene* **2007**, *402*, 35–39. [CrossRef]
32. Maruyama, S.; Suzaki, T.; Weber, A.P.; Archibald, J.M.; Nozaki, H. Eukaryote-to-eukaryote gene transfer gives rise to genome mosaicism in euglenids. *BMC Evol. Biol.* **2011**, *11*, 105. [CrossRef] [PubMed]
33. Cramer, M.; Myers, J. Growth and photosynthetic characteristics of Euglena gracilis. *Archiv. Für Mikrobiol.* **1952**, *17*, 384–402. [CrossRef]
34. Wilson, B.W.; Danforth, W.F. The extent of acetate and ethanol oxidation by Euglena gracilis. *J. Gen. Microbiol.* **1958**, *18*, 535–542. [CrossRef]
35. Buetow, D. Ethanol stimulation of oxidative metabolism in Euglena gracilis. *Nature* **1961**, *190*, 1196. [CrossRef]
36. Mego, J.L.; Farb, R.M. Alcohol dehydrogenases of Euglena gracilis, strain Z. *Biochim. Biophys. Acta* **1974**, *350*, 237–239. [CrossRef]
37. Sharpless, T.K.; Butow, R.A. An inducible alternate terminal oxidase in Euglena gracilis mitochondria. *J. Biol. Chem.* **1970**, *245*, 58–70. [CrossRef]
38. App, A.A.; Jagendorf, A.T. Repression of chloroplast development in Euglena gracilis by substrates. *J. Protozool.* **1963**, *10*, 340–343. [CrossRef]
39. Buetow, D.E. Acetate repression of chlorophyll synthesis in Euglena gracilis. *Nature* **1967**, *213*, 1127–1128. [CrossRef]
40. Vannini, G.L. Degeneration and regeneration of chloroplasts in Euglena gracilis grown in the presence of acetate: Ultrastructural evidence. *J. Cell Sci.* **1983**, *61*, 413–422. [CrossRef]
41. Calvayrac, R.; Laval-Martin, D.; Briand, J.; Farineau, J. Paramylon synthesis by Euglena gracilis photoheterotrophically grown under low O2 pressure: Description of a mitochloroplast complex. *Planta* **1981**, *153*, 6–13. [CrossRef]
42. Monfils, A.K.; Triemer, R.E.; Bellairs, E.F. Characterization of paramylon morphological diversity in photosynthetic euglenoids (Euglenales, Euglenophyta). *Phycologia* **2011**, *50*, 156–169. [CrossRef]
43. Shibakami, M.; Tsubouchi, G.; Hayashi, M. Thermoplasticization of euglenoid beta-1,3-glucans by mixed esterification. *Carbohydr. Polym.* **2014**, *105*, 90–96. [CrossRef]
44. Watanabe, T.; Shimada, R.; Matsuyama, A.; Yuasa, M.; Sawamura, H.; Yoshida, E.; Suzuki, K. Antitumor activity of the beta-glucan paramylon from Euglena against preneoplastic colonic aberrant crypt foci in mice. *Food Funct.* **2013**, *4*, 1685–1690. [CrossRef]
45. Matsuda, F.; Hayashi, M.; Kondo, A. Comparative profiling analysis of central metabolites in Euglena gracilis under various cultivation conditions. *Biosci. Biotechnol. Biochem.* **2011**, *75*, 2253–2256. [CrossRef]

46. Furuhashi, T.; Ogawa, T.; Nakai, R.; Nakazawa, M.; Okazawa, A.; Padermschoke, A.; Nishio, K.; Hirai, M.Y.; Arita, M.; Ohta, D. Wax ester and lipophilic compound profiling of Euglena gracilis by gas chromatography-mass spectrometry: Toward understanding of wax ester fermentation under hypoxia. *Metabolomics* **2015**, *11*, 175–183. [CrossRef]

47. Inui, H.; Ishikawa, T.; Tamoi, M. Wax Ester Fermentation and Its Application for Biofuel Production. *Adv. Exp. Med. Biol.* **2017**, *979*, 269–283. [CrossRef] [PubMed]

48. Ogbonna, J.C.; Tomiyamal, S.; Tanaka, H. Heterotrophic cultivation of Euglena gracilis Z for efficient production of α-tocopherol. *J. Appl. Phycol.* **1998**, *10*, 67–74. [CrossRef]

49. Hallick, R.B.; Hong, L.; Drager, R.G.; Favreau, M.R.; Monfort, A.; Orsat, B.; Spielmann, A.; Stutz, E. Complete sequence of Euglena gracilis chloroplast DNA. *Nucleic Acids Res.* **1993**, *21*, 3537–3544. [CrossRef] [PubMed]

50. Spencer, D.F.; Gray, M.W. Ribosomal RNA genes in Euglena gracilis mitochondrial DNA: Fragmented genes in a seemingly fragmented genome. *Mol. Genet. Genom.* **2011**, *285*, 19–31. [CrossRef] [PubMed]

51. Dobakova, E.; Flegontov, P.; Skalicky, T.; Lukes, J. Unexpectedly streamlined mitochondrial genome of the euglenozoan Euglena gracilis. *Genome Biol. Evol.* **2015**. [CrossRef]

52. O'Neill, E.C.; Trick, M.; Hill, L.; Rejzek, M.; Dusi, R.G.; Hamilton, C.J.; Zimba, P.V.; Henrissat, B.; Field, R.A. The transcriptome of Euglena gracilis reveals unexpected metabolic capabilities for carbohydrate and natural product biochemistry. *Mol. Biosyst.* **2015**. [CrossRef]

53. Yoshida, Y.; Tomiyama, T.; Maruta, T.; Tomita, M.; Ishikawa, T.; Arakawa, K. De novo assembly and comparative transcriptome analysis of Euglena gracilis in response to anaerobic conditions. *BMC Genom.* **2016**, *17*. [CrossRef] [PubMed]

54. Ebenezer, T.E.; Carrington, M.; Lebert, M.; Kelly, S.; Field, M.C. Euglena gracilis Genome and Transcriptome: Organelles, Nuclear Genome Assembly Strategies and Initial Features. *Adv. Exp. Med. Biol.* **2017**, *979*, 125–140. [CrossRef] [PubMed]

55. Loppes, R.; Radoux, M. Identification of short promoter regions involved in the transcriptional expression of the nitrate reductase gene in Chlamydomonas reinhardtii. *Plant Mol. Biol.* **2001**, *45*, 215–227. [CrossRef] [PubMed]

56. Andrews, S. *FastQC: A Quality Control Tool for High Throughput Sequence Data*; Babraham Bioinformatics, Babraham Institute: Cambridge, UK, 2010.

57. Schmieder, R.; Edwards, R. Quality control and preprocessing of metagenomic datasets. *Bioinformatics* **2011**, *27*, 863–864. [CrossRef]

58. Bolger, A.M.; Lohse, M.; Usadel, B. Trimmomatic: A flexible trimmer for Illumina sequence data. *Bioinformatics* **2014**, *30*, 2114–2120. [CrossRef]

59. Haas, B.J.; Papanicolaou, A.; Yassour, M.; Grabherr, M.; Blood, P.D.; Bowden, J.; Couger, M.B.; Eccles, D.; Li, B.; Lieber, M.; et al. De novo transcript sequence reconstruction from RNA-seq using the Trinity platform for reference generation and analysis. *Nat. Protoc.* **2013**, *8*, 1494–1512. [CrossRef]

60. Gilbert, D. Gene-omes built from mRNA-seq not genome DNA. *F1000Research* **2016**, *5*, 1695.

61. Gilbert, D.G. Genes of the pig, Sus scrofa, reconstructed with EvidentialGene. *PeerJ* **2019**, *7*, e6374. [CrossRef]

62. Camacho, C.; Coulouris, G.; Avagyan, V.; Ma, N.; Papadopoulos, J.; Bealer, K.; Madden, T.L. BLAST+: Architecture and applications. *BMC Bioinform.* **2009**, *10*, 421. [CrossRef]

63. Sayers, E.W.; Beck, J.; Bolton, E.E.; Bourexis, D.; Brister, J.R.; Canese, K.; Comeau, D.C.; Funk, K.; Kim, S.; Klimke, W.; et al. Database resources of the National Center for Biotechnology Information. *Nucleic Acids Res.* **2021**, *49*, D10–D17. [CrossRef] [PubMed]

64. Howe, K.L.; Achuthan, P.; Allen, J.; Allen, J.; Alvarez-Jarreta, J.; Amode, M.R.; Armean, I.M.; Azov, A.G.; Bennett, R.; Bhai, J.; et al. Ensembl 2021. *Nucleic Acids Res.* **2021**, *49*, D884–D891. [CrossRef] [PubMed]

65. Langmead, B.; Salzberg, S.L. Fast gapped-read alignment with Bowtie 2. *Nat. Methods* **2012**, *9*, 357–359. [CrossRef]

66. Langmead, B.; Wilks, C.; Antonescu, V.; Charles, R. Scaling read aligners to hundreds of threads on general-purpose processors. *Bioinformatics* **2019**, *35*, 421–432. [CrossRef] [PubMed]

67. Li, W.; Godzik, A. Cd-hit: A fast program for clustering and comparing large sets of protein or nucleotide sequences. *Bioinformatics* **2006**, *22*, 1658–1659. [CrossRef]

68. Fu, L.; Niu, B.; Zhu, Z.; Wu, S.; Li, W. CD-HIT: Accelerated for clustering the next-generation sequencing data. *Bioinformatics* **2012**, *28*, 3150–3152. [CrossRef]

69. Li, B.; Fillmore, N.; Bai, Y.; Collins, M.; Thomson, J.A.; Stewart, R.; Dewey, C.N. Evaluation of de novo transcriptome assemblies from RNA-Seq data. *Genome Biol.* **2014**, *15*, 553. [CrossRef] [PubMed]

70. Smith-Unna, R.; Boursnell, C.; Patro, R.; Hibberd, J.M.; Kelly, S. TransRate: Reference-free quality assessment of de novo transcriptome assemblies. *Genome Res.* **2016**, *26*, 1134–1144. [CrossRef]

71. Rice, P.; Longden, I.; Bleasby, A. EMBOSS: The European Molecular Biology Open Software Suite. *Trends Genet.* **2000**, *16*, 276–277. [CrossRef]

72. Lagesen, K.; Hallin, P.; Rodland, E.A.; Staerfeldt, H.H.; Rognes, T.; Ussery, D.W. RNAmmer: Consistent and rapid annotation of ribosomal RNA genes. *Nucleic Acids Res.* **2007**, *35*, 3100–3108. [CrossRef] [PubMed]

73. Tang, S.; Lomsadze, A.; Borodovsky, M. Identification of protein coding regions in RNA transcripts. *Nucleic Acids Res.* **2015**, *43*, e78. [CrossRef] [PubMed]

74. Simao, F.A.; Waterhouse, R.M.; Ioannidis, P.; Kriventseva, E.V.; Zdobnov, E.M. BUSCO: Assessing genome assembly and annotation completeness with single-copy orthologs. *Bioinformatics* **2015**, *31*, 3210–3212. [CrossRef] [PubMed]

75. Waterhouse, R.M.; Seppey, M.; Simao, F.A.; Manni, M.; Ioannidis, P.; Klioutchnikov, G.; Kriventseva, E.V.; Zdobnov, E.M. BUSCO Applications from Quality Assessments to Gene Prediction and Phylogenomics. *Mol. Biol. Evol.* **2018**, *35*, 543–548. [CrossRef] [PubMed]

76. Huerta-Cepas, J.; Forslund, K.; Coelho, L.P.; Szklarczyk, D.; Jensen, L.J.; von Mering, C.; Bork, P. Fast Genome-Wide Functional Annotation through Orthology Assignment by eggNOG-Mapper. *Mol. Biol. Evol.* **2017**, *34*, 2115–2122. [CrossRef]

77. Huerta-Cepas, J.; Szklarczyk, D.; Heller, D.; Hernandez-Plaza, A.; Forslund, S.K.; Cook, H.; Mende, D.R.; Letunic, I.; Rattei, T.; Jensen, L.J.; et al. eggNOG 5.0: A hierarchical, functionally and phylogenetically annotated orthology resource based on 5090 organisms and 2502 viruses. *Nucleic Acids Res.* **2019**, *47*, D309–D314. [CrossRef]

78. UniProt, C. UniProt: The universal protein knowledgebase in 2021. *Nucleic Acids Res.* **2021**, *49*, D480–D489. [CrossRef]

79. The Gene Ontology Consortium. The Gene Ontology resource: Enriching a GOld mine. *Nucleic Acids Res.* **2021**, *49*, D325–D334. [CrossRef]

80. Kanehisa, M.; Furumichi, M.; Sato, Y.; Ishiguro-Watanabe, M.; Tanabe, M. KEGG: Integrating viruses and cellular organisms. *Nucleic Acids Res.* **2021**, *49*, D545–D551. [CrossRef] [PubMed]

81. Yadav, K.N.S.; Miranda-Astudillo, H.V.; Colina-Tenorio, L.; Bouillenne, F.; Degand, H.; Morsomme, P.; Gonzalez-Halphen, D.; Boekema, E.J.; Cardol, P. Atypical composition and structure of the mitochondrial dimeric ATP synthase from Euglena gracilis. *Biochim. Biophys. Acta Bioenerg.* **2017**, *1858*, 267–275. [CrossRef]

82. Miranda-Astudillo, H.V.; Yadav, K.N.S.; Colina-Tenorio, L.; Bouillenne, F.; Degand, H.; Morsomme, P.; Boekema, E.J.; Cardol, P. The atypical subunit composition of respiratory complexes I and IV is associated with original extra structural domains in Euglena gracilis. *Sci. Rep.* **2018**, *8*, 9698. [CrossRef] [PubMed]

83. Sobotka, R.; Esson, H.J.; Konik, P.; Trskova, E.; Moravcova, L.; Horak, A.; Dufkova, P.; Obornik, M. Extensive gain and loss of photosystem I subunits in chromerid algae, photosynthetic relatives of apicomplexans. *Sci. Rep.* **2017**, *7*, 13214. [CrossRef] [PubMed]

84. Koziol, A.G.; Durnford, D.G. Euglena Light-Harvesting Complexes Are Encoded by Multifarious Polyprotein mRNAs that Evolve in Concert. *Mol. Biol. Evol.* **2008**, *25*, 92–100. [CrossRef] [PubMed]

85. Van Vlierberghe, M.; Philippe, H.; Baurain, D. Broadly sampled orthologous groups of eukaryotic proteins for the phylogenetic study of plastid-bearing lineages. *BMC Res. Notes* **2021**, *14*. [CrossRef] [PubMed]

86. Cornet, L.; Meunier, L.; Van Vlierberghe, M.; Leonard, R.R.; Durieu, B.; Lara, Y.; Misztak, A.; Sirjacobs, D.; Javaux, E.J.; Philippe, H.; et al. Consensus assessment of the contamination level of publicly available cyanobacterial genomes. *PLoS ONE* **2018**, *13*, e0200323. [CrossRef]

87. Huson, D.H.; Auch, A.F.; Qi, J.; Schuster, S.C. MEGAN analysis of metagenomic data. *Genome Res.* **2007**, *17*, 377–386. [CrossRef]

88. R Core Team. *R: A Language and Environment for Statistical Computing*; R Foundation for Statistical Computing: Vienna, Austria, 2017.

89. Li, B.; Dewey, C.N. RSEM: Accurate transcript quantification from RNA-Seq data with or without a reference genome. *BMC Bioinform.* **2011**, *12*, 323. [CrossRef] [PubMed]

90. Soneson, C.; Love, M.I.; Robinson, M.D. Differential analyses for RNA-seq: Transcript-level estimates improve gene-level inferences. *F1000Research* **2015**, *4*, 1521. [CrossRef]

91. Leek, J.T. Svaseq: Removing batch effects and other unwanted noise from sequencing data. *Nucleic Acids Res.* **2014**, *42*. [CrossRef]

92. Maechler, M.; Rousseeuw, P.; Struyf, A.; Hubert, M.; Hornik, K. Cluster: Cluster Analysis Basics and Extensions, R Package Version 2.0.7-1. 2018. Available online: https://cran.r-project.org/web/packages/cluster/index.html (accessed on 28 May 2021).

93. Dunn, J.C. Well-separated clusters and optimal fuzzy partitions. *J. Cybern.* **1974**, *4*, 95–104. [CrossRef]

94. Kolde, R.; Kolde, M.R. Pheatmap: Pretty Heatmaps, R Package Version 1.0.12. 2019. Available online: https://rdrr.io/cran/pheatmap/ (accessed on 28 May 2021).

95. Bindea, G.; Mlecnik, B.; Hackl, H.; Charoentong, P.; Tosolini, M.; Kirilovsky, A.; Fridman, W.H.; Pages, F.; Trajanoski, Z.; Galon, J. ClueGO: A Cytoscape plug-in to decipher functionally grouped gene ontology and pathway annotation networks. *Bioinformatics* **2009**, *25*, 1091–1093. [CrossRef]

96. Shannon, P.; Markiel, A.; Ozier, O.; Baliga, N.S.; Wang, J.T.; Ramage, D.; Amin, N.; Schwikowski, B.; Ideker, T. Cytoscape: A software environment for integrated models of biomolecular interaction networks. *Genome Res.* **2003**, *13*, 2498–2504. [CrossRef]

97. Bader, G.D.; Hogue, C.W. An automated method for finding molecular complexes in large protein interaction networks. *BMC Bioinform.* **2003**, *4*, 2. [CrossRef]

98. Lusk, R.W. Diverse and widespread contamination evident in the unmapped depths of high throughput sequencing data. *PLoS ONE* **2014**, *9*, e110808. [CrossRef] [PubMed]

99. Strong, M.J.; Xu, G.; Morici, L.; Splinter Bon-Durant, S.; Baddoo, M.; Lin, Z.; Fewell, C.; Taylor, C.M.; Flemington, E.K. Microbial contamination in next generation sequencing: Implications for sequence-based analysis of clinical samples. *PLoS Pathog.* **2014**, *10*, e1004437. [CrossRef] [PubMed]

100. Simion, P.; Philippe, H.; Baurain, D.; Jager, M.; Richter, D.J.; Di Franco, A.; Roure, B.; Satoh, N.; Queinnec, E.; Ereskovsky, A.; et al. A Large and Consistent Phylogenomic Dataset Supports Sponges as the Sister Group to All Other Animals. *Curr. Biol.* **2017**, *27*, 958–967. [CrossRef] [PubMed]

101. Tae, H.; Karunasena, E.; Bavarva, J.H.; McIver, L.J.; Garner, H.R. Large scale comparison of non-human sequences in human sequencing data. *Genomics* **2014**, *104*, 453–458. [CrossRef]

102. Nakazawa, M.; Minami, T.; Teramura, K.; Kumamoto, S.; Hanato, S.; Takenaka, S.; Ueda, M.; Inui, H.; Nakano, Y.; Miyatake, K. Molecular characterization of a bifunctional glyoxylate cycle enzyme, malate synthase/isocitrate lyase, in Euglena gracilis. *Comp. Biochem. Physiol. B Biochem. Mol. Biol.* **2005**, *141*, 445–452. [CrossRef] [PubMed]

103. Nakazawa, M.; Nishimura, M.; Inoue, K.; Ueda, M.; Inui, H.; Nakano, Y.; Miyatake, K. Characterization of a bifunctional glyoxylate cycle enzyme, malate synthase/isocitrate lyase, of Euglena gracilis. *J. Eukaryot. Microbiol.* **2011**, *58*, 128–133. [CrossRef]

104. Liu, F.; Thatcher, J.D.; Barral, J.M.; Epstein, H.F. Bifunctional glyoxylate cycle protein of Caenorhabditis elegans: A developmentally regulated protein of intestine and muscle. *Dev. Biol.* **1995**, *169*, 399–414. [CrossRef]

105. Gilbert, D. Evigene Versus Transdecoder for Proteins from Transcripts. Available online: https://sourceforge.net/p/evidentialgene/blog/2017/11/-evigene-versus-transdecoder-for-proteins-from-transcripts/ (accessed on 3 April 2021).

106. Kitaoka, M.; Sasaki, T.; Taniguchi, H. Purification and properties of laminaribiose phosphorylase (EC 2.4 1.31) from Euglena gracilis Z. *Arch. Biochem. Biophys.* **1993**, *304*, 508–514. [CrossRef]

107. Kuhaudomlarp, S.; Patron, N.J.; Henrissat, B.; Rejzek, M.; Saalbach, G.; Field, R.A. Identification of Euglena gracilis beta-1,3-glucan phosphorylase and establishment of a new glycoside hydrolase (GH) family GH149. *J. Biol. Chem.* **2018**, *293*, 2865–2876. [CrossRef] [PubMed]

108. Barsanti, L.; Evangelista, V.; Passarelli, V.; Frassanito, A.M.; Gualtieri, P. Fundamental questions and concepts about photoreception and the case of Euglena gracilis. *Integr. Biol.* **2012**, *4*, 22–36. [CrossRef] [PubMed]

109. Gallo, J.M.; Schrevel, J. Homologies between paraflagellar rod proteins from trypanosomes and euglenoids revealed by a monoclonal antibody. *Eur. J. Cell Biol.* **1985**, *36*, 163–168. [PubMed]

110. Cachon, J.; Cachon, M.; Cosson, M.-P.; Cosson, J. The paraflagellar rod: A structure in search of a function. *Biol. Cell* **1988**, *63*, 169–181. [CrossRef]

111. Maharana, B.R.; Tewari, A.K.; Singh, V. An overview on kinetoplastid paraflagellar rod. *J. Parasit. Dis.* **2015**, *39*, 589–595. [CrossRef] [PubMed]

112. Verni, F.; Rosati, G.; Lenzi, P.; Barsanti, L.; Passarelli, V.; Gualtieri, P. Morphological relationship between paraflagellar swelling and paraxial rod in Euglena gracilis. *Micron Microsc. Acta* **1992**, *23*, 37–44. [CrossRef]

113. Iseki, M.; Matsunaga, S.; Murakami, A.; Ohno, K.; Shiga, K.; Yoshida, K.; Sugai, M.; Takahashi, T.; Hori, T.; Watanabe, M. A blue-light-activated adenylyl cyclase mediates photoavoidance in Euglena gracilis. *Nature* **2002**, *415*, 1047–1051. [CrossRef] [PubMed]

114. Koumura, Y.; Suzuki, T.; Yoshikawa, S.; Watanabe, M.; Iseki, M. The origin of photoactivated adenylyl cyclase (PAC), the Euglena blue-light receptor: Phylogenetic analysis of orthologues of PAC subunits from several euglenoids and trypanosome-type adenylyl cyclases from Euglena gracilis. *Photochem. Photobiol. Sci.* **2004**, *3*, 580–586. [CrossRef]

115. Cherry, J.M.; Hong, E.L.; Amundsen, C.; Balakrishnan, R.; Binkley, G.; Chan, E.T.; Christie, K.R.; Costanzo, M.C.; Dwight, S.S.; Engel, S.R.; et al. Saccharomyces Genome Database: The genomics resource of budding yeast. *Nucleic Acids Res.* **2012**, *40*, D700–D705. [CrossRef]

116. Markunas, C.M.; Triemer, R.E. Evolutionary History of the Enzymes Involved in the Calvin-Benson Cycle in Euglenids. *J. Eukaryot. Microbiol.* **2016**, *63*, 326–339. [CrossRef]

117. Lakey, B.; Triemer, R.; Müller, K. The tetrapyrrole synthesis pathway as a model of horizontal gene transfer in euglenoids. *J. Phycol.* **2017**, *53*, 198–217. [CrossRef]

118. Ponce-Toledo, R.I.; Moreira, D.; Lopez-Garcia, P.; Deschamps, P. Secondary Plastids of Euglenids and Chlorarachniophytes Function with a Mix of Genes of Red and Green Algal Ancestry. *Mol. Biol. Evol.* **2018**. [CrossRef]

119. Novak Vanclova, A.M.G.; Zoltner, M.; Kelly, S.; Soukal, P.; Zahonova, K.; Fussy, Z.; Ebenezer, T.E.; Lacova Dobakova, E.; Elias, M.; Lukes, J.; et al. Metabolic quirks and the colourful history of the Euglena gracilis secondary plastid. *New Phytol.* **2020**, *225*, 1578–1592. [CrossRef]

120. Cavalier-Smith, T. Principles of protein and lipid targeting in secondary symbiogenesis: Euglenoid, dinoflagellate, and sporozoan plastid origins and the eukaryote family tree. *J. Eukaryot. Microbiol.* **1999**, *46*, 347–366. [CrossRef]

121. Cavalier-Smith, T. Genomic reduction and evolution of novel genetic membranes and protein-targeting machinery in eukaryote-eukaryote chimaeras (meta-algae). *Philos. Trans. R. Soc. Lond. B Biol. Sci.* **2003**, *358*, 109–133. [CrossRef] [PubMed]

122. Cenci, U.; Bhattacharya, D.; Weber, A.P.M.; Colleoni, C.; Subtil, A.; Ball, S.G. Biotic Host-Pathogen Interactions as Major Drivers of Plastid Endosymbiosis. *Trends Plant Sci.* **2017**, *22*, 316–328. [CrossRef] [PubMed]

123. Curtis, B.A.; Tanifuji, G.; Burki, F.; Gruber, A.; Irimia, M.; Maruyama, S.; Arias, M.C.; Ball, S.G.; Gile, G.H.; Hirakawa, Y.; et al. Algal genomes reveal evolutionary mosaicism and the fate of nucleomorphs. *Nature* **2012**, *6*, 59–65. [CrossRef] [PubMed]

124. Read, B.A.; Kegel, J.; Klute, M.J.; Kuo, A.; Lefebvre, S.C.; Maumus, F.; Mayer, C.; Miller, J.; Monier, A.; Salamov, A.; et al. Pan genome of the phytoplankton Emiliania underpins its global distribution. *Nature* **2013**, *499*, 209–213. [CrossRef] [PubMed]

125. Dorrell, R.G.; Gile, G.; McCallum, G.; Meheust, R.; Bapteste, E.P.; Klinger, C.M.; Brillet-Gueguen, L.; Freeman, K.D.; Richter, D.J.; Bowler, C. Chimeric origins of ochrophytes and haptophytes revealed through an ancient plastid proteome. *eLife* **2017**, *6*. [CrossRef] [PubMed]

126. Simion, P.; Belkhir, K.; Francois, C.; Veyssier, J.; Rink, J.C.; Manuel, M.; Philippe, H.; Telford, M.J. A software tool 'CroCo' detects pervasive cross-species contamination in next generation sequencing data. *BMC Biol.* **2018**, *16*, 28. [CrossRef]

127. Kumar, S.; Jones, M.; Koutsovoulos, G.; Clarke, M.; Blaxter, M. Blobology: Exploring raw genome data for contaminants, symbionts and parasites using taxon-annotated GC-coverage plots. *Front. Genet.* **2013**, *4*, 237. [CrossRef]

128. Lukes, J.; Leander, B.S.; Keeling, P.J. Cascades of convergent evolution: The corresponding evolutionary histories of euglenozoans and dinoflagellates. *Proc. Natl. Acad. Sci. USA* **2009**, *106* (Suppl. 1), 9963–9970. [CrossRef]

129. Goh, W.W.B.; Wang, W.; Wong, L. Why Batch Effects Matter in Omics Data, and How to Avoid Them. *Trends Biotechnol.* **2017**, *35*, 498–507. [CrossRef] [PubMed]

130. Almasi, A.; Pescod, M.B. Wastewater treatment mechanisms in anoxic stabilization ponds. *Water Sci. Technol.* **1996**, *33*, 125–132. [CrossRef]

131. Hauslage, J.; Strauch, S.M.; Eßmann, O.; Haag, F.W.M.; Richter, P.; Krüger, J.; Stoltze, J.; Becker, I.; Nasir, A.; Bornemann, G.; et al. Eu:CROPIS—"Euglena gracilis: Combined Regenerative Organic-food Production in Space"—A Space Experiment Testing Biological Life Support Systems Under Lunar And Martian Gravity. *Microgravity Sci. Technol.* **2018**, *30*, 933–942. [CrossRef]

132. Perez, E. *Analyses Biochimiques, Protéomiques et Transcriptomiques du Métabolisme Énergétique chez L'algue Secondaire verte Euglena gracilis (Euglenozoa, Excavata)*; Université de Liège: Liège, Belgique, 2015.

133. Simpson, A.G.; Stevens, J.R.; Lukes, J. The evolution and diversity of kinetoplastid flagellates. *Trends Parasitol.* **2006**, *22*, 168–174. [CrossRef] [PubMed]

134. Schwartzbach, S.D. Photo and nutritional regulation of Euglena organelle development. *Euglena Biochem. Cell Mol. Biol.* **2017**, 159–182. [CrossRef]

135. Gain, G.; Vega de Luna, F.; Cordoba, J.; Perez, E.; Degand, H.; Morsomme, P.; Thiry, M.; Baurain, D.; Pierangelini, M.; Cardol, P. Trophic state alters the mechanism whereby energetic coupling between photosynthesis and respiration occurs in Euglena gracilis. *New Phytol.* **2021**. (in revision).

136. Koski, L.B.; Golding, G.B. The closest BLAST hit is often not the nearest neighbor. *J. Mol. Evol.* **2001**, *52*, 540–542. [CrossRef]

137. Triemer, R.E. Feeding in Peranema trichophorum revisited (Euglenophyta) 1. *J. Phycol.* **1997**, *33*, 649–654. [CrossRef]

138. Schnepf, E.; Deichgräber, G. «Myzocytosis», a kind of endocytosis with implications to compartmentation in endosymbiosis. *Naturwissenschaften* **1984**, *71*, 218–219. [CrossRef]

139. Hehenberger, E.; Gast, R.J.; Keeling, P.J. A kleptoplastidic dinoflagellate and the tipping point between transient and fully integrated plastid endosymbiosis. *Proc. Natl. Acad. Sci. USA* **2019**. [CrossRef] [PubMed]

140. Bodyl, A. Did some red alga-derived plastids evolve via kleptoplastidy? A hypothesis. *Biol. Rev. Camb. Philos. Soc.* **2018**, *93*, 201–222. [CrossRef] [PubMed]

141. Yamaguchi, A.; Yubuki, N.; Leander, B.S. Morphostasis in a novel eukaryote illuminates the evolutionary transition from phagotrophy to phototrophy: Description of Rapaza viridis n. gen. et sp. (Euglenozoa, Euglenida). *BMC Evol. Biol.* **2012**, *12*, 29. [CrossRef] [PubMed]

MDPI
St. Alban-Anlage 66
4052 Basel
Switzerland
Tel. +41 61 683 77 34
Fax +41 61 302 89 18
www.mdpi.com

Genes Editorial Office
E-mail: genes@mdpi.com
www.mdpi.com/journal/genes